Mechanisms and Machine Science

Volume 27

Series Editor
Marco Ceccarelli
Univ Cassino & South Latium Lab of Robotics & Mechatronics
Cassino
Italy

The aim of the Book Series is to establish a well defined forum for monographs, edited Books, and proceedings on mechanical engineering with particular emphasis on MMS (Mechanism and Machine Science). The final goal is the publication of works that can show developments of mechanical engineering and particularly MMS in all their technical aspects, even in very recent assessments. The works will have a technical approach by which technical details and formulation are discussed and formulate in a modern formalism with the aim to circulate research and technical achievements for use in professional, research, academic, and teaching activities. This technical approach is an essential characteristic of the series. By discussing technical details and formulations in terms of modern formalisms, the possibility is created not only to show technical developments but also to explain achievements for technical teaching and research activity today and for the future. The book series is intended to collect technical views on developments of the broad field of MMS in a unique frame that can be seen in its totality as an Encyclopaedia of MMS but with the additional purpose of archiving and teaching MMS achievements. Therefore the book series is intended not only for researchers and teachers in Mechanical Engineering but also for professionals and students for their formation and future work.

More information about this series at http://www.springer.com/series/8779

Vigen Arakelian • Sébastien Briot

Balancing of Linkages and Robot Manipulators

Advanced Methods with Illustrative Examples

Springer

Vigen Arakelian
Institut National des Sciences
Appliquées (INSA), Rennes, France,
and Institut de Recherche en
Communications et Cybernétique
de Nantes (IRCCyN), Nantes, France
Rennes
France

Sébastien Briot
Institut de Recherche en Com.
et Cybernétique de Nantes
CNRS
Nantes
France

ISSN 2211-0984 ISSN 2211-0992 (electronic)
Mechanisms and Machine Science
ISBN 978-3-319-12489-6 ISBN 978-3-319-12490-2 (eBook)
DOI 10.1007/978-3-319-12490-2

Library of Congress Control Number: 2014958013

Springer Cham Heidelberg New York Dordrecht London

Printed on acid-free paper

Springer is part of Springer Science+Business Media (www.springer.com)

Vigen Arakelian dedicates this work to his charming wife Tatiana and beloved son David.

Sébastien Briot dedicates this work to his beloved wife and sons, Sylvie, Élouan and Guénaël.

Preface

The balancing of linkages is an integral part of the mechanism design. The challenge of reducing vibrations of the frame on which the mechanism is mounted is nothing new. Despite its long history, mechanism balancing theory continues to be developed and new approaches and solutions are constantly being reported. Hence, the balancing problems are of continued interest to researchers. Several laboratories around the world are very active in this area and new results are published regularly. In recent decades, new challenges have presented themselves, particularly, the balancing of robots for fast manipulation.

The authors believe that this is an appropriate moment to present the state of the art of the studies devoted to balancing and to summarize their research results. This monograph is based on the material published by the first author over the last twenty years and the doctoral dissertation of the second author defended in 2007 and rewarded by the Research Group in Robotics of the French National Center for Scientific Research (GDR Robotique, CNRS, 2008), the French Section of the ASME (2011) and the French Région Bretagne in the category "Sciences, Technologies and Interdisciplinarities" (2011).

Some results given in the book were reached in collaboration with Mike Smith, Clément Gosselin, Ilian Bonev, Simon Lessard and Cédric Baradat. The authors acknowledge for their contributions, as well as the "Mechanical Center" of the National Institute of Applied Sciences of Rennes for the development of the prototypes permitting the validation and improvement of the obtained theoretical results.

The authors will be also genuinely grateful to the readers for any critical remarks on the material presented in the book and for any suggestion for its improvement.

Rennes/Nantes, France
June, 2014

Vigen Arakelian
Sébastien Briot

Contents

Part I Introduction to Balancing

1 Introduction 3

2 An Overview of Balancing Methods 7
2.1 Shaking Force and Shaking Moment Balancing of Linkages 7
2.1.1 Shaking Force Balancing of Linkages 8
2.1.2 Shaking Moment Balancing of Linkages 12
2.2 Shaking Force and Shaking Moment Balancing of Robots and Manipulators 19
2.2.1 Shaking Force Balancing 19
2.2.2 Shaking Moment Balancing 22
2.3 Gravity Balancing in Robotics 27
2.3.1 Gravity Compensation in Automatic Robot-Manipulators ... 27
2.3.2 Gravity Compensation in Hand-Operated Balanced Manipulators (HOBM) 44
2.3.3 Gravity Compensation in Rehabilitation Systems of Human Extremities, Exoskeletons and Walking Assist Devices 46

Part II Balancing of Linkages

3 Partial Shaking Force and Shaking Moment Balancing of Linkages .. 55
3.1 Shaking Moment Minimization of Fully Force-balanced Planar Linkages by Displacing One Counterweight 56
3.1.1 Complete Shaking Force and Partial Shaking Moment Balancing of Planar Linkages 56
3.1.2 Numerical Example and Comparative Analysis 59
3.2 Shaking Moment Minimization of Fully Force-balanced Planar Linkages by Displacing Several Counterweights 60
3.2.1 Minimization of the Shaking Moment by Parallel Displacements of Counterweights Mounted on the Frame .. 60
3.2.2 Example: Balancing of a Six-Bar Linkage 62
3.2.3 Numerical Example 66

3.3 Shaking Moment Minimization of Fully Force-balanced Spatial Linkages 67
3.3.1 Complete Shaking Force and Partial Shaking Moment Balancing of Spatial Linkages 67
3.3.2 Numerical Example and Comparative Analysis 70
3.4 An Approximate Method of Calculating a Counterweight for the Optimum Shaking Force and Shaking Moment Balancing of Linkages 72
3.4.1 Shaking Force Balancing 72
3.4.2 Shaking Moment Balancing 73
3.4.3 Numerical Example 74

4 Complete Shaking Force and Shaking Moment Balancing of Linkages 77
4.1 Complete Shaking Force and Shaking Moment Balancing of In-Line Four-Bar Linkages by Adding a Class-Two *RRR* or *RRP* Assur Group 78
4.1.1 Complete Shaking Force and Shaking Moment Balancing by Adding a Class-Two *RRR* Assur Group 78
4.1.2 Complete Shaking Force and Shaking Moment Balancing by Adding a Class-Two *RRP* Assur Group 84
4.1.3 Illustrative Examples and Numerical Simulations 87
4.2 Complete Shaking Force and Shaking Moment Balancing of Planar Linkages by Adding the Articulated Dyads 90
4.2.1 Complete Shaking Force and Shaking Moment Balancing of Sub-linkages 90
4.2.2 Application of the Methods for Complete Shaking Force and Shaking Moment Balancing of Multilink Mechanisms 100
4.3 Complete Shaking Force and Shaking Moment Balancing of *RSS'R* Spatial Linkages 101
4.3.1 Statement of the Problem 101
4.3.2 Coupler Shape Design 102
4.3.3 Numerical Example 105
4.3.4 Input Torque of the Balanced Linkage 106
4.4 Design of Self-balanced Mechanical Systems 111
4.4.1 Shaking Force Balancing 111
4.4.2 Shaking Moment Balancing 113
4.4.3 Numerical Example and Simulation Results 114

5 Balancing of Slider-Crank Mechanisms 117
5.1 Generalized Lanchester Balancer 117
5.1.1 Shaking Force Balancing of Off-set Crank-Slider mechanism 117
5.1.2 Numerical Simulations 120

5.2 Balancing via the Properties of the Watt Gear-Slider Mechanism ... 121
5.2.1 Watt Gear-Slider Mechanism ... 121
5.2.2 Shaking Force and Shaking Moment of the Slider-Crank Mechanism ... 122
5.2.3 Shaking Force and Shaking Moment Balancing ... 124
5.2.4 Numerical Example ... 126
5.3 Shaking Moment Cancellation of Self-balanced Slider-Crank Mechanical Systems by Means of Optimum Mass Redistribution ... 129
5.3.1 Shaking Force and Shaking Moment Balancing ... 129
5.3.2 Numerical Example ... 131
5.4 Simultaneous Inertia Force/Moment Balancing and Torque Compensation of Slider-Crank Mechanisms ... 133
5.4.1 Design of the Inertia Force/Moment Balanced and Torque Compensated Slider-Crank Mechanism ... 133
5.4.2 Illustrative Example ... 136
5.5 Shaking Force and Shaking Moment Balancing of Slider-Crank Mechanisms via Optimal Generation of the Input Crank Rotation ... 139
5.5.1 Problem Statement ... 139
5.5.2 Shaking Force and Shaking Moment Minimization ... 140
5.5.3 Illustrative Example ... 142

Part III Balancing of Robot Manipulators

6 Balancing of Manipulators by Using the Copying Properties of Pantograph Mechanisms ... 147
6.1 Design of Balancing Mechanisms for Spatial Parallel Manipulators: Application to the Delta Robot ... 147
6.1.1 Description of the Balancing Mechanism ... 148
6.1.2 Minimization of the Torque by a Constant Force Applied to the Robot Platform ... 151
6.1.3 Minimization of the Input Torques by a Variable Force Applied to the Platform of the Robot ... 157
6.1.4 Prototype and Experimental Validation ... 159
6.2 Design of Self-Balanced Parallel Manipulators: PAMINSA with 4-*dof* ... 163
6.2.1 A New Concept for the Design of Partially Decoupled Parallel Manipulators ... 164
6.2.2 Static Analysis of the PAMINSA with 4-*dof* ... 175
6.2.3 Prototype and Experimental Validations ... 179
6.3 Design and Balancing of Hand-operated Manipulators ... 182
6.3.1 Methodology ... 184
6.3.2 Application ... 185

7 Shaking Force and Shaking Moment Balancing of Robot Manipulators 189
7.1 Complete Shaking Force and Shaking Moment Balancing of 3-*dof* 3-*RRR* Parallel Manipulators 190
7.1.1 3-*dof* 3-*RRR* Planar Parallel Manipulator and Dynamic Model with Concentrated Masses 190
7.1.2 Balancing of Legs 191
7.1.3 Balancing of the 3-*RRR* Robot by Using an Inertia Flywheel 196
7.2 Complete Shaking Force and Shaking Moment Balancing of Planar Parallel Manipulators with Prismatic Pairs 199
7.2.1 Complete Shaking Moment and Shaking Force Balancing by Adding an Idler Loop Between the Base and the Platform 199
7.2.2 Complete Shaking Force and Shaking Moment Balancing Using Scott-Russell Mechanism 203
7.3 Shaking Force Minimization of High-speed Robots via Centre of Mass Acceleration Control 212
7.3.1 Minimization of the Shaking Forces via an Optimal Motion Planning of the Total Mass Centre of Moving Links 212
7.3.2 Illustrative Examples 215
7.4 Balancing of Robot Manipulators via Optimal Motion Control 224
7.4.1 Dynamic Balancing of the SCARA Robot 224
7.4.2 Dynamic Balancing of a Position/Orientation Decoupled PAMINSA Robot 230

8 Gravitational Force Balancing of Robotic Systems 241
8.1 Balancing of Pantograph Mechanisms 241
8.2 Optimal Balancing of the Parallel Robot for Medical 3D-ultrasound Imagining 243
8.2.1 Complete Static Balancing 244
8.2.2 Input Torques 245
8.2.3 Minimization of the Root-mean-square Values of the Input Torques 247
8.2.4 Results 251
8.3 Improvement of Balancing Accuracy of Robot-manipulators Taking into Account the Spring Mass 252
8.3.1 Improvement of Balancing Accuracy by Taking into Account the Spring Mass 252
8.3.2 Numerical Examples and Error Analysis 257
8.3.3 Application to the Balancing of Leg Orthosis for Rehabilitation Devices 260

8.4 Optimal Balancing of Serial Manipulators with Decoupled Dynamics . . . 262
8.4.1 Complexity and the Nonlinearity of Robot Arm Dynamics: Basic Notions . . . 262
8.4.2 Design of Decoupled 2-*dof* Planar Serial Manipulator . . . 264
8.4.3 Design of Decoupled 3-*dof* Spatial Serial Manipulator . . . 266
8.4.4 Illustrative Examples . . . 268

References . . . 271

List of Symbols and Abbreviations

In the whole book, vectors are represented by bold lowercase symbols and matrices by bold uppercase symbols, except for greek symbols.

List of Symbols

$\alpha, \beta, \gamma, \phi, \varphi, \theta, \psi$	angles used for geometry description of a mechanism
$\mathbf{f}$	a vector of force
$\mathbf{f}^{sh}$	a vector of shaking force
F	a force
F^{sh}	a shaking force
$\mathbf{h}$	an angular momentum
H_x, H_y, H_z	the components of the angular momentum $\mathbf{h}$ around $\mathbf{x}$, $\mathbf{y}$ and $\mathbf{z}$ axes, respectively
$\mathbf{I}_j$	the inertia matrix for body j, expressed at the *com* in the local frame attached to this body
I_{cr_j}	the axial moment of inertia of the counter-rotation j
I_{S_j}	the axial moment of inertia of the link j expressed at the *com* when link j is considered to have a planar motion
$I_{xx}^{(j)}, I_{yy}^{(j)}, I_{zz}^{(j)}$	the axial moments of inertia around $\mathbf{x}$, $\mathbf{y}$ and $\mathbf{z}$ axes, respectively, for body j, expressed at the *com* in the local frame attached to this body
$I_{xy}^{(j)}, I_{yz}^{(j)}, I_{yz}^{(j)}$	the inertial cross-moments around $\mathbf{z}$, $\mathbf{y}$ and $\mathbf{x}$ axes, respectively, for body j, expressed at the *com* in the local frame attached to this body
L	the Lagrangian of a system
l_{PQ}	the length of the segment PQ
$\mathbf{m}$	a vector of moment
M	a moment
$\mathbf{m}^{sh}$	a vector of shaking moment
M^{sh}	a shaking moment
m_j	the mass of the body j
m_{cw_j}	the mass of the counterweight j
ω_j	the rotational velocity of body j expressed in the base frame

$\omega_{j_x}, \omega_{j_y}, \omega_{j_z}$	the components of the vector ω_j around **x**, **y** and **z** axes, respectively
$\dot{\omega}_j$	the rotational acceleration of body j expressed in the base frame
$\dot{\omega}_{j_x}, \dot{\omega}_{j_y}, \dot{\omega}_{j_z}$	the components of the vector $\dot{\omega}_j$ around **x**, **y** and **z** axes, respectively
$\mathbf{p}$	a linear momentum
P_x, P_y, P_z	the components of the linear momentum **p** along **x**, **y** and **z** axes, respectively
$\mathbf{q}, \dot{\mathbf{q}}, \ddot{\mathbf{q}}$	vectors of actuated coordinates, velocities and accelerations, respectively
$q, \dot{q}, \ddot{q}$	some actuated coordinates, velocities and accelerations, respectively
${}^i\mathbf{R}_j$	the rotation matrix from the frame i to the frame j
$\mathbf{r}_P$	the position of point P expressed in the base frame
$\dot{\mathbf{r}}_P$	the velocity of point P expressed in the base frame
$\ddot{\mathbf{r}}_P$	the acceleration of point P expressed in the base frame
S_j	the *com* of the body j
t	the time variable
T	the kinetic ernergy of a system
$\mathbf{x}$	for a manipulator, the Cartesian position of its end-effector
x_P, y_P, z_P	the components of the vector $\mathbf{r}_P$ along **x**, **y** and **z** axes, respectively
$\dot{x}_P, \dot{y}_P, \dot{z}_P$	the components of the vector $\dot{\mathbf{r}}_P$ along **x**, **y** and **z** axes, respectively
$\ddot{x}_P, \ddot{y}_P, \ddot{z}_P$	the components of the vector $\ddot{\mathbf{r}}_P$ along **x**, **y** and **z** axes, respectively
τ	an input torque
V	the potential ernergy of a system
$\mathbf{w}$	a wrench

List of Abbreviations

com	centre of mass
dof	degree of freedom
HOBM	hand-operated balanced manipulators
PAMINSA	parallel manipulator of the INSA
PKM	parallel kinematic machine
PPM	planar parallel mechanism
rms	root mean square
SPM	spatial parallel mechanism

Part I
Introduction to Balancing

Chapter 1
Introduction

It is known that fast-moving machinery with rotating and reciprocating masses is a significant source of vibration excitation. The high-speed linkages can generate significant fluctuating forces with even small amounts of unbalance. In general, two types of forces must be considered: the externally applied forces and the inertial forces. Inertial forces arise when links of a mechanism are subjected to large accelerations. The inertial force system acting on a given link can be represented as an inertia force acting on a line through the center of mass and an inertia torque about the center of mass. The determination of the inertial forces and torques is well known and it has been disclosed in various hand books. With regard to the external forces, which are associated with the useful function that the mechanism is to perform, these are often smaller than inertia forces with a much lower variation. On the other hand, when formulating balancing conditions of a mechanism, it is necessary to recognize that, in many cases, external active forces applied to mechanism links constitute internal forces with respect to the mechanism as a whole. Thus, if all external active forces applied to the links of a mechanism are internal forces for the mechanism as whole, then the balance of the mechanism will be ensured under the fulfillment of inertia forces and inertia torque cancellation. Therefore, the balancing of shaking force and shaking moment due to the inertial forces of links acquires a specific importance. The quality of balancing of the moving masses has the influence not only on the level of vibrations but also on the resource, reliability and accuracy of mechanisms. Besides the mentioned negative effects, vibrations bring to the environments pollution and the loss of energy, and can also provoke various health issues. Consequently, the quality improvement of the mass balancing has not only technical, technological and economical aspects but also social.

A new field for balancing methods applications is the design of mechanical systems for fast manipulation, which is a typical problem in advanced robotics. Here also we have similar problems relating to the cancelation or reduction of inertia forces. However, the mechanical systems with multi degrees of freedom lead to new solutions, such as the shaking force and shaking moment reduction by optimal motions of links, by adding flywheels with prescribed motions, or with the design of new self-balanced manipulators.

V. Arakelian, S. Briot, *Balancing of Linkages and Robot Manipulators*,
Mechanisms and Machine Science 27, DOI 10.1007/978-3-319-12490-2_1

It should also be noted that many robotic systems are operated at low speed to ensure the different tasks. In this situation, gravitational torques generated by the masses of links are often much greater than dynamic torques. Thus, gravity compensation is beneficial where a robotic system can be operated with relatively small actuators. Therefore, the development of gravitational force balancing methods is still current.

In this book, the advanced balancing methods for planar and spatial linkages, hand operated and automatic robot manipulators are presented. It is organized into three main parts and eight chapters. The main parts are the introduction to balancing, the balancing of linkages and the balancing of robot manipulators. The suggested balancing methods are illustrated by numerous examples.

Chapter 2 is devoted to an overview of balancing methods, which is presented in three main parts: shaking force and shaking moment balancing of linkages; shaking force and shaking moment balancing of robots and manipulators, as well as gravity balancing used in robotics. We considered that such participation reflects the particularities of the reviewed balancing methods and their specific characteristics.

It is known that the complete shaking force and shaking moment balancing of linkages can only be reached by a considerably more complicated design of the initial linkage and by an unavoidable increase of the total mass. This is the reason why in most cases, the partial balancing is used in the machinery. The methods of partial balancing are discussed in Chap. 3. However, the complete shaking force and shaking moment balancing methods are often indispensable. In Chap. 4, new methods for full shaking force and shaking moment balancing of linkages are considered. The balancing methods are carried out by adding articulated dyads permitting an optimal redistribution of moving masses, as well as by optimal design providing the conditions for a complete shaking force and shaking moment balancing of linkages with a relatively small increase of the total mass of movable links. It is achieved by mounting the gear inertia counterweights on the base of the mechanism. The balancing of spatial linkages and the design of self-balanced mechanical systems are also studied.

Special attention is given to the balancing of slider-crank mechanisms. The methods of shaking force and shaking moment balancing of axial and off-set slider-crank mechanisms are disclosed in Chap. 5, which completes the second part.

The copying properties of pantograph mechanisms are used in Chap. 5 in order to balance or to design robot manipulators. The proposed auxiliary balancing linkage, which can be added into the base and the platform of the Delta robot, allows a significant unloading of the robot's actuators. The design and properties of the PAMINSA manipulator based on the three legs, which are pantograph linkages, are also considered. The last subchapter deals with the problem of the balancing of hand-operated manipulators of the pantograph types.

Chapter 7 is devoted to the shaking force and shaking moment balancing of robot manipulators. The development of reactionless 3-*RRR* planar parallel manipulators, which apply no reaction forces or moments to the mounting base during motion, is discussed. The total angular momentum of the manipulator is reduced to zero using two approaches: (i) on the basis of counter-rotations and (ii) using an inertia flywheel rotating with a prescribed angular velocity. The complete shaking force and shaking moment balancing of planar parallel manipulators with prismatic pairs is also disclosed. Then, a simple and effective balancing method, which allows the considerable reduction of the shaking force of non-redundant manipulators without adding counterweights, is studied. It is based on the optimal control of the acceleration of the total center of mass (*com*) of moving links. The full shaking force and shaking moment balancing of robots using an optimal motion control is also used in the last subchapter.

The balancing methods of gravitation forces of robot manipulators are given in the last Chapter. The problems relating to the balancing with reduced number of springs, as well as the balancing of mechanical systems by considering the spring mass are discussed.

At the end of this short introduction, we would like to point out that in this book advanced methods of balancing are presented. In order to properly understand the content, the reader should possess a certain level of knowledge in the field of theoretical mechanics and balancing of mechanisms.

Chapter 2
An Overview of Balancing Methods

Abstract The review of state-of-the-art literature including more than 500 references is given in this chapter. The balancing methods illustrated via various kinematic schemes are presented in three main parts: shaking force and shaking moment balancing of linkages; shaking force and shaking moment balancing of robots and manipulators, as well as gravity balancing used in robotics. We consider that such participation reflects the particularities of the reviewed balancing methods and their specific characteristics.

2.1 Shaking Force and Shaking Moment Balancing of Linkages

The balancing of mechanisms is a well-known problem in the field of mechanical engineering because the variable dynamic loads cause noise, wear and fatigue of the machines. The resolution of this problem consists in the balancing of the shaking force and shaking moment, fully or partially, by internal mass redistribution or by adding auxiliary links.

From very ancient times, with building works that were widely carried out, different auxiliary technical means appeared in which various simple mechanisms were used. The practical experience of the creators of such mechanisms showed that in many cases, during the displacement of heavy objects, the necessity arose for compensation of moving masses by additional means. Since for a long time the driving force of such mechanical systems was human physical force, the creation of additional balancing means was considered to be a significant technical problem that would increase the hoisting capacity of mechanisms. At that time, the speeds of the objects to be displaced were very low and the inventors simply confined themselves to balancing gravitational forces of mechanism links. The design methods of such mechanisms were based on intuition and the simplest arithmetical computations. The situation began to change at the beginning of the last century. With the emergence of the first steam machines and, particularly, of internal combustion engines, it became evident that the fast moving elements of machines brought about undesirable effects, such as vibration, noise and rapid wearing. The explosive growth in the production of high speed mechanisms presented scientists with the problem of creating the theoretical principles for the balancing of mechanisms. The problem of balancing

V. Arakelian, S. Briot, *Balancing of Linkages and Robot Manipulators*,
Mechanisms and Machine Science 27, DOI 10.1007/978-3-319-12490-2_2

gravitational forces ceased to be critical and was transformed into the problem of balancing the inertia forces of mechanisms. This problem may be formulated as follows: determination of parameters, redistribution of the rapidly moving mechanism masses that will provide small dynamic loads onto the mechanism foundation. Two main types of balancing have emerged: static—when the shaking force is cancelled, and dynamic—when the shaking force is cancelled together with the shaking moment.

Here, we point out that in the theory of balancing, the term "static balancing" should be understood arbitrarily and has nothing in common with the well-known mechanical phenomenon of "static character" (i.e. when there is no motion). By its character, "balancing of mechanism" is a dynamic phenomenon and any imbalance is the result of an accelerated motion of mechanism links. However, the mode of balancing the shaking force was called "static", as imbalance of shaking force can be detected in static conditions, i.e. imbalance of shaking force in any mechanism can be demonstrated experimentally in the static state, without the links having to be driven, while imbalance of the principal moment of inertia may be revealed during mechanism motion only, i.e. in the dynamic behavior.

The term "static balancing" has almost fallen out of use now in the theory of balancing of mechanisms. Now, the term "shaking force balancing" is well known. The term "static balancing" is most often applied when considering the balancing problems of rotating bodies, for example rotors, turbines, etc.

First, let us consider the methods of shaking force balancing of linkages.

2.1.1 *Shaking Force Balancing of Linkages*

One of the first publications in this field may be considered to be the work of O. Fischer (Fischer 1902) in which a method called the method of "principal vectors" was suggested. The aim of this approach was to study the balancing of the mechanism relative to each link and in the determination of those points on the links relative to which a static balance was reached. These points were called "principal points". Then, from the condition of similarity of the vector loop of the principal points and the structural loop of the mechanism, the necessary conditions of balancing were derived. It was thereby shown that the necessary and sufficient condition for balancing the shaking force is the fixation of the common centre of masses of the moving links of the mechanism. This method was used in the works of V. P. Goryachkin (Goryachkin 1914), Kreutzinger (Kreutzinger 1942), V. A. Yudin (Yudin 1941). At that time, it was of a particular importance as it served to create several auxiliary devices intended for studying the motion of the centres of mechanism masses. This method was also used for determination of the mass centers of mechanisms (Shchepetilnikov 1968), for balancing of mechanisms with unsymmetrical links (Shchepetilnikov 1975) and for shaking moment balancing of three elements in series (van der Wijk 2013; van der Wijk and Herder 2012, 2013).

Another well known method for balancing which was one of the first that was developed, was the "method of static substitution of masses". Its aim was to statically substitute the mass of the coupler by concentrated masses, which are balanced thereafter together with the rotating links. Such an approach allows changing the problem of mechanism balancing into a simpler problem of balancing rotating links. It was used in the works of F. R. Grossley (Grossley 1954), R. L. Maxwell (Maxwell 1960), M. R. Smith and L. Maunder (Smith and Maunder 1967), G. J. Talbourdet and P. R. Shepler (Talbourdet and Shepler 1941).

From the beginning of the 1920s, special attention was paid to balancing of engines (Cormac 1923; Dalby 1923; Delagne 1938; Doucet 1946; Kobayashi 1931; Lanchester 1914; Root 1932) and mechanisms in agricultural machines (Artobolevsky and Edelshtein 1935; Artobolevsky 1938). Engineers successfully used the "Lanchester balancer" (Lanchester 1914). It should be noted here that the principle proposed by Lanchester remains classic and practical even today. In modern cars, to balance the inertia forces in four-stroke engines, opposed balancing shafts are used in four-cylinder in-line engines, these shafts being synchronized with the crankshaft by means of a geared belt drive. These balancing shafts for balancing the second harmonic are designed in the same way as in the "Lanchester balancer". This approach has been investigated in (Chiou and Davies 1994) in order to minimize the shaking moment and in (Arakelian and Makhsudyan 2010) for shaking force minimization in offset slider-crank mechanisms.

Another trend in the balancing theory was developed by means of the "duplicated mechanism" (Arakelian 2006; Artobolevsky 1977; Davies 1968; Kamenski 1968b). The addition of an axially symmetric duplicate mechanism to any given mechanism will make the new combined centre of mass stationary. This approach resulted in the building of self-balancing mechanical systems. The principle of construction of self-balanced mechanical systems is to have two identical mechanisms executing similar but opposite movements. The opposite motion for shaking force balancing has also been used in (Berkof 1979a; Doronin and Pospelov 1991; Dresig 2001; Dresig and Holzweißig 2004; Filonov and Petrikovetz 1987; Frolov 1987; Turbin et al. 1978; van der Wijk and Herder 2010b).

The known kinematic diagrams of self-balanced systems are shown in Fig. 2.1. They can be arranged into three groups: (a) the systems built by adding an axially symmetric duplicate mechanism with separated input cranks (a1–a3); (b) the systems built by adding an axially symmetric duplicate mechanism with common input crank (b1–b6); (c) the systems built via an asymmetric model of duplicate mechanisms (c1–c3). Such mechanical systems were used successfully in agricultural machines, mills and in various automatic machines.

V. A. Kamenski (Kamenski 1968a) first used the cam mechanism for the balancing of linkages. In his work, the variation of inertia forces was performed by means of a cam bearing a counterweight and it was shown how cam-driven masses may be used to keep the total centre of mass of a mechanism stationary. This approach was further developed in (Arakelian and Briot 2010), in which a design concept permitting the simultaneous shaking force/shaking moment balancing and torque compensation in slider-crank mechanisms has been proposed. First, the shaking force and shaking

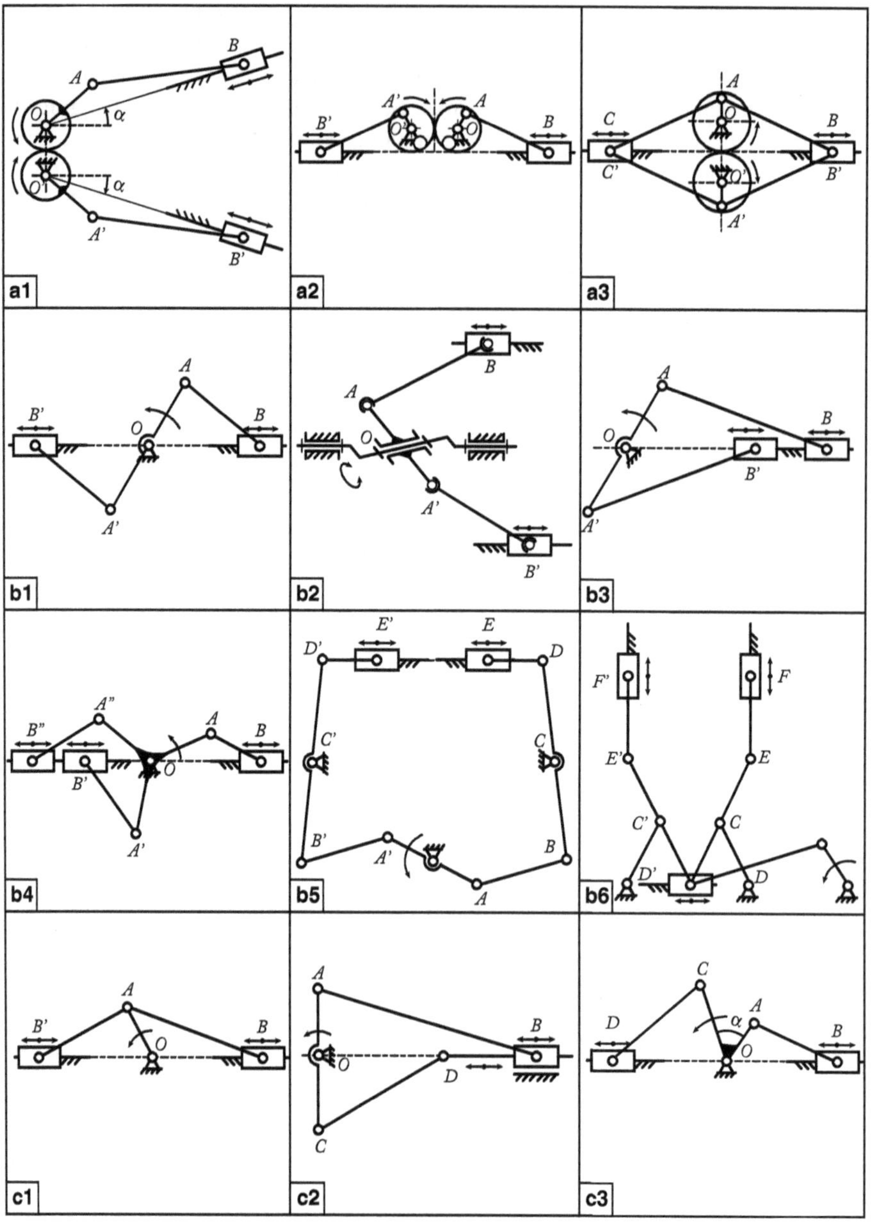

Fig. 2.1 Kinematic diagrams of self-balanced systems

moment have been cancelled via a cam mechanism carrying a counterweight. Then, the spring designed for maintaining contact in this balancing cam mechanism is used for torque minimization. The designs of cam mechanisms for shaking force minimization in press machines have been investigated in (Chiou and Davies 1997).

Among several works, the study of H. Hilpert (Hilpert 1968) in which a pantograph mechanism is used for the displacement of the counterweight may also be distinguished. This approach was further developed in works (Arakelian 1993, 1998b; Arakelian and Smith 2005c) in which the duplicating properties of the pantograph are used by connecting to the balancing mechanism a two-link group forming a parallelogram pantograph with the initial links. For example, for the balancing of a slider-crank mechanism, the additional two-link group forms a pantograph with the crank and coupler of the initial mechanism. The formed pantograph system executes a rectilinear translation that is opposite to the movement of the slider. Thus a new solution of a self-balanced mechanical system without any additional slider (prismatic) pair is proposed. The pantograph system may be formed by gears or by toothed-belt transmission carrying a counterweight. Such an approach permits the balancing of mechanisms with a smaller increase of link mass compared to earlier methods.

In the 1940's, partial balancing methods based on function approximation were successfully developed. Such a solution was proposed by Y. L. Gheronimus (Gheronimus 1968a, b). In these works, the balancing conditions are formulated by the minimization of root-mean-square (*rms*) or maximal values (Chebichev approach) of shaking force and they are called "best uniform balancing" of mechanisms. This approach has been used in (Arakelian 1995) and (Arakelian 2004a). A similar study has been developed in (Han 1967).

The use of the slider-crank mechanism in internal combustion engines brought about the rapid development of methods based on harmonic analysis. The reduction of inertia effects is primarily accomplished by the balancing of certain harmonics of the forces and moments. Unbalanced forces and moments are divided into Fourier series (or Gaussian least-square formulation) and then studied by parts. This solution found a large application as it may be realized by means of rotating balancing elements connected to the crank.

The force harmonics of slider-crank mechanisms of various types were examined and a large quantity of works concerning the problem of balancing of engines and linkages was published. We would like to note certain references (Emöd 1967; Gappoev 1979; Gappoev and Tabouev 1980; Gappoev and Salamonov 1983; Innocenti 2007; Semenov 1968b; Stevensen 1973 ; Tsai and Maki 1989; Urba 1978, 1980). The properties of the Watt-gear slider-crank mechanism which are similar to harmonics has also been used in order to solve the balancing problem (Arakelian and Smith 2005a).

In (Tsai 1984), it was shown that by a proper arrangement of two Oldham couplings, a balancer can be obtained for the elimination of second-harmonic shaking forces or second-harmonic shaking moments or a combination of both shaking forces and moments. The advantage of this balancer is that it runs at the primary speed of the machine to be balanced whereas the Lanchester-type balancer must run at twice the primary speed to achieve the same balancing effect. The harmonic balancing has also been applied in (Davies and Niu 1994) in order to find that there are boundaries to the regions where additional shafts can be located.

In 1968, R. S. Berkof and G. G. Lowen (Berkof and Lowen 1969) proposed a new solution for shaking force balancing of mechanisms that is called the method of "linearly independent vectors". In this method, the vector equation describing the position of the centre of total mass of the mechanism is treated in conjunction with the closed equation of its kinematic chain. The result is an equation of static moments of moving link masses containing single linearly independent vectors. They follow the conditions for balancing the mechanism by reducing the coefficients to zero which are time-dependent. This method found further development and applications in works (Bagci 1979; Balasubramanian and Bagci 1978; Berkof et al. 1977; Elliot and Tesar 1982; Smith 1975; Tepper and Lowen 1972a; Walker and Oldham 1978; Yao and Smith 1993).

Particularly, in (Smith 1975), an interactive computer program is developed which allows the design of fully force balanced four-bar linkages by the method of "linearly independent vectors". The increase in the shaking moment of these linkages is controlled by designing the counterweight such that the total moment of inertia of the associated links is made as small as possible.

2.1.2 Shaking Moment Balancing of Linkages

In the 1970's, great attention was given to the development of dynamic balancing methods. The principal schemes for complete shaking force and shaking moment balancing of four-bar linkages are presented in Fig. 2.2. In Berkof's approach (Berkof 1973; Fig. 2.2a), the mass of the connecting coupler 3 is substituted dynamically by concentrated masses located at joints B and C. Thus, the dynamic model of the coupler represents a weightless link with two concentrated masses. This allows for the transformation of the problem of four-bar linkage dynamic balancing (shaking force and shaking moment) into a problem of balancing rotating links carrying concentrated masses.

The parallelogram structure (Fig. 2.2b) has also been applied for complete shaking force and shaking moment balancing of four-bar linkages (Arakelian et al. 1992; Bagci 1982).

Ye and Smith (Ye and Smith 1991), Arakelian and Smith (Arakelian and Smith 1999), Gao (Gao 1989, 1990, 1991) and Berestov (Berestov 1975, 1977a; Fig. 2.2c, d) have proposed methods for complete shaking force and shaking moment balancing by counterweights with planetary gear trains. Esat and Bahai (Esat and Bahai 1999; Fig. 2.2e) used a toothed-belt transmission to rotate counterweights 5 and 6 intended for shaking force balancing which also allowed shaking moment balancing.

Another approach applied by Kochev (Kochev 1992a; Fig. 2.2f) was to balance the shaking moment (in the force balanced mechanism) by a prescribed input speed

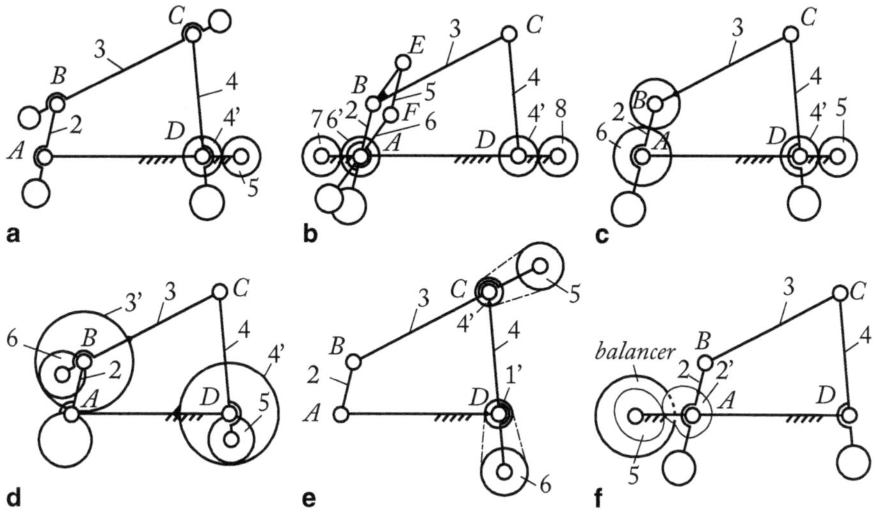

Fig. 2.2 Principal schemes for complete shaking force and shaking moment balancing of four-bar linkages with constant input speed

fluctuation achieved by non-circular gears or by a microprocessor speed-controlled motor.

In practice, all the known methods for complete shaking force and shaking moment balancing of four-bar linkages face serious technical problems. The schemes presented in Fig. 2.2a–d have a common disadvantage which is the connection of gears to the rocker. The resulting oscillations of the rocker create considerable noise unless expensive anti-backlash gears are used. Thus, in high-speed systems it is inadvisable to use gears connected to oscillating links. In the solution presented in Fig. 2.2e, this problem is solved partially by the use of toothed-belt transmission but the oscillations still cause serious technical problems. The method of non-circular gears balancing (Fig. 2.2f) always presents great engineering difficulty requiring the development of a special type of driver-generators.

Moore, Schicho and Gosselin have proposed all possible sets of design parameters for which a planar four-bar linkage is dynamically balanced without counter-rotations (Moore et al. 2009). This approach has been used in (Briot and Arakelian 2012) for the complete shaking force and shaking moment balancing of any four-bar linkage.

Figure 2.3 shows the schemes of complete shaking force and shaking moment balancing of four-bar linkages via copying properties of pantograph systems formed by gears (Arakelian and Dahan 2002; Arakelian and Smith 2005c). They will be further detailed in Chaps. 4 and 5.

Dresig and Nguyen proposed the shaking force and shaking moment balancing of mechanisms using a single rigid body called "balancing body" (Dresig and Nguyen 2011). By motion control of the balancing body, any resultant inertia forces and moments of several mechanisms can be fully compensated. The desired motion of

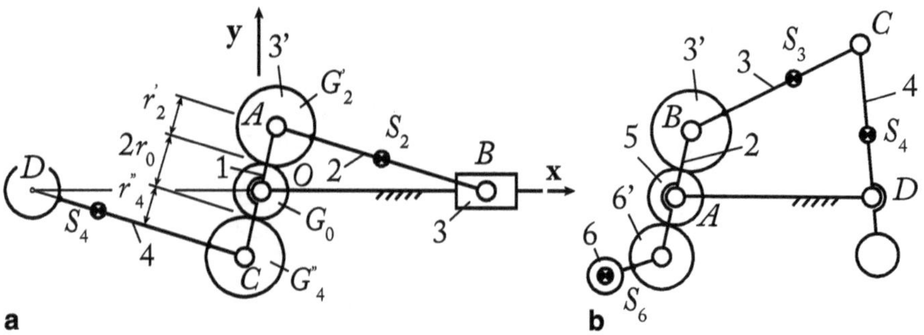

Fig. 2.3 Complete shaking force and shaking moment balancing of four-bar linkages based on the copying properties of pantograph systems

the balancing body is calculated in order that the sum of inertia forces and moments of the mechanisms and the balancing body will be zero.

However, it is evident that the complete dynamic balancing of mechanisms can only be reached by a considerably complicated design of the initial mechanism and by an unavoidable increase of the total mass. This is the reason why methods of partial dynamic balancing of mechanisms underwent a further development.

In the works (Berkof and Lowen 1971; Carson and Stephens 1978; Freudenstein 1973; Jacobi 1969; Lowen and Berkof 1970, 1971; Sconfeld 1974; Tricamo and Lowen 1983a, b), different modes of minimization of the shaking moment are suggested and are of interest. F. Freudenstein, J. P. Macey, E. R. Maki (Freudenstein et al. 1981) derive the equations for minimizing any order of combined pitching and yawing moments by counterweighting the driveshaft or a shaft geared to the driveshaft. The equations are given directly as a function of the harmonic coefficients of pitch and yaw and apply to any plane machine configuration. J. L. Wiederrich and B. Roth (Wiederrich and Roth 1976) proposed simple and general conditions for determination of the inertial properties of a four-bar linkage that allow partial momentum balancing. Dresig et al. (Dresig et al. 1994; Dresig and Schönfeld 1971, 1976a, b; Dresig and Jacobi 1974) examined the optimum balancing conditions for various structural forms of planar six and eight-bar linkages. A least-square theory for the optimization of the shaking moment of fully force-balanced inline four-bar linkages, running at constant input angular velocity, is developed in the studies of J. L. Elliot and D. Tesar (Elliot and Tesar 1977) and (Haines 1981).

V. A. Shchepetilnikov (Shchepetilnikov 1968, 1982) suggested the minimization of the unbalance of shaking moment by transferring the rotation axis of the counterweight mounted on the input crank. In his works, the first harmonic of the shaking moment is eliminated by attaching the required input link counterweight, not to the input shaft itself, but to a suitable offset one which rotates with the same angular velocity. This approach is original in that, while maintaining the shaking force balance of the mechanism, it is possible to create an additional balancing moment, reducing thereby the shaking moment. This approach has been developed in (Arakelian and Dahan 2000 a, b, 2001a, b; Arakelian and Smith 2004).

The particularities of the studies (Tepper and Lowen 1973; Urba 1981) resides in that a method is suggested permitting the comparison of the efficiency of balancing methods by the criterion of the minimum value of the shaking moment. F. R. Tepper and G. G. Lowen (Tepper and Lowen 1973) showed that in shaking force balanced mechanisms, the root-mean-square value of the principal inertia moment is constant relative to some ellipses located in the mechanism plane. By decreasing the dimensions of the ellipses, the root-mean-square value decreases and reaches a global minimum in the centre of this family of ellipses. This theory of isomomental ellipses was developed by A. L. Urba (Urba 1981) for the case of three-dimensional mechanisms. It was shown that the ellipses are transformed into ellipsoids and the properties mentioned are maintained.

Optimization algorithms based on programming are also widely used in balancing theory. The following studies are of interest: the studies of J. P. Sadler et al. (Conte et al. 1975; Porter and Sandler 1973; Sadler and Mayne 1973; Sadler 1975), H. Dresig and S. Schönfeld (Dresig and Schönfeld 1976 a, b), P. Jacobi (Jacobi 1972), J. M. O'Leary and G. W. Gatecliff (O'Leary and Gatecliff 1989), N. M. Qi and E. Pennestri (Qi and Pennestri 1991), M. J. Walker and R. S. Haines (Walker and Haines 1982a), as well as the studies (Demeulenaere et al. 2004b; Lee and Cheng 1984; Smith and Walker 1976; Smith et al. 1977a, b; Tepper and Lowen 1972b; Yan and Soong 2001). Among the recent studies based on various optimization techniques, it should be noted (Chaudhary and Saha 2007, 2008a, b, 2009; Chiou et al. 1998; Demeulenaere 2004; Demeulenaere et al. 2004a, b, 2006, 2008; Emdadi et al. 2013; Erkaya 2013; Ettefagh et al. 2011; Farmani and Jaamiolahmadi 2009; Ilia and Sinatra 2009; Li and Tso 2006; Verschuure et al. 2007, 2008a; Yan and Soong 2001).

M. A. K. Zobairi, S. S. Rao and B. Sahay (Rao 1977; Zobairi et al. 1986a, b) studied the problems of balancing taking into consideration the elasticity of links. The acceleration field resulting from the vibration of the links develops additional inertia forces called kineto-elastodynamic inertia forces. These works take into account the contribution of the kineto-elastodynamic inertia forces towards the shaking force and shaking moment while balancing planar mechanisms. Combining kinematic design and dynamic stress considerations, an optimal kinematic design of the mechanism satisfying the given aim and optimal cross-sectional areas of the links were determined such that the shaking force transmitted to the foundations due to the combined effect of rigid-body inertia forces and kineto-elastodynamic inertia forces is a minimum. The effect of the inclusion of kineto-elastodynamic inertia forces has been demonstrated by taking an example problem in which the maximum shaking force produced during the complete cycle of motion of mechanism has been minimized using nonlinear programming techniques.

The elastic behavior of a counterweighted four-bar linkage was first investigated theoretically and experimentally by Jandrasits and Lowen (Jandrasits and Lowen 1979a, b). The effect of link shape on the dynamic response of flexible mechanisms has also been studied (Yu and Smith 1995). The shaking force and shaking moment balancing of flexible mechanisms using redundant drives has been investigated theoretically and experimentally in (Yu and Jiang 2007). Experimental study on the elastodynamic behavior of the unbalanced and the counterweighted four-bar mechanisms has been considered in (Raghu and Balasubramonian 1990). The dynamic operation of a four-bar linkage, taking into account elastodynamic aspects, has also been studied in (Martini et al. 2013).

A novel method has been developed in (Lin 2000; Yu and Lin 2003) for the shaking force and shaking moment balancing of flexible mechanisms. The theoretical analysis and numerical results of a flexible four-bar linkage illustrated that the redundant actuators are useful for the optimum balancing of flexible linkages.

A. P. Bessonov (Bessonov 1967, 1968), for the first time, formulated and solved the balancing problem of mechanisms with variable masses of links. To obtain the optimal balancing of such mechanisms, he successfully applied the root-mean-square and mini-max methods of minimization.

The studies (Jacobi and Rose 1972; Offt 1974; Walker and Haines 1982a) are noticeable from the point of view of the experimental study of balancing of mechanisms. P. Jacobi and W. Rose (Jacobi and Rose 1972) conducted an experimental investigation of a theoretically fully force-balanced four-bar linkage. This study shows that the agreement between experimental and computed results was generally satisfactory. F. R. Tricamo and G. G. Lowen, in (Tricamo and Lowen 1981a, b) described a new concept for force balancing machines for four-bar linkages. On the base of the theoretical study, they proposed a device for the experimental application of this technique to a four-bar linkage. For the examined four-bar linkage the reduction of the shaking moment was more than 50 %.

Interesting results are also available in the field of balancing of spatial mechanisms. One of the first, M. V. Semenov (Semenov 1968a) was able to show that the kth harmonic of the shaking force for any spatial mechanism may be balanced by three counterweights disposed in mutually perpendicular planes. In (Gill and Freudenstein 1983a, b), computer-aided design procedures have been developed for the optimum mass distribution of the links of high-speed spherical four-bar linkages. R. E. Kaufman and G. N. Sandor (Kaufman and Sandor 1971) developed the method of linearly independent vectors for spatial mechanisms. The general approach is illustrated by the balancing of *RSSR* and *RSSP* spatial mechanisms. T. T. Gappoev developed the method (Gappoev and Tabouev 1979; Gappoev and Salamonov 1983) generalizing the Shchepetilnikov approach (Shchepetilnikov 1968, 1982) for the spatial version. He eliminated the first harmonic of the shaking moment by attaching the required input link counterweight, not to the input shaft proper, but to a suitably offset one which rotates with the same angular velocity. Balancing of the Bennett mechanism and *RCCC* spatial mechanism are studied in the works of N. Chen and Q. Zang (Chen and Zhang 1983; Chen 1984a, b). Y. Q. Yu (Yu 1987a, b, 1988) develops the method for balancing mechanisms by connecting additional dyads to the initial mechanism. The "method of static substitution of masses" has been used successfully in (Arakelian 2007) for complete shaking force and shaking moment balancing of *RSS'R* spatial linkage.

The studies of Wawrzecki (Wawrzecki 1998, 1999) relate to spatial mechanisms moving the needle of sewing machines. In the works of I. D. Belonovskaya, F. M. Dimentberg, L. B. Maysuk (Belonovskaya et al. 1987) the principles for building self-balanced spatial mechanisms are proposed. I. S. Chiou, M.-G. Shieh and R. J. Tsai examined the balancing of spatial mechanisms by means of two (Chiou et al. 1997) or three rotating counterweights (Chiou and Tsai 1995). The shaking force is eliminated by equal and opposite forces exerted by counterweights mounted on the shafts. The shaking moment is eliminated even though the location of every third shaft is chosen arbitrarily, the locations of the other two shafts are then determined. Demonstration of such a balancing is illustrated by a seven-link $7R$ spatial linkage. Shaking force balancing of spherical $4R$ linkages was discussed in (Moore et al. 2010; Moore 2009). It was shown that in spherical four-bar linkages, the shaking force and shaking moment cannot be completely balanced without introducing additional mechanical components, such as counter-rotations or added loops. The shaking force balancing of Bennett linkage has also been studied by Moore and Schicho (Moore and Schicho 2009). Genetic algorithm for shaking force and shaking moment balancing of spatial *RSSR* mechanism has also been proposed (Feng et al. 2000).

The "Finite position method" that can solve the shaking force balancing for spatial mechanism as well as planar ones has been proposed in (Zhu et al. 2009). The proposed method uses only discrete motion positions. As mentioned in the study, the linear equations of force balancing can be automatically generated, and therefore it is especially suitable for spatial mechanisms.

Nguyen (Nguyen and Nguyen 2007) proposed a method to algebraically derive the balancing conditions for shaking force and shaking moment of spatial one-degree-of-freedom mechanisms, which provides a helpful tool to obtain the exact balancing conditions of spatial mechanisms.

Zhang and Chen (Zhang 1994; Zhang and Chen 1995) introduced the theory of mechanical vibration to the balance of the shaking moment of linkages. They consider that the frame is a three-degree-of-freedom vibration system. The excitation of this system is caused by the components of shaking force and shaking moment. On the basis of such a model, they propose a new compromise for the optimum balance method. The calculated results show that the new method is very efficient in reducing the system vibrational response. Similar studies have been developed in (Ishida and Matsuda 1977, 1979). In (Xi and Sinatra 1997), the authors investigated the effect of dynamic balancing on four-bar linkage vibrations. Results of the simulation have indicated that the dominant factor on the vibrations of a dynamically balanced four-bar linkage may be the moment of inertia of the counterweight attached to the input link.

A six-degrees-of-freedom vibration model for any mechanical vibration system, in which a linkage is mounted, and the relationship between the response and the counterweight mass parameters of the linkage has been studied in (Zhang et al. 2000). It was shown that by the proposed technique, the vibration response can be reduced.

Verschuure, Demeulenaere, Swevers and De Schutter (Verschuure et al. 2008b) have proposed two optimization criteria: the frame vibration itself and the dynamic force transmitted to the machine floor. Numerical comparison with known algorithms for nonlinear optimization shows that the mentioned approach results in a substantial reduction of the required computational time.

As was shown in (Zhe and Shixian 1992), the shaking characteristics of mechanisms with clearances differ greatly from those of mechanisms without clearances. The system performance will be seriously deteriorated if the clearance effects are not controlled properly. The methods presented in the mentioned paper can effectively prevent the occurrence of contact loss and impacts between pairing elements and can reduce the shaking force and the shaking moment. An optimal design formulation is developed in (Park and Kwak 1987) to reduce undesirable dynamic effects due to clearance in a joint. The design variables in this study are the magnitude and the location of the added counterweight. Numerical examples for an offset slider-crank mechanism were considered. The analysis result given in (Feng et al. 2002) showed that the high frequency vibrations occur when the joint forces change sharply. By optimizing the mass distribution of the moving links, it was shown in the mentioned study that the amplitude and the direction of the joint forces can be controlled. As a result, the vibration can be reduced effectively.

In (Zhe 1998), the sensitivity formulae of the shaking force and shaking moment of general planar articulating mechanisms to the link mass parameters has been derived. A planar four-bar mechanism has been taken as an example to illustrate the balancing process, the sensitivity analysis and the robust balancing method. The numerical results demonstrated the necessity of the sensitivity analysis to the balancing results and the effectiveness of the proposed robust balancing method.

I. S. Kochev has worked successfully on the development of balancing theory (Kochev 1987, 1988, 1989, 1990a, b, 1991a, b, 1992a, b, c, d, e). In his work, the theory of isomomental ellipses, the linearly independent complex method, and studies concerning mechanical systems with double cranks having symmetrical properties are generalized. He developed a method of active shaking moment balancing of linkages realized by a prescribed input speed fluctuation of the balancers and the design of self-balanced systems based on the optimal assemblies and the angular positions of their sub-linkages. The active balancing has also been used in studies (Angeles et al. 1992; van der Wijk and Herder 2010a; Zhang et al. 2007).

The two surveys by G. G. Lowen and R. S. Berkof (Lowen and Berkof 1968) in 1968 and by G. G. Lowen, F. R. Tepper, R. S. Berkof (Lowen et al. 1983) in 1983 are of great interest. The overviews (Arakelian et al. 2000; Arakelian and Smith 2005b, c) should also be noted. These works present a very valuable discussion and systematization of the methods of shaking force and shaking moment balancing. The present authors believe that it is desirable to present a new survey with a complete analysis of the balancing literature taking into account the large number of recent publications.

Despite its long history, mechanism balancing theory continues to be developed and new approaches and solutions are constantly being reported. Methods which take into account physical aspects such as the elasticity of links and the clearance and impacts in the joints of the mechanisms, using finite element methods are of increasing relevance. The synchronous development of high-speed computing methods creates a basis for a more realistic representation of those physical effects which occur in high-speed mechanical systems.

2.2 Shaking Force and Shaking Moment Balancing of Robots and Manipulators

It is known that a mechanical system with unbalance shaking force/moment transmits substantial vibration to the frame. Thus, a primary objective of the balancing is to cancel or reduce the variable dynamic loads transmitted to the frame and surrounding structures. The reduction of vibrations leads to the increased accuracy of manipulators (Foucault and Gosselin 2002), which is one of the positive consequences of the balancing. As was mentioned in (van der Wijk et al. 2012), balancing brings other advantages such as a reduced cycle time (Raaijmakers 2007), reduced noise, wear and fatigue (Lowen and Berkof 1968), as well as improved ergonomics (Ishida and Matsuda 1979).

Different approaches and solutions devoted to the shaking force and shaking moment balancing have been developed and documented for one-degree-of-freedom mechanisms (Lowen et al. 1983; Arakelian et al. 2000; Arakelian and Smith 2005c). A new field for their applications is the design of mechanical systems for fast manipulation, which is a typical problem in advanced robotics.

The balancing of manipulators is generally carried out in two steps: (i) the cancellation (or reduction) of the shaking force and (ii) the cancellation (or reduction) of the shaking moment.

First, let us consider the methods for the shaking force balancing.

2.2.1 Shaking Force Balancing

The review of methods devoted to the shaking force balancing of manipulators has shown that three principal approaches can be distinguished.

2.2.1.1 Shaking Force Balancing by Adding Counterweights in Order to Keep the Total Centre of Mass of Moving Links Stationary

In the case of open-chain manipulators, the balancing of shaking force starts from the final link and a counterweight is added in order to locate the centre of mass (*com*) position of this link on the preceeding joint axis. Such a balancing process must be repeated sequentially until the *com* of the whole chain is fixed on the base pivot (Filaretov and Vukobratovic 1993; Agrawal and Fattah 2004b).

It is obvious that the adding of the supplementary mass due to the counterweights is not desirable because it leads to the increase of the total mass, of the overall size of the robot-manipulator and of the efforts in joints. That is why in many designs of industrial robots, the masses of the motors are often used as counterweights (Bayer and Merk 2011).

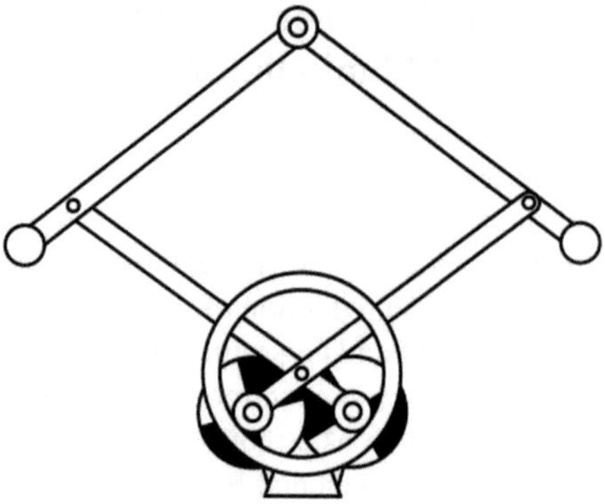

Fig. 2.4 Counterweight balancing of a planar parallel manipulator. (Menschaar et al. 2006)

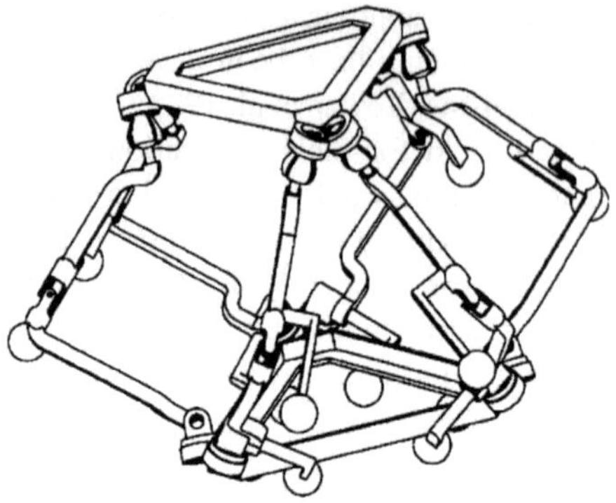

Fig. 2.5 Counterweight balancing of a spatial parallel manipulator. (Gosselin 2008)

With regard to the parallel manipulators, the approach is the same: adding counterweights to keep the total *com* of moving links stationary. However, the approach is simpler to carry out in planar parallel manipulators (Fig. 2.4) than in spatial parallel manipulators (Fig. 2.5).

2.2.1.2 Shaking Force Balancing by Adding Auxiliary Structures

Different approaches have been developed in order to keep the total centre of mass of moving links stationary by adding auxiliary structures.

In (Agrawal and Fattah 2004b; Fattah and Agrawal 2003, 2005a), the parallelograms were used as auxiliary structures in order to create balanced manipulators. As shown in Fig. 2.6, three scaled links are added to form parallelograms and are then used to identify the center of mass C. For the 3-link mechanism, the system consists of parallelograms in two layers: the first layer has two parallelograms while the second layer has one. As is mentioned in the cited papers, this procedure can be extended to n-link mechanisms.

The pantograph has also been used in order to balance the shaking force. Different solutions were proposed for Delta robot (van der Wijk and Herder 2009a; Herder and

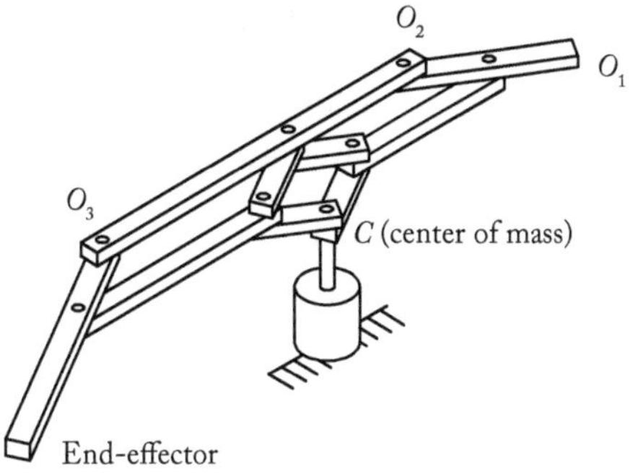

Fig. 2.6 Manipulator with auxiliary parallelograms to locate the center of mass

van der Wijk 2010): by adding a pantograph to each leg or by adding a pantograph connected with the center of mass localized by using the parallelograms, the shaking force can be cancelled.

2.2.1.3 Shaking Force Balancing by Elastic Components

The studies (Alici and Shirinzadeh 2003a, b) are focused on optimum force balancing of a five-bar parallel manipulator by a combination of a proper distribution of link masses with springs connected to the driving links. The force balancing is formulated as an optimization problem in such a way that the root-mean-square of the values of bearing and spring forces are minimized. However, it should be noted that the springs connected to the driving links produce elastic forces which are internal forces and the added springs cannot have an influence on the shaking force minimization due to the inertia of the moving links. They influence the gravitational forces and the input torques which are also included in the objective function. In the mentioned studies, the authors overlook this fact.

2.2.1.4 Shaking Force Balancing by Adjustment of Kinematic Parameters

The studies (Ouyang and Zhang 2002, 2005) deal with the synthesis of the balanced five-bar mechanism via changing the geometric and kinematic parameters of the mechanical structure. The shaking force balancing leads to the conditions which are traditionally satisfied by the redistribution of moving masses. In the mentioned studies, the mass of the link is considered unchanged and the length and the mass center of the links are determined in order to carry out the shaking force balancing. Thus, a new kinematic chain is obtained which is fully force balanced. With regard to the trajectory planning, the authors propose to estimate the given positions of the end effector of the mechanism by the controllers of servomotors. As is rightly mentioned in these studies, the proposed design approach will change the workspace, so some regions of the original workspace may not be reachable. The drawback of this approach is that the project designers: (i) set the structural and kinematic tasks, and (ii) then the dynamic optimization, sequentially. Fixing the values of moving masses and then finding the kinematic parameters of the mechanism is quite unusual.

This approach was also applied on the design of a spatial three degree-of-freedom (*dof*) parallel manipulator (Zang et al. 2011). Theoretical results were obtained, but cannot be easily used for real application. Therefore, for the shaking force balancing of the proposed spatial three-degree-of-freedom parallel manipulator, another method was used (Russo et al. 2005).

It seems that the combined optimization including mass and geometric parameters is more attractive for a wide range of applications of this technique.

2.2.1.5 Shaking Force Minimization via Centre of Mass Acceleration Control

An innovative solution was developed in (Briot et al. 2010, 2012), which is based on the optimal control of the robot centre of masses. The aim of the suggested method consists in the fact that the manipulator is controlled not by applying end-effector trajectories but by planning the displacements of the total mass centre of moving links. The trajectories of the total mass centre of moving links are defined as straight lines and are parameterized with "bang-bang" motion profiles. Such a control approach allows the reduction of the maximal value of the centre of mass acceleration and, consequently, leads to the reduction in the shaking force. It should be mentioned that such a solution is also very favourable for reduction of input torques because it is carried out without adding counterweights.

2.2.2 *Shaking Moment Balancing*

With regard to the shaking moment balancing of manipulators, the following approaches have been developed.

2.2.2.1 Shaking Moment Balancing by Counter-Rotation

The concept of the shaking moment balancing by counter-rotation was studied for the first time in (Berestov 1975, 1977a). This approach was developed further in the various studies devoted to the balancing of 1-*dof* mechanisms and later in (Menschaar et al. 2006; Herder and Gosselin 2004a; Hess-Coelho et al. 2004; Fattah and Agrawal 2006c; Arakelian and Smith 2008; Acevedo et al. 2012) to multi-*dof* mechanisms (Fig. 2.7).

As is rightly pointed out in (Kochev 2000), this technique leads to the unavoidable increase in the initial mass, as well as in the mechanism dimensions. Moreover, the price paid for complete shaking moment balancing is usually unjustifiably high.

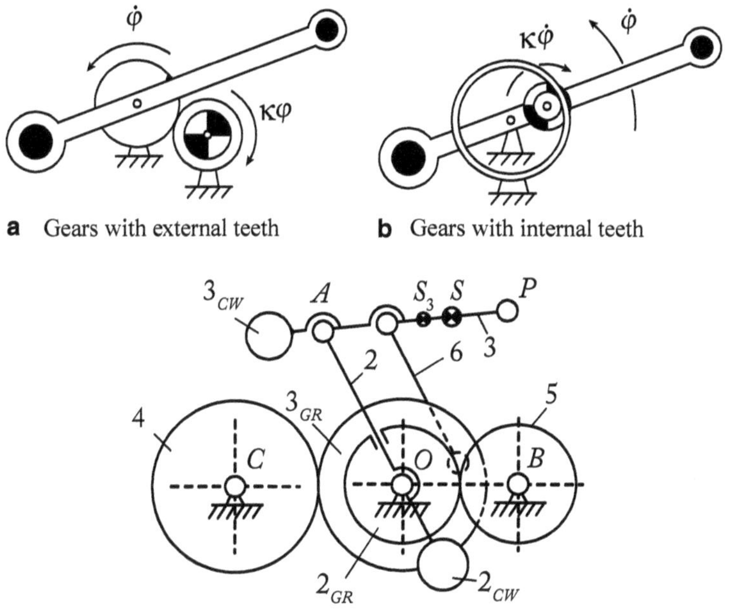

a Gears with external teeth **b** Gears with internal teeth

c Balancing an articulated dyad by gears

Fig. 2.7 Shaking moment balancing by counter-rotation

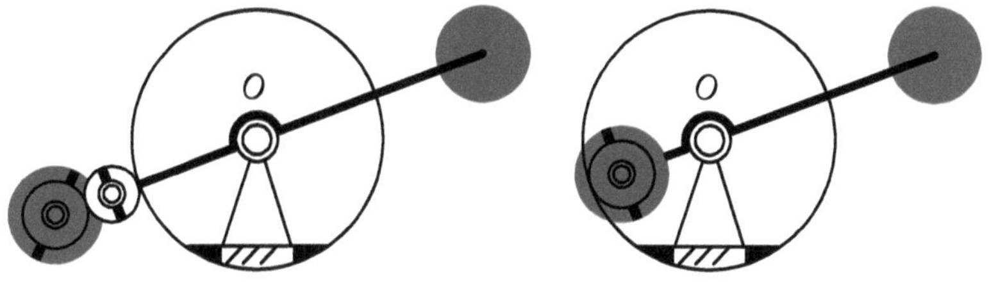

a External gears **b** Internal gears

Fig. 2.8 Counter-rotary counter-mass

In (Herder and Gosselin 2004a; van der Wijk and Herder 2008a, b, 2009b; van der Wijk et al. 2009, 2012), a new design concept was proposed, studied and optimized for light-weight shaking moment balancing by gears. The aim of this concept is to combine both the functions of counter-rotation and counterweight in the same mechanism (Fig. 2.8), which helps to reduce the mass of the resulting system.

The major disadvantage of this technique is the need for the connection of gears to the oscillating links. The oscillations of the links of the manipulator will create noise unless expensive anti-backlash gears are used. Anti-backlash gears are devices that pre-load the gear always to favor one side of the tooth through spring action. Regardless of the direction of movement, they should always "push" up against the same side of the tooth. They are basically comprised of two gears that are spring-loaded in opposite directions. One gear is attached to the mechanism being moved, while the other simply "floats" to provide the pre-loading.

2.2.2.2 Shaking Moment Balancing with Modules Based on Dynamically Balanced Four-Bar Linkages

In (Ricard and Gosselin 2000; Gosselin et al. 2004; Wu and Gosselin 2004; Lecours and Gosselin 2010), the complete shaking force and shaking moment balancing is carried out without any separate counter-rotation. It becomes possible thanks to the synthesis of fully balanced four-bar linkages. It was shown that a four-bar linkage having specific geometric parameters and assuming some ratio between the lengths of links can be fully balanced only by optimal choice of mass and inertia parameters of moving links. This principle is also practicable when the input angular velocity of the four-bar linkage is variable. Thus, various structures of manipulators were designed by special legs constructed with modules based on dynamically balanced four-bar linkages (Fig. 2.9).

2.2.2.3 Shaking Moment Balancing by Generating Optimal Trajectories of Moving Links

In (Papadopoulos and Abu-Abed 1994), a redundant 3-*dof* manipulator is designed in which the system center of mass is fixed by an optimal redistribution of masses. Moreover, the dynamics of the system is decoupled. The latter feature simplified the planning of optimal motions in order to balance the shaking moment of the manipulator. A similar study is carried out in (He and Lu 2006).

Shaking moment balancing by prescribed rotation of the end-effector was proposed in (Fattah and Agrawal 2006c; Arakelian and Briot 2008; Briot and Arakelian 2009). The shaking moment of 3-*dof* planar parallel manipulator (Fattah and Agrawal 2006c) was cancelled using two approaches: through a proper choice of inertia and geometric parameters and by using appropriate motion planning. The shaking moment on the frame of the SCARA-type robots with 4-*dof* has been eliminated by a prescribed velocity of the end-effector (Arakelian and Briot 2008). Taking into account that the two angles of the linear positioning do not depend on the orientation angle, it was proposed to rotate the end-effector during the linear displacements of the end-effector and to balance in such a manner the shaking moment of the robot. The advantage of such a balancing is its simplicity because the complete balancing of the shaking moment is achieved without significant design modifications. The major drawback is the increase of the inertia moment of the end-effector in order to

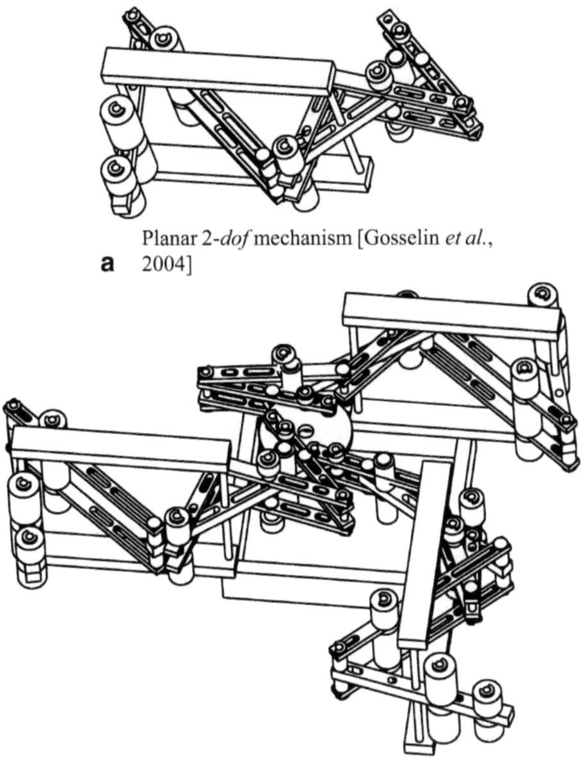

Fig. 2.9 Balancing by adding four-bar linkages

a Planar 2-*dof* mechanism [Gosselin *et al.*, 2004]

b Planar 3-*dof* mechanism [Gosselin *et al.*, 2004]

compensate the inertia moment of the other rotating links. A similar approach has been applied on the PAMINSA manipulator in (Briot and Arakelian 2009).

2.2.2.4 Shaking Moment Balancing by Adding an Inertia Flywheel Rotating with a Prescribed Angular Velocity

It is well known that after shaking force balancing, the shaking moment applied on the base is constant relative to any point, i.e. for a given position of the manipulator it has the same value for any point of the base. Taking into account this property, the shaking moment of any planar manipulator can be balanced adding an inertia flywheel rotating with a prescribed angular velocity (Arakelian and Smith 2008). A similar approach based on the active balancing of the shaking moment of the Delta robot by three additional rotating inertia was discussed in (van der Wijk and Herder 2009a, 2010a). Active balancing of the Hummingbird minipositioner with three axis servo mechanisms was discussed in (Karidis et al. 1992).

2.2.2.5 Other Techniques

It should be noted that new balanced structures have also been developed. In (Foucault and Gosselin 2004), a dynamically balanced 3-*dof* planar parallel manipulator was presented and tested. The manipulator is composed of two independently force-balanced five-bar linkages mounted in opposition on the base and coupled with the end-effector link. In this manipulator, each leg was balanced separately, which was made possible by distributing the inertia of the platform on each of its attachment points (Arakelian and Smith 2008; Wu and Gosselin 2007).

In (Wu and Gosselin 2005), a novel 3-*dof* parallel mechanism referred to as a parallelepiped mechanism was developed. Counterweights and counter-rotations were used to dynamically balance the proposed mechanism.

The design of a dynamically balanced redundant planar 4-*RRR* parallel manipulator was also presented together with design approaches for adapting a given kinematic architecture and obtaining it from known balanced architectures (van der Wijk et al. 2011, 2013).

The complete shaking force and shaking moment balancing of planar parallel manipulators with prismatic pairs (Briot et al. 2009b) and with variable payload (Hess-Coelho et al. 2004; Lecours and Gosselin 2010; van der Wijk and Herder 2010a; van der and Herder 2009; Chungand et al. 1984) were also studied.

In the field of free-floating space robots, the design of reactionless robots was also studied. The formalism called "Reaction Null Space" was initially introduced in (Nenchev et al. 1988) (see also (Nenchev 2013)). The idea is to optimally plan the trajectory to cancel some components of the shaking force or shaking moment. Later, it was applied to reactionless motion generation and vibration control with flexible base robots (Nenchev et al. 1996, 1999; Yoshida et al. 1996a, b).

The study (Agrawal and Shirumalla 1995) deals with a novel scheme for motion planning of a dual-arm free-floating planar manipulator where one arm must perform desired tasks while the other provides compensating motions to keep the base inertially fixed.

The use of kinematic redundancy for robot base reaction reduction was explored in (Chung and Desa 1989; Quinn et al. 1994). The given numerical examples demonstrated that the developed approach is effective for reducing base reactions for planar and spatial robots.

The study (Longman et al. 1987) demonstrated that three orthogonally mounted wheels in the attitude-control system can compensate the total moment about the system mass center. They further show that induced translational motion of the base can be counteracted by using a set of augmented inverse-kinematic relations when calculating the commanded joint variables.

In (Carpenter and Peck 2009), control-moment gyroscopes are proposed as actuators for a spacecraft-mounted robotic arm to reduce reaction forces and torques on the spacecraft base.

Finally, it should be noted that the various optimization methods were also applied in order to reduce the dynamic loads due to the shaking force and shaking moment of manipulators (Xi 1999; Alici and Shirinzadeh 2004, 2006; Ilia et al. 2007; Ilia and Sinatra 2009; Buganza and Acevedo 2011).

2.3 Gravity Balancing in Robotics

Let us now consider the methods of gravity balancing in robotics. In this Section, three typical main applications are presented: automatic robot-manipulators, hand-operated balanced manipulators, and rehabilitation systems of human extremities, exoskeletons and walking assist devices. The advantages/drawbacks of the compensation methods are presented and the design particularities of the gravity compensation of each category are reviewed taking into account the nature of the compensation force.

2.3.1 Gravity Compensation in Automatic Robot-Manipulators

The gravity compensation methods for automatic robot-manipulators can be systematized taking into account the particularity of the design concept, as well as the nature of the compensation force: counterweight (group A), spring (group B), pneumatic or hydraulic cylinder, electromagnetic device, etc. (group C).

2.3.1.1 Goup A: Gravity Compensation by Counterweights

The use of counterweights has been applied to the design of mechanical systems for a long time (Arakelian et al. 2000; Ciupitu et al. 2010a; Lowen et al. 1983). The classical approach consists in adding counterweights in order to keep the total centre of mass of moving links stationary. With regard to the several approaches employed for the redistribution of movable masses, the developed design concepts can be divided into two principal subgroups.

Goup A1: Gravity Compensation by Counterweighs Mounted on the Links of the Initial System Mechanisms presented in (Bayer and Merk 2011, Bolotin 1982; Dunlop and Jones 1996; Gosselin 2008; Gosselin and Wang 1998; Kazerooni 1989; Kazerooni and Kim 1988; Laliberté et al. 1999; Newman and Hogan 1986; Wang and Gosselin 1999, 2000) belong to that category. Examples of such mechanisms are presented in Fig. 2.10.

It is obvious that the adding of the supplementary mass as counterweight is not desirable because it leads to the increase of the total mass, of the overall size of the robot-manipulator and the efforts in joints. That is why in many industrial robots, the masses of the motors are often used for gravity compensation (Fig. 2.10b, Fig. 2.11; Bayer and Merk 2011; Bolotin 1982).

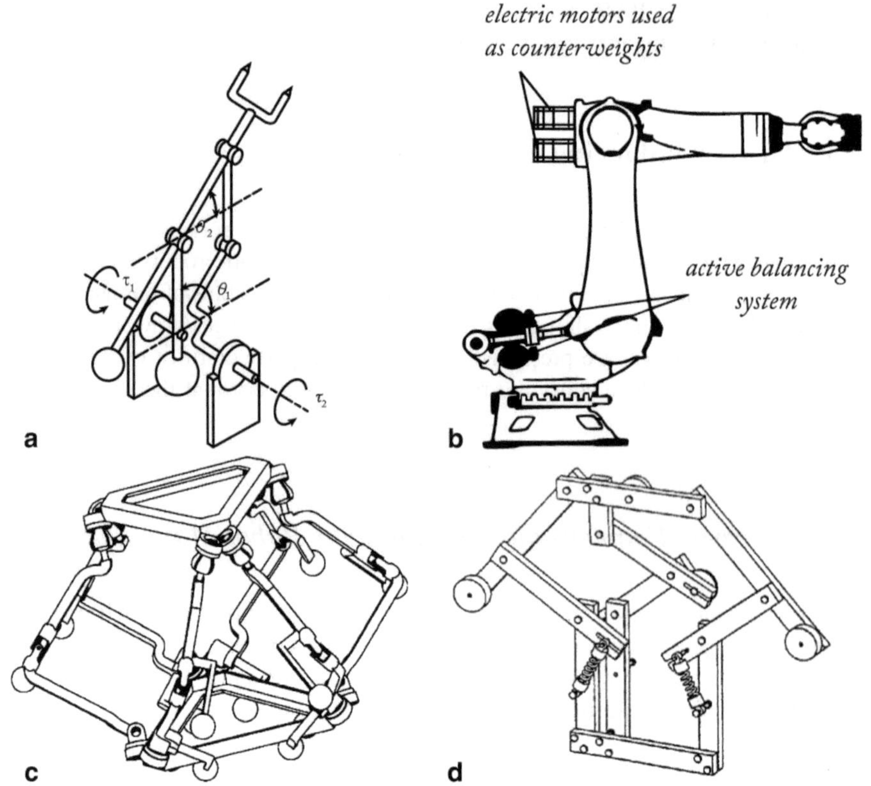

Fig. 2.10 Gravity compensation by counterweighs mounted on the links: serial **a**, **b** (Newman and Hogan 1986; Bayer and Merk 2011) and parallel **c**, **d** manipulators. (Gosselin 2008; Laliberté et al. 1999)

Fig. 2.11 Gravity compensation carried out by using the weights of motors: KUKA R360 (**a**) and PUMA 200 (**b**)

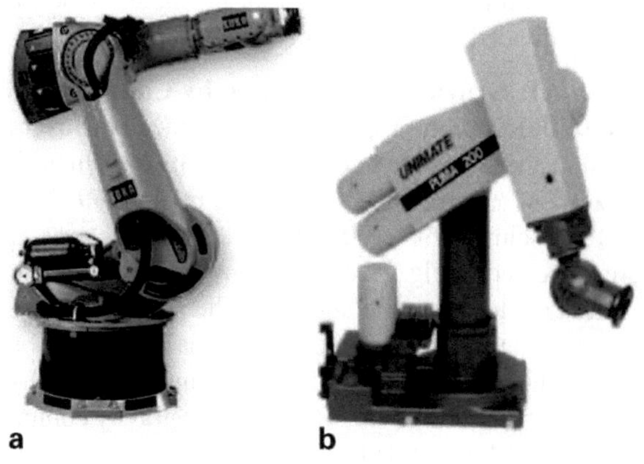

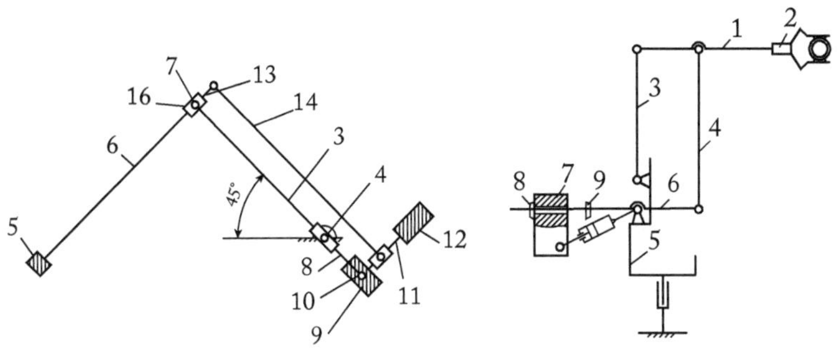

Fig. 2.12 Gravity compensation by counterweights mounted on the auxiliary linkage connected with the initial system. (Fujikoshi 1976; Arakelian et al. 1989)

Goup A2: Gravity Compensation by Counterweights Mounted on an Auxiliary Linkage Connected with the Initial System Mechanisms presented in (Arakelian 1987, 1988; Arakelian et al. 1989; Fujikoshi 1976; Fukushima et al. 2005; Glachet et al. 1985; Russo et al. 2005; Reiner et al. 1991; Zang et al. 2011; Zayzev 1981) belong to that category. Their review showed that the gravity compensation by counterweights mounted on the links is more appropriate for serial and planar parallel manipulators. It is much more difficult for spatial parallel manipulators.

The examples given in Fig. 2.12 show serial manipulators comprising auxiliary systems equipped with the counterweights. (Arakelian et al. 1989) also proposed to cancel the weight of the payload via moving counterweights. Such an approach has also been used in (Arakelian 1987, 1988; Zayzev 1981).

The counterweight balancing of a mine detection vehicle through the use of a pantograph manipulator has been studied in (Fukushima et al. 2005). It has been shown that the robot arm with properly dimensioned counterweights can be efficiently actuated with very low power and energy consumption.

The study (Cheng and Cheng 2011) provides the methodology and an index to evaluate the influence of gravity compensation on the dynamic performance of manipulators. On the base of the PUMA 560 robot, it is shown that the application of the auxiliary linkages is better than the use of counterweights mounted on the moving links. A similar study has been carried out in (Kolarski et al. 1994). The advantages and disadvantages of using counterweights or springs for the balancing are also discussed in (Mahalingam and Sharan 1986). Their effects on the system's natural frequencies are illustrated using numerical examples based on the use of a three-dimensional finite element analysis as a mathematical tool for finding the natural frequencies.

In (Bruzzone and Bozzini 2011), the balancing of the SCARA robot by means of a counterweight or a spring was proposed. The obtained simulation results showed that for low-speed motions, the counterweight balancing is more efficient, while for high-speed motions, the elastic balancing is advantageous.

Many schemes illustrate the parallel manipulators comprising auxiliary systems equipped with counterweights. However, the industrial applications of such approaches are often quite complicated because of the limitation of the overall size of the manipulators and because of the collision possibilities of extended moving links carrying counterweights.

It may be noted that there are also studies devoted to the reactionless manipulators, i.e. high-speed manipulators which apply no reaction forces or moments to the fixed base during the robot motion. It can be reached when the shaking force and shaking moment of the manipulator are cancelled. This goal is usually achieved by adding counterweights or auxiliary linkages in order to keep the total mass centre of moving links stationary. Thus, it becomes obvious that the cancellation of the shaking force by the redistribution of the movable masses leads to the design of a manipulator with a constant potential energy and, as a result, to the compensation of the effects of gravity. However, it should be emphasized that the aim of the shaking force balancing is the cancellation of the variable dynamic loads on the frame of high-speed manipulators and not the minimization of input torques. For this raison, in the studies concerning the reactionless manipulators, the gravity compensation is not a goal but only a result due to the balancing of inertia forces. However, the increase of the accelerations of moving links leads to the increase of the inertia forces. Thus, the complete gravity compensation by adding counterweights in a dynamic mode of operation leads to to the increase of the input torques. So, given that the aim of this investigation of this section is the review of the design concepts permitting the reduction of actuator efforts in static operation, the studies devoted to the reactionless manipulators are not included. They have been considered in Sec. 2.2.

2.3.1.2 Goup B: Gravity Compensation by Springs

First, let us disclose the properties of two types of springs which are used for gravity compensation in robotic systems: zero-free length (Fig. 2.13a) and non zero-free length springs (Fig. 2.13b).

Zero-free length spring is a term for a specially designed coil spring that would exert zero force if it had zero length. That is, in a line graph of the spring's force versus its length, the line passes through the origin (Fig. 2.13a; Carwardine 1932, 1934).

Obviously, a coil spring cannot contract to zero length because at some point the coils will touch each other and the spring will not be able to shorten any more. Zero-free length springs are made by manufacturing a coil spring with built-in tension. So if it could contract further, the equilibrium point of the spring, i.e. the point at which its restoring force is zero, should be for a spring length of zero. In practice, zero-free length springs are made by combining a "negative length" spring, made with even more tension so that its equilibrium point would be at a "negative length", with a piece of inelastic material of the proper length in order that the zero force point would occur at zero length (Lacoste 1935; Melton 1971).

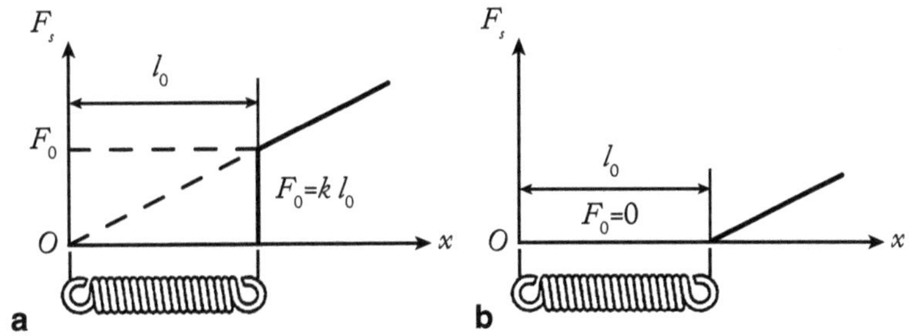

Fig. 2.13 Force-length characteristics of zero free length (**a**) and non-zero free length springs (**b**)

Fig. 2.14 Gravity compensation of a rotating link

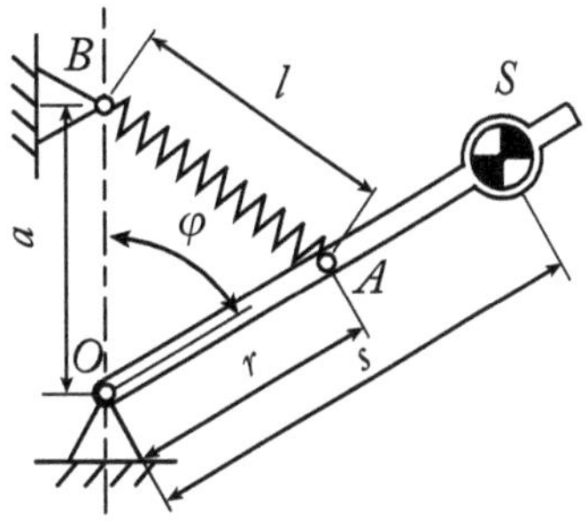

In order to better understand the difference between the zero-free and non zero-free length springs, let us consider the gravity compensation of a rotating link (Fig. 2.14).

It is obvious that the potential energy of this system will be constant if the moment of the gravitational forces is fully balanced by the moment of the elastic force of the spring, i.e.

$$mgs \sin\varphi = (F_{sp}\, a\, r/l) \sin\varphi \tag{2.1}$$

where m is the mass of the rotating link, $s = l_{OS}$ is the distance of gravity center S from axis O, φ is the angle between the vertical axis and the link axis, $F_{sp} = F_0 + k(l - l_0)$ is the elastic force of the spring, F_0 is the initial force of the spring (the initial force is the internal force that holds the coils tightly together), k is the stiffness coefficient of the spring, l_0 is the initial length of the spring, $a = l_{OB}$ is the distance of point B from axis O, $r = l_{OA}$ is the distance of point A from axis O, $l = l_{AB}$ is the length of the spring at current angle φ.

One can see from Eq. (2.1) that a full gravity compensation can be achieved when $F_0 = kl_0$, i.e. when a zero-free length spring is used. In the case of applying of a non zero-free length spring with $F_0 = 0$ or $F_0 \neq kl_0$, only partial gravity compensation of a rotating link can be achieved.

It is important to emphasize that the use of a zero-free length spring for complete gravity compensation is inevitable when the spring is connected directly with the robot links and such a necessity disappears when the spring is connected with the robot links via a cable or an auxiliary mechanism.

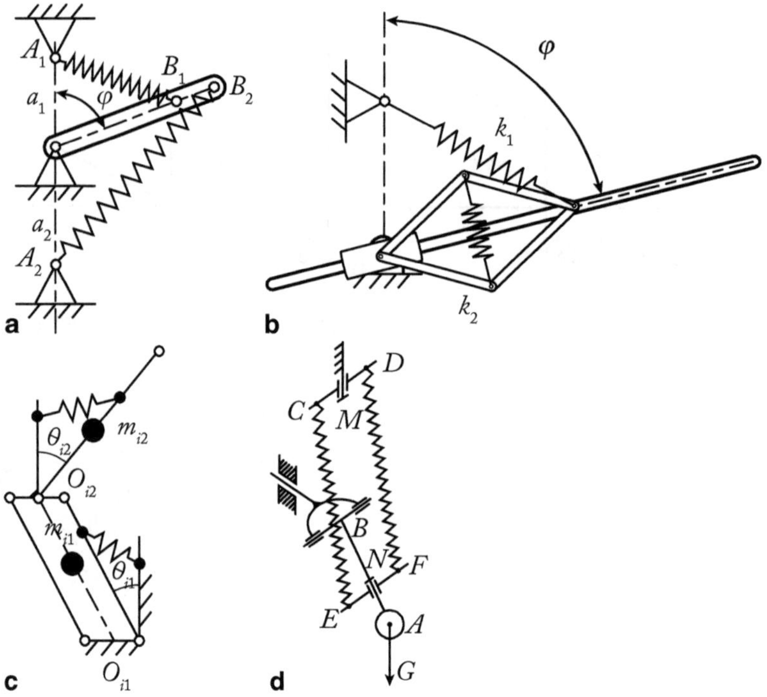

Fig. 2.15 Gravity compensation by springs jointed directly with the manipulator links: **a** (Herder 2001); **b** (Streit and Shin 1993); **c** (Ebert-Uphoff et al. 2000); **d** (Maksimov and Moiseenkov 1978)

To preserve the structure of the systematization via design particularities, which has been adopted above, let us classify the spring compensators in three subgroups as follows.

Goup B1: Balancing by Springs Jointed Directly with Manipulator Links Mechanisms presented in (Carwardine 1940; Tuda 1983; Gopalswamy et al. 1992; Streit and Shin 1990; Shin and Streit 1991, 1993; Rahman et al. 1995; Pons et al. 1998; Gosselin 1998; Herder 1998, 2001; Ebert-Uphoff et al. 2000; Maksimov and Moiseenkov 1978; Riele and Herder 2001; Gunnarsson 2002; Tuijthof and Herder 2000; Ono and Morita 2004; Perrot 2004; Soethoudt and Herder 2007; Lin et al. 2009, 2010b, 2012; Cho and Kang 2011; Wisse et al. 2007; Lu et al. 2011; Deepak and Ananthasuresh 2009, 2012a, b) belong to that category.

Some examples of the gravity compensation by springs jointed directly with manipulator links are shown in Fig. 2.15. Such an approach has also been applied to the spatial manipulators (Walsh et al. 1991; Morita et al. 2003; Wongratanaphisan and Chew 2002; Agrawal and Fattah 2004a).

In order to create springs with adjustable stiffness, the "Jack spring" concept has been proposed (Hollander and Sugar 2004; Hollander et al. 2005). It is based upon the principle of adding and subtracting coils from a spring. Thus, with this method,

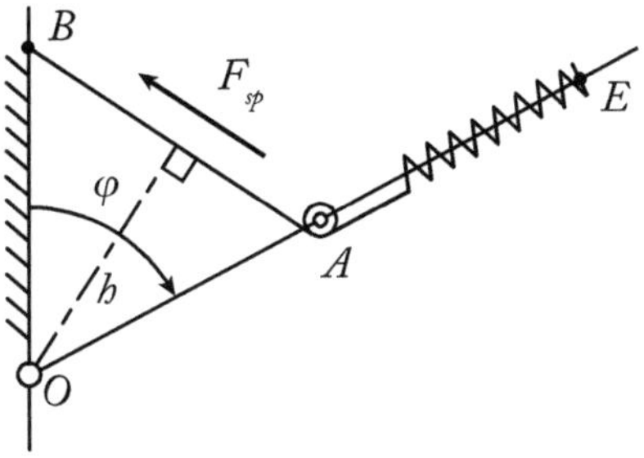

Fig. 2.16 Simple scheme of the gravity compensation by a spring, a cable and a pulley

by changing the number of coils in a spring, the actual or intrinsic stiffness of the spring is structurally changed. A simple and practical method to adjust the number of coils was proposed in (Sugar and Hollander 2012).

In (Klimchik et al. 2014), an approach for the stiffness modeling of robots with the spring equilibrators, which are located between the manipulator links has been presented. The aim of this approach is to replace the gravity equilibrator by an equivalent non-linear virtual spring integrated in the corresponding actuated joint. Efficiency of the developed approach and its industrial value have been confirmed by an application example.

In (Eckenstein and Yim 2013) the design and analysis of the modular gravity compensated manipulators have been presented. Modular advantages and kinematic decoupling have been disclosed. It has been shown that the decoupling simplifies the kinematic equations reducing the order of complexity of calculation.

Various design concepts have also been developed for adjustment of gravity equilibrators (Wisse et al. 2007; Nathan 1983, 1985; van Dorsser et al. 2007).

Goup B2: Balancing by Using a Cable and Pulley Arrangement Mechanisms presented in (van Dorsser et al. 2008; Vrijlandt and Herder 2002; Vladov and Danilevskij 1981; Tuda and Mizuguchi 1983; Ebert-Uphoff and Johnson 2002; Endo and Hirose 2008; Endo et al. 2010; Kobayashi 2001b; Schmit and Okada 2012; Tomas 1995; Bailey 2012; Barents et al. 2011) belong to that category.

The adding of the cable and pulley allows the full gravity compensation by using a non-zero free length spring.

Let us consider a simple example in order to understand the effect of the additional cable in the gravity compensation of a rotating link (Fig. 2.16).

The condition of the gravity compensation (2.1) can be rewritten as:

$$m\,g\,s\,\sin\varphi = F_{sp}\,h \tag{2.2}$$

where $h = a\,r/\,l_{AB}\sin\varphi$. Thus, expression (2.2) is similar to (2.1) when the length l of the spring is equal to l_{AB}. However, in this case, thanks to the cable, it is possible to consider that $l_{AB} = l - l_0$, which leads to the condition $m\,g\,s = k\,a\,r$, with $F_0 = 0$. So, the rotating link can be balanced with a non-zero free length spring.

Figure 2.17 shows several examples of gravity compensation using a cable and pulley arrangement.

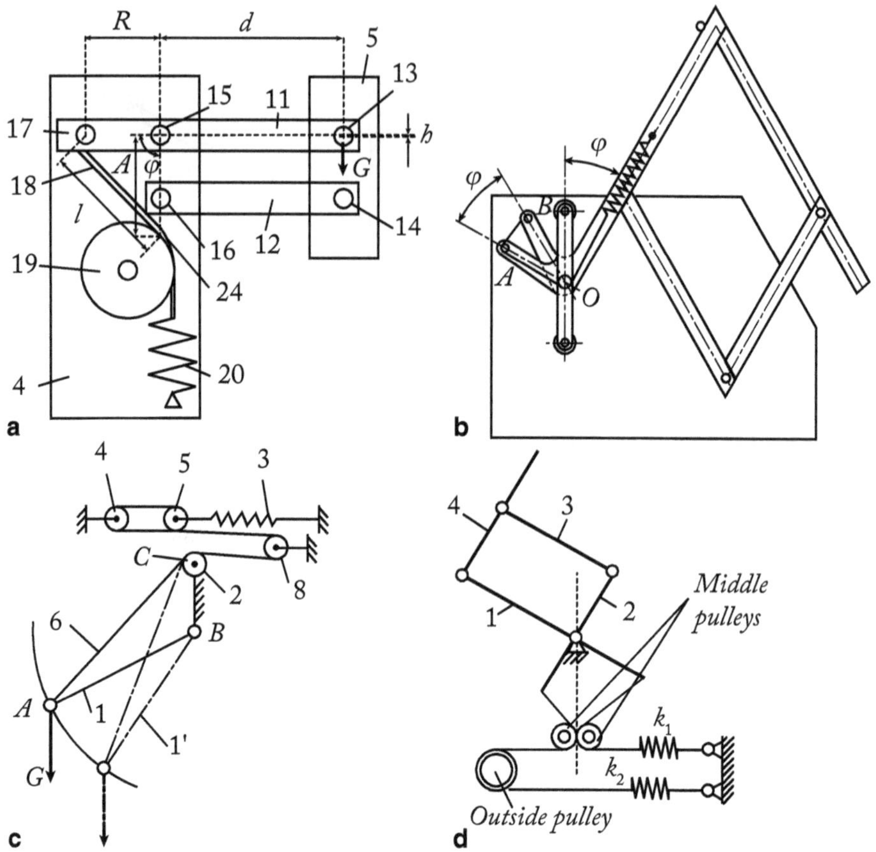

Fig. 2.17 Gravity compensation by using a cable and pulley arrangement: **a** (van Dorsser et al. 2008); **b** (Vladov and Danilevskij 1981); **c** (Tuda and Mizuguchi 1983); **d** (Ebert-Uphoff and Johnson 2002)

Gravity compensation with the non-circular pulleys and the springs has been examined in (Endo and Hirose 2008; Endo et al. 2010). After preliminary verification of the design methodology for a single pendulum system, the authors extend the weight compensation mechanism to the two degrees of freedom parallel five-bar linkage arm. It has been shown that the introduction of the weight compensation mechanism reduces the maximum static torque by 50–80 %. The spiral pulley with spring has also been used in (Kobayashi 2001b; Schmit and Okada 2012).

As mentioned in (Ebert-Uphoff and Johnson 2002), several error sources in the pratical implementations decrease the efficiency of the gravity compensation with springs and pulleys. Errors are mainly caused by the non-linearity of the springs due to the manufacturing tolerance. Often the nominal values of the calculated springs are different to the real values. Therefore the values of spring stiffness must be ajusted. Another error source is the radius of the pulleys.

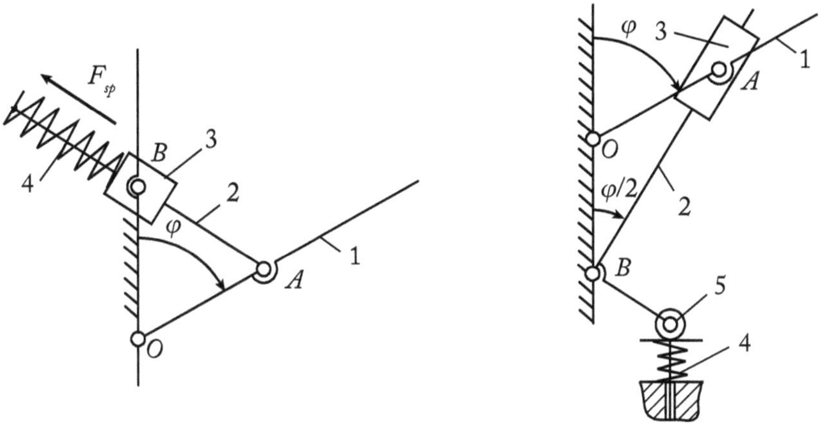

Fig. 2.18 Gravity compensation by the auxiliary mechanisms. (Popov and Tyurin 1988; Popov et al. 1984)

Goup B3: Balancing by Using an Auxiliary Mechanism Mechanisms presented in (Leblond and Gosselin 1998; Segla et al. 1998; Saravanana et al. 2008; Herder 2002; Wang and Gosselin 2004; Segla 2004; Fattah and Agrawal 2006b; Popov and Tyurin 1983, 1988; Kobrinskii and Kobrinskii 1989; Simionescu and Ciupitu 2000a, b; Minotti and Pracht 1988; Dzhavakhyan and Dzhavakhyan 1989; Hervé 1985; Kolotenkov 1984; Arakelian and Dahan 1995; Kim and Song 2013; Bartlett et al. 1988; Popov et al. 1984; Lakota and Petrov 1985; Pracht et al. 1987; Kondrin et al. 1990; Petrov and Polishchuk 1979; Ulrich and Kumar 1991; Koser 2009; Akeel 1987; Gvozdev 1992, 1990, 1987; Greene 1995; Cho et al. 2010, 2011, 2012; Frolov and Vorobev 1988) belong to that category.

The auxiliary mechanisms have the same effect as the cables and the pulleys. They allow the gravity compensation by using non-zero free length springs. Let us consider two illustrative examples.

Figure 2.18a shows an equilibrator in which rotating link 1 is connected with coulisse 2 and slider 3. The added links of the mechanism allows the complete compensation of the gravity of the rotating link 1 by using a non-zero free length compression spring.

Another similar solution is given in Fig. 2.18b. In this case, the lengths of links of the mechanism must satisfy to the condition $l_{OA} = l_{OB}$ leading to a displacement of the spring proportional to $\sin(\varphi/2)$, which ensures the complete gravity compensation of rotating link 1. This condition was also applied to the design of the gravity equilibrator by using a gear train (Fig. 2.19).

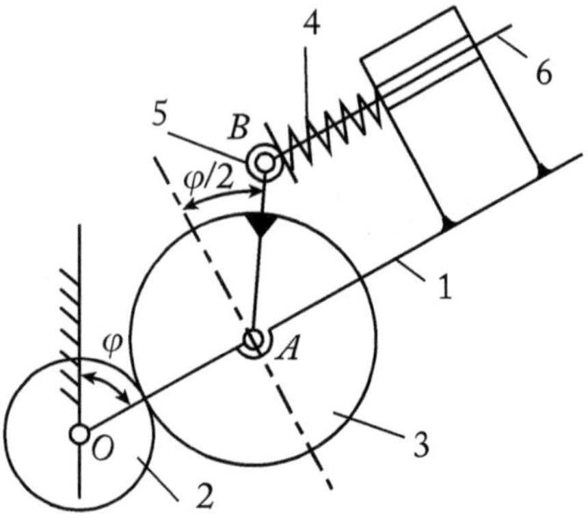

Fig. 2.19 Gravity compensation by using gears. (Frolov and Vorobev 1988)

The design solutions via adding an auxiliary mechanism can be classified in three subgroups:

- *B3.1: Balancing by using an auxiliary linkage* (Leblond and Gosselin 1998; Segla et al. 1998; Saravanana et al. 2008; Herder 2002; Wang and Gosselin 2004; Segla 2004; Fattah and Agrawal 2006b; Popov and Tyurin 1988; Kobrinskii and Kobrinskii 1989; Simionescu and Ciupitu 2000a; Minotti and Pracht 1988; Dzhavakhyan and Dzhavakhyan 1989; Hervé 1985; Kolotenkov 1984; Arakelian and Dahan 1995; Kim and Song 2013; Bartlett et al. 1988; Popov et al. 1984);
- *B3.2: Balancing by using a cam mechanism* (Simionescu and Ciupitu 2000b; Lakota and Petrov 1985; Pracht et al. 1987; Kondrin et al. 1990; Petrov and Polishchuk 1979; Ulrich and Kumar 1991; Koser 2009);
- *B3.3: Balancing by using gear train* (Popov and Tyurin 1983; Akeel 1987; Gvozdev 1987, 1992, 1990; Greene 1995; Cho et al. 2010, 2011, 2012; Frolov and Vorobev 1988).

Some examples of the design concepts developped by adding auxiliary mechanisms with corresponding references are given in Figs. 2.20–2.22.

The following two schemes shown in Fig. 2.21a illustrate the gravity compensation by using compression springs mounted on the guides. The compression springs are chosen to have a force-deflection characteristics to account for the gravitational moment and the geometry of the mechanism.

The advantages of these equilibrators consists in the high rigidity of the system because they do not contain auxiliary elements that have a tendency to reduce the balancing accuracy. However, as discussed earlier, the errors due to the manufacturing tolerances of links can decrease the quality of the gravity compensation.

Note that the two springs of the equalibrator shown in Fig. 2.21a have the same stiffness. The combined interaction of two springs provides the variation of the compensation moment by a sine law.

The advantage of the addition of an auxiliary mechanism consists in the increase of the free design parameters of the system. This allows optimization of the gravity compensation by applying linkage synthesis methods.

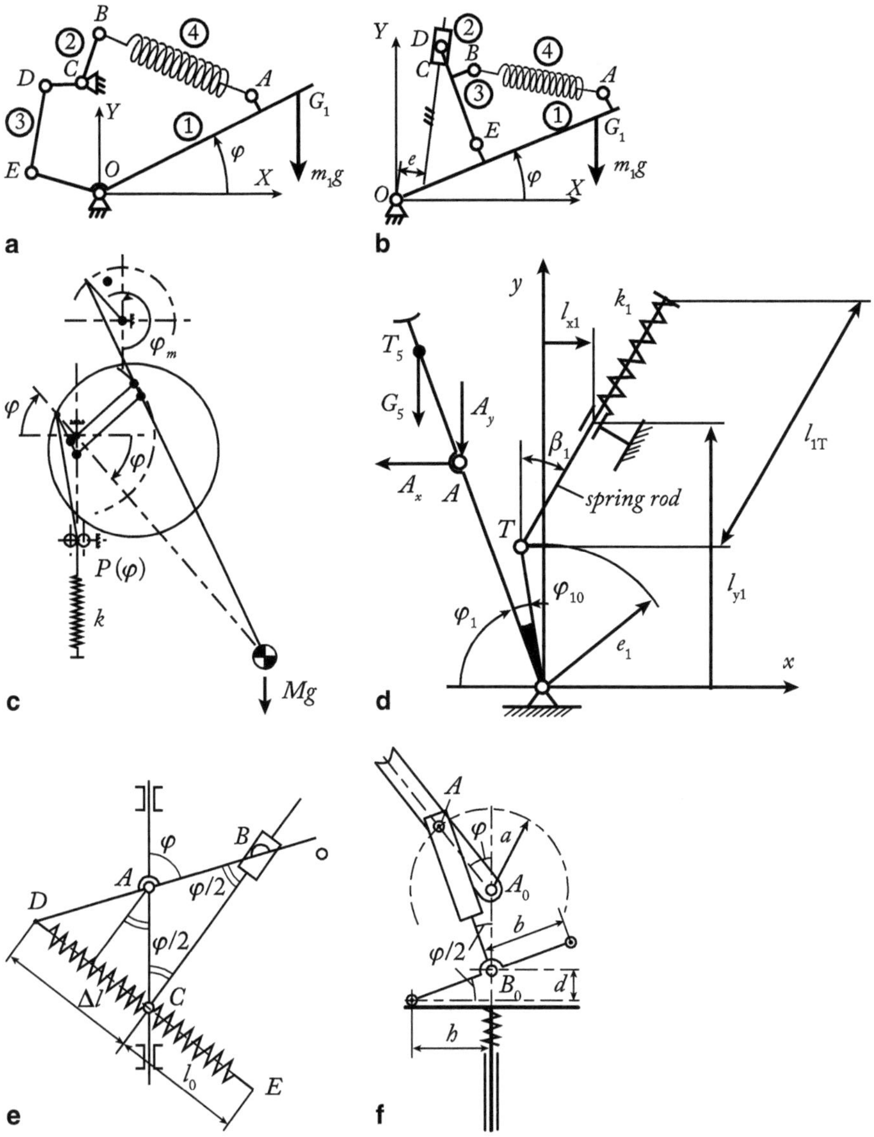

Fig. 2.20 Gravity compensation by using auxiliary linkages: **a** and **b** (Simionescu and Ciupitu 2000a); **c** (Minotti and Pracht 1988); **d** (Segla et al. 1998); **e** (Dzhavakhyan and Dzhavakhyan 1989); **f** (Hervé 1985)

Gravity compensation of links that have translational and vertical motions has also been studied. Four illustrative examples are given in Fig. 2.24.

The primary equations of static equilibrium between the vertical load P (Fig. 2.24) and the elastic forces of springs (F_{sp}) lead to the conditions of complete gravity compensation.

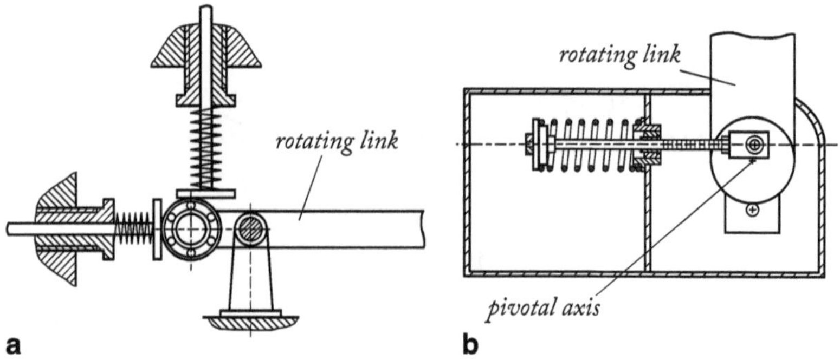

Fig. 2.21 Gravity compensation by using an auxiliary linkage with compressing springs: **a** (Kolotenkov 1984) and **b** (Bartlett et al. 1988)

In (Morozovski 1993), the gravity compensation of a gantry system has been proposed (Fig. 2.25). Arm 1 of the robot carried out the horizontal motion by using a carriage 4 which is mounted on the frame 2 and equipped with pulleys 5, 6, 10 and a cable. When the carriage moves in the horizontal direction, the rotation of the device 3 ensures a constant tension in cable and link 11 keeps a stable position. When arm 1 moves in the vertical direction, device 3 locks its rotation, which produces a transmission carried out by using links 7, 8, 9 and 11. As a result, the rocker 11 rotates about its axis and drives the spring equilibrator. A similar work has also been carried out in (Brown and Dolan 1994).

The drawback of the systems designed for balancing of links with vertical motion consists in obtaining relatively small vertical displacements. In order to eliminate this drawback, a supplementary transmission mechanism can be used, for example pantographs, polyspasts or gear trains.

A general equation of inertia force for both a gravity balancer by counterweight and spring balancer has been derived in (Kobayashi 2001a). The two equations were compared and the conditions that make the spring balancer more attractive were investigated.

It should be added that the spring compensation has also been studied for the spatial robotic systems (Agrawal and Fattah 2004 a; Simionescu and Ciupitu 2000a; Simionescu and Ciupitu 2000b; Wongratanaphisan and Chew 2002; Tuijthof and Herder 2000; Brown and Di Guilio 1980; Gosselin 2006; Arsenault and Gosselin 2007; Schenk et al. 2006; Lin et al. 2010a; Dehkordi et al. 2012).

2.3.1.3 Goup C: Gravity Compensation by Using Pneumatic or Hydraulic Cylinders

Mechanisms presented in (Belyanin 1988a; Fahim and Fernandez 1988; Yamamoto et al. 2010; Wildenberg 2002; Dzhavakhyan and Dzhavakhyan 1987; Segawa et al. 2000; Lauzier et al. 2009; Lacasse et al. 2013; Laliberté et al. 2010; Baradat et al. 2005; Baradat et al. 2008) belong to that category.

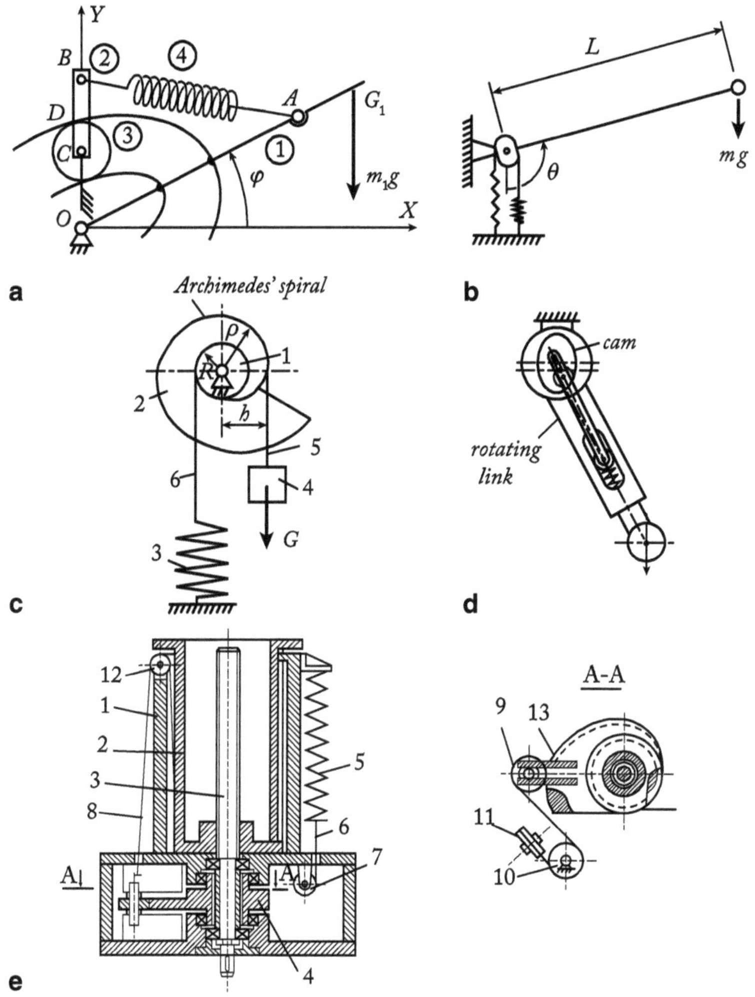

Fig. 2.22 Gravity compensation by using cam mechanisms: **a** (Simionescu and Ciupitu 2000b); **b** (Ulrich and Kumar 1991); **c** (Lakota and Petrov 1985); **d** (Kondrin et al. 1990); **e** (Petrov and Polishchuk 1979)

In this case, a pneumatic or hydraulic cylinder is connected with some manipulator links (Bayer and Merk 2011; Belyanin 1988a; Fahim and Fernandez 1988; Yamamoto et al. 2010) or directly with the moving platform (Wildenberg 2002). There are also some approaches based on special counterweights, which are fluid reservoirs. Continuous gravity compensation is achieved by the pumping of fluid from the first reservoir-counterweight to the second (Dzhavakhyan and Dzhavakhyan 1987) (Fig. 2.26).

Electromagnetic effects have also been used (Segawa et al. 2000).

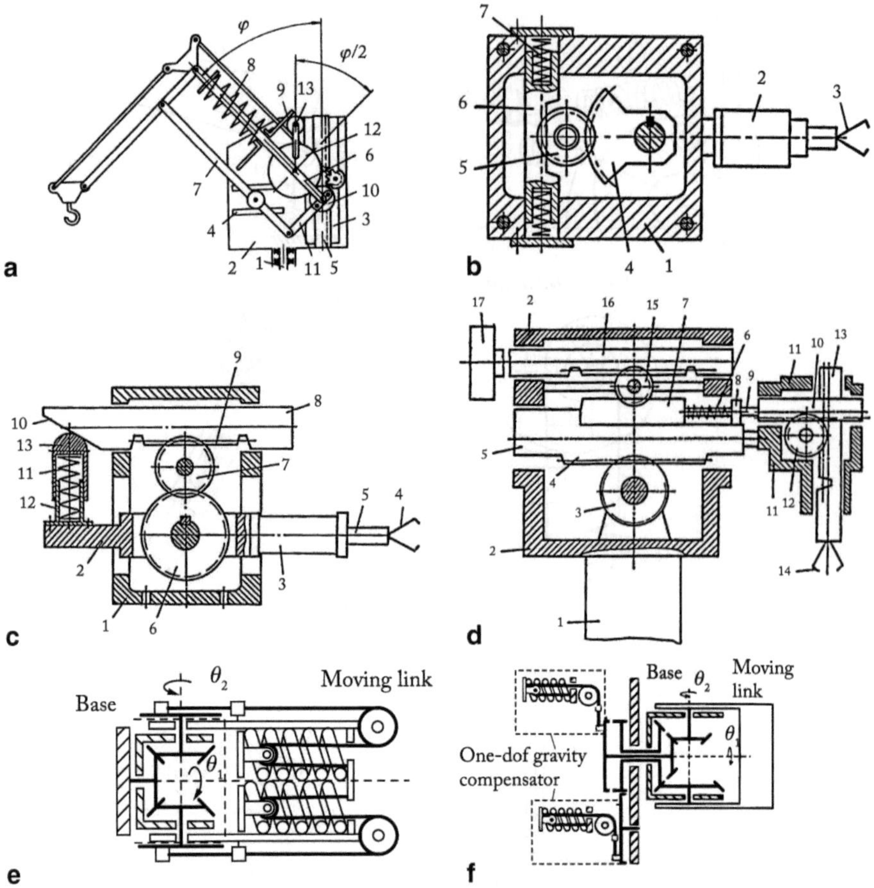

Fig. 2.23 Gravity compensation by using gears: **a** (Popov and Tyurin 1983); **b** (Gvozdev 1992); **c** (Gvozdev 1990); **d** (Gvozdev 1987); **e** (Cho et al. 2011); **f** (Cho et al. 2012)

The gravity compensation technique developed in (Lauzier et al. 2009; Lacasse et al. 2013; Laliberté et al. 2010) uses remote counterweights connected to the robot via a hydraulic transmission. As has been shown in (Lacasse et al. 2013) the built prototype of the 7-*dof* robot is able to adapt its balancing counterweights to a payload of up to 10 kg, which was a maximal payload for the tested prototype.

However, it should be noted that many gravity compensation methods are applicable only for serial or planar parallel manipulators. The gravity compensation of spatial parallel architectures is a complicated problem because it can only be achieved either by an unavoidable increase of the total mass of moving links or by a considerably more complicated design of the initial parallel mechanism.

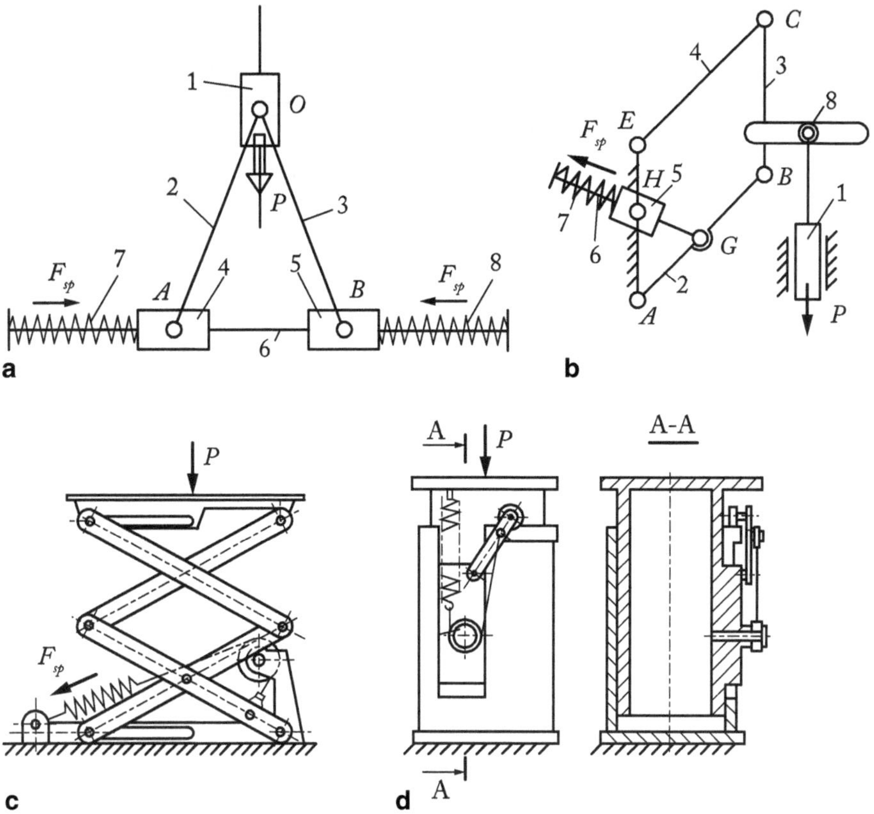

Fig. 2.24 Gravity compensation of the link with the vertical translational motion: **a** and **b** (Frolov and Vorobev 1988); **c** (Lakota 1980); **d** (Verkhovcki 1980)

It seems that an optimal approach is to combine an auxiliary linkage with pneumatic or hydraulic cylinders.

An illustrative example is shown in Fig. 2.27.

The suggested approach involves connecting an auxiliary mechanism to the initial structure, which generates a vertical force applied to the manipulator platform (Baradat et al. 2005, 2008). The details of this approach will be disclosed in Section 6.1.

2.3.1.4 Other Works

It is necessary to note that the studies concerning gravity balancing are generally devoted to gravity compensation for a constant weight. In (Takesue et al. 2011), a variable gravity compensation mechanism is proposed. It uses two types of linear springs and changes the equilibrium position of one of these. It should also be noted that the complete compensation of gravity often requires numerous complex

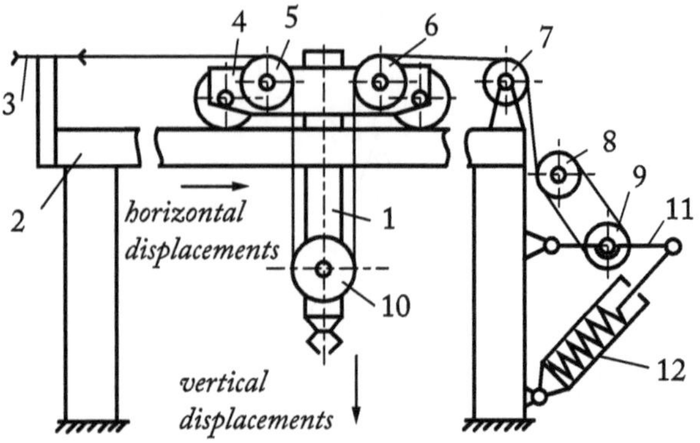

Fig. 2.25 Gravity compensation of the robots with the vertical and horizontal translational motions. (Morozovski 1993)

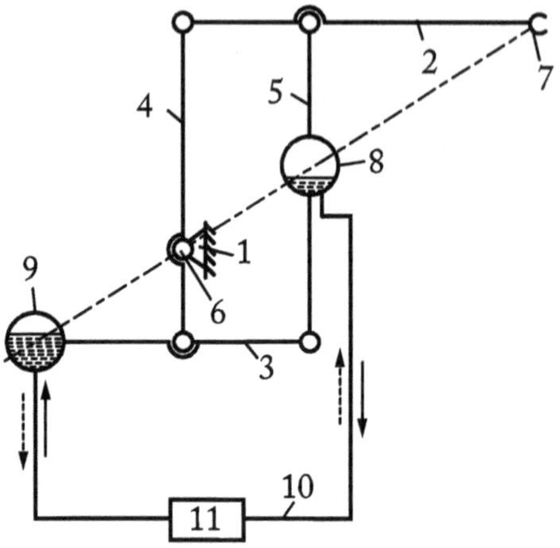

Fig. 2.26 Continuous gravity compensation accomplished by the pumping of the fluid from the first reservoir-counterweight to the second. (Dzhavakhyan and Dzhavakhyan 1987)

mechanical add-ons or unavoidable addition of mass. This is the reason why methods of partial gravity compensation have also been developed (Lessard et al. 2007a, b).

The gravity compensation of the parallel cable-driven mechanisms has also been studied (Perreault and Cardou 2009). It has been shown that by using non-linear springs in parallel with motors, it is possible to maintain the minimum torsions in cables and as a result to minimize the static loads over its entire workspace.

In (Brudniok et al. 2012) a method for the automated movement of a gravity-compensated payload and an automated handling system for gravity compensation of the payload were presented (Fig. 2.28). It involves supporting a payload by a holding unit that is connected with an end-effector flange of the robot for automatically moving of the load body.

Some recent works have dealt with the gravity compensation for the manipulators with Schoenflies motions (three translations plus one rotation around the vertical axis) in which the vertical motion is decoupled from other degrees of freedom. In the latter situation, only the mechanical system that generates the vertical motion needs to be gravity compensated in order to eliminate actuator torque caused by the weight of the moving parts and the payload (Zabalza et al. 2002; Briot et al. 2008a, 2009a).

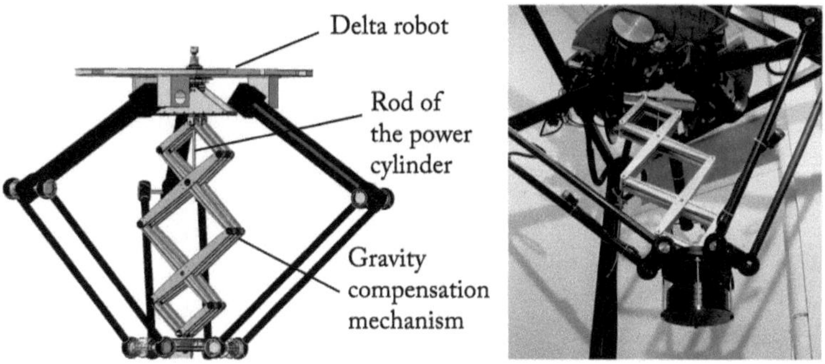

Fig. 2.27 Gravity compensation of the Delta robot

Fig. 2.28 Automated movement of a gravity-compensated payload. (Brudniok et al. 2012)

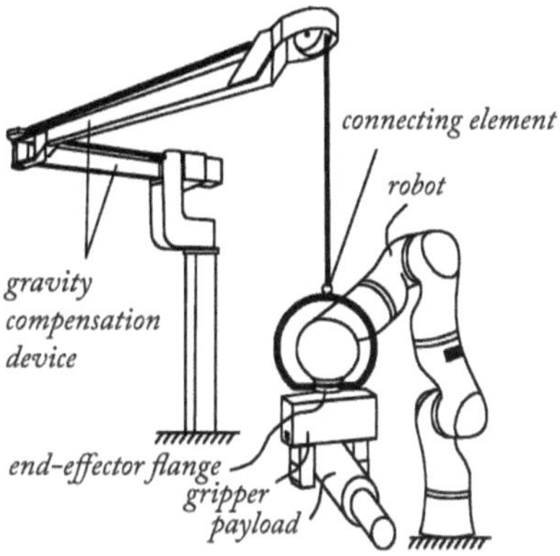

This technique has been applied for the design of PAMINSA manipulators (Fig. 2.29; Briot et al. 2008a, 2009a).

The particularity of this architecture is the decoupling of the displacements of the platform in the horizontal plane from its translation along the vertical axis. Such a decoupling allows the cancellation of the gravity loads on the actuators which displace the platform in the horizontal plane. A prototype of the proposed manipulator with four degrees of freedom has been built (Briot et al. 2008a). Two cases have been examined on the built prototype: the manipulator with payload and without it.

The input torques with and without payload for the three rotating actuators that allow the planar motions of the platform are similar, i.e. the gravity loads caused by the payload on these actuators are cancelled. Regarding the actuator carrying out the vertical translations of the platform, it carries the payload and the increase of the input force is significant. More detailed information about the properties of the manipulator PAMINSA will be presented in Sect. 6.2.

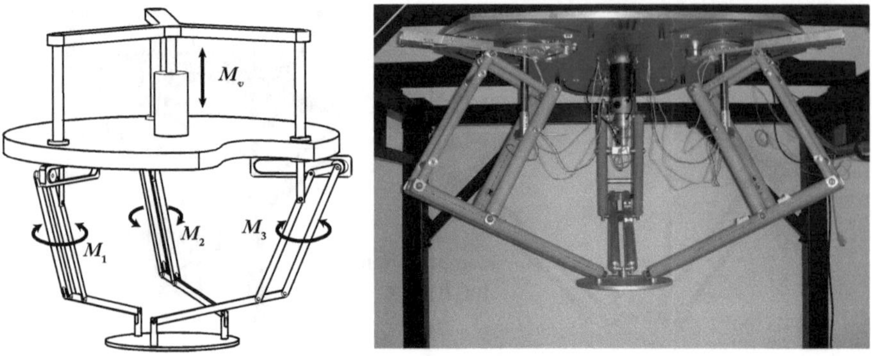

Fig. 2.29 3D view and the prototype of the PAMINSA. (Briot et al. 2008a)

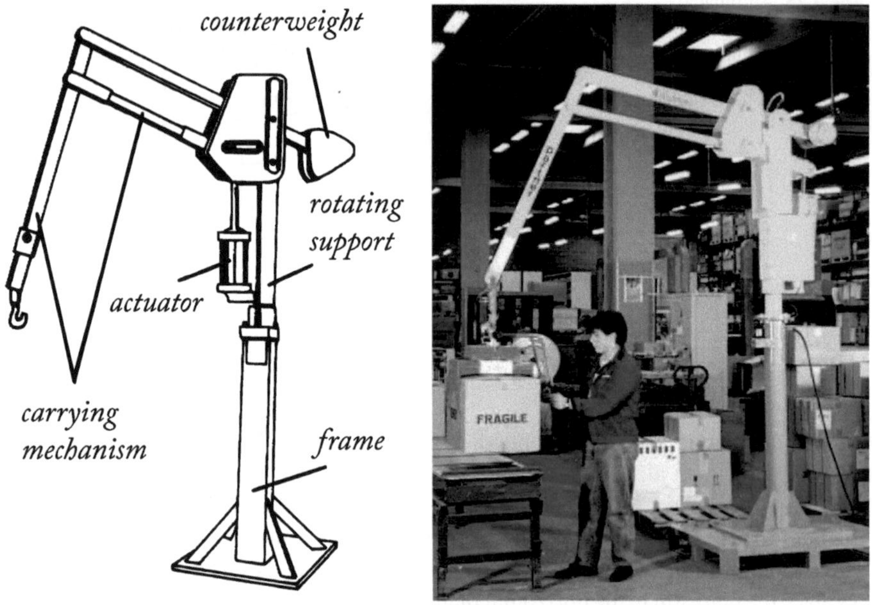

Fig. 2.30 Hand-operated balanced manipulator (Dalmec, Italy)

2.3.2 *Gravity Compensation in Hand-Operated Balanced Manipulators (HOBM)*

The HOBM (Fig. 2.30) is a handling system with a simple mechanical actuator in which the manipulated object in any position of the workspace is balanced (Arakelian 2004b). Such a state of constant gravity cancellation allows the manual displacement of heavy objects. In the technical literature, these devices are also called "self-balancing manual manipulators". Such manipulators have found a broad application in several fields of industry where it is necessary to carry out mechanization of heavy manual work.

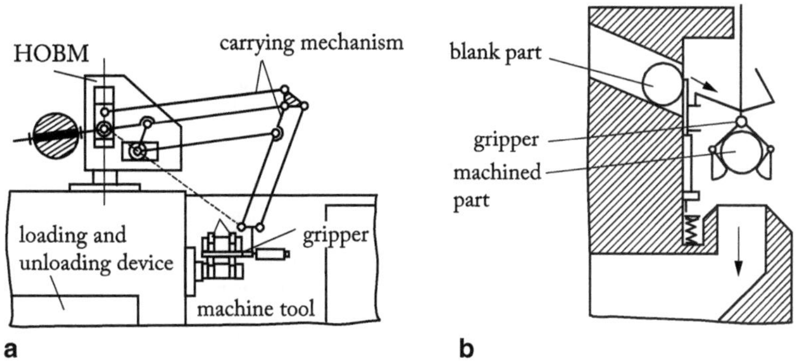

Fig. 2.31 HOBM without driving system (**a**) and the loading and the unloading device (**b**) of the manipulator. (Arakelian 1986)

The advantages of HOBM with respect to automatic robot-manipulators are the simplicity of their construction and their low cost, as well as a great weight-carrying capacity and a very large worskspace.

The term "balanced manipulator" shows that in the operating procedure of these systems, it is very important to achieve an accurate compensation of gravity.

Many studies and design concepts have been devoted to the gravity compensation of HOBM (Olsen 1964, 1966; Matsumoto 1975; Hank 1976, 1987; Krsnak and Howe 1976; Previdia et al. 2012; Arakelian 1986, 1998a; Arakelian and al. 1991; Djavakhian 1987, 1990; Djavakhian and Djavakhian 1988; Moor and Akouna 2003; Haaker 1968; Kiyoaki 1999; Bittenbinder 1995; Gvozdev 1981; Popov and Tyurin 1987, 1989; Vladov and Danilevskij, 1981, 1999; Mishkind 1981; Derkach and Umantsev 1990; Kfoury 2013; Heiko 1998; Derkach 1993). Two principal ways for gravity compensation of the HOBM exist which depend on the structural architecture of the manipulator and the type of actuation: either the movements of the carrying mechanical system are decoupled and the force of gravity of the payload is canceled during its movement in the horizontal plane (in this case, a compensation must be provided only during the vertical movements), or the actuators provide permanent compensation for any position of the payload. In the last case, the load weight measurement for performing the movement control is required. The general approach for the determination of the gravity compensation is based on the study of the motion of the centre of mass of the mechanical system carrying the payload and on the optimal control of the drivers.

Figure 2.31a shows a type of hand-operated balanced manipulators without a driving system designed to assist the loading and unloading of a machine tool (Arakelian 1986). The need for a driving system is removed by the gravity compensation method, in which the system parameters take into account the mass of the effective load equal to one half of the masses of the blank part and the machined part. In other words, the manipulator is always loaded either by the blank part or the machined part. Figure 2.31b shows the loading and the unloading device of the manipulator in addition to the input of the blank part and the ejection of the machined piece.

Let us now consider the gravity compensation in rehabilitation systems.

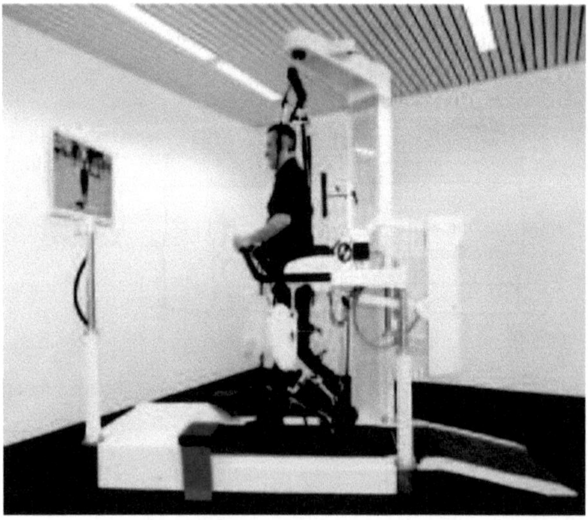

Fig. 2.32 Lokomat system

2.3.3 Gravity Compensation in Rehabilitation Systems of Human Extremities, Exoskeletons and Walking Assist Devices

It is known that weight support can facilitate human extremities movements and patients may do more and more meaningful exercises earlier in the rehabilitation process (Ellis et al. 2007; Agrawal and Agrawal 2004,). Several overviews of balancing methods of this field have been published (Ciupitu et al. 2005, 2010b; van Ninhuijs et al. 2013; Duning and Herder 2013).

First of all, before discussing the methods of gravity compensation, it is necessary to note that for the rehabilitation systems, two levels of the balancing problem should be distinguished: gravity compensation of bodyweight and gravity compensation of the extremities.

In stationary rehabilitation devices, the passive compensation of gravity is often combined with the active compensation executed via the motors. Gravity-compensating suspension systems are frequently used for bodyweight support. They can take different forms with different suspension elements such as ribbons, cables or linkages.

Let us consider some examples of gravity-compensating suspension systems.

The Lokomat (Hocoma AG, Volketswil, Switzerland) consists of a robotic gait orthosis and an advanced body weight support system (Fig. 2.32; LOKOMAT 2014). This system includes the ability to measure the patient's activity by a way of force transducers fitted directly on the motors and offers the possibility to adjust the level of gait assistance for each leg between full and zero guidance force.

The String-Man is a wire robot for manipulating the torso of patients based on the string-puppet principle (Fig. 2.33; Surdilovic and Bernhardt 2004). Seven wires connected to the patient's trunk achieve weight bearing and 6-*dof* manipulation.

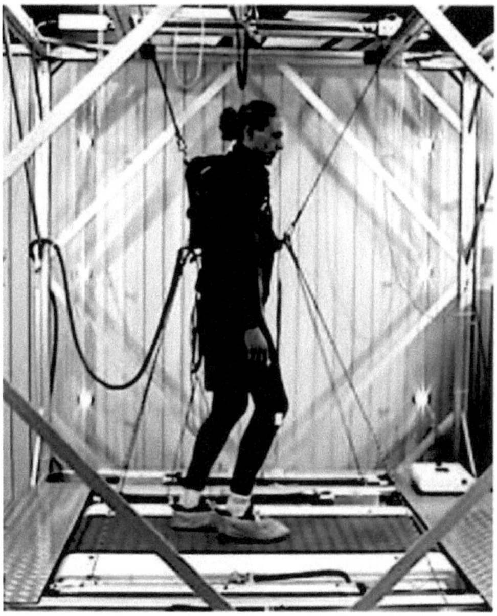

Fig. 2.33 String-Man rehablilitation robot

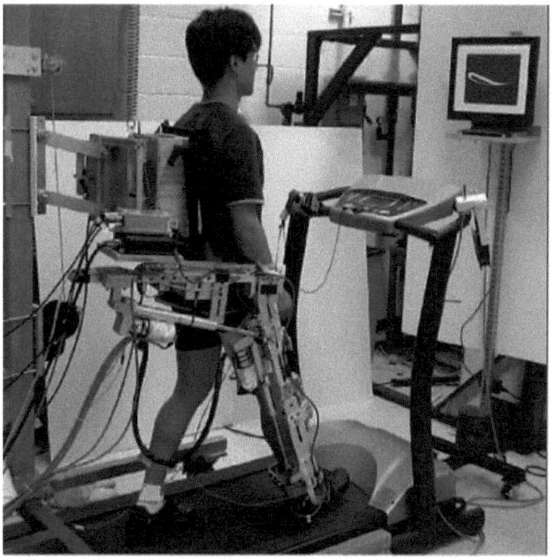

Fig. 2.34 Motorized orthosis ALEX

Force sensors between the wires and the actuators allow force or impedance control: this permits adjustment of the interaction control from totally passive to completely active and allows patients to test their balancing capabilities. A similar study devoted to the pelvic device for assistance during locomotion was carried out in (Mokhtarian et al. 2010), in which six springs were used for the suspension system.

Recently, the orthosis ALEX with force-field controller has been developed for gait rehabilitation of stroke survivors. In this case, the bodyweight suspension system is carried out via articulated linkage (Fig. 2.34; Banala et al. 2009).

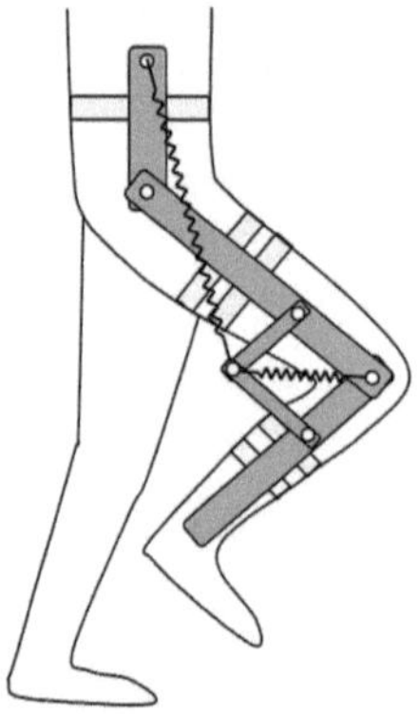

Fig. 2.35 Basic components of gravity balancing mechanism. (Banala et al. 2004)

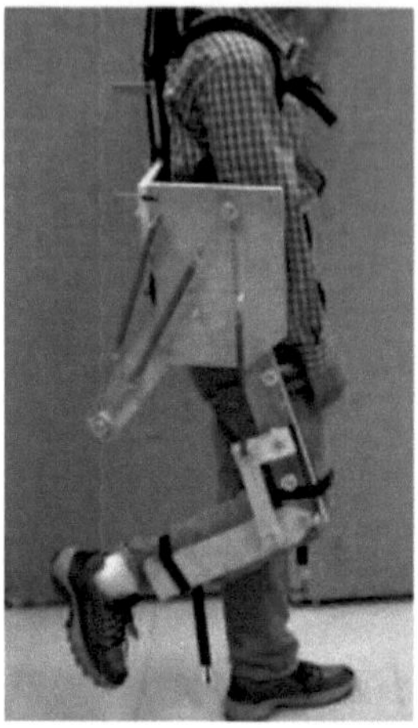

Fig. 2.36 Simple prototype. (Banala et al. 2004)

A similar suspension system for gravity compensation was proposed in (Ma and Wang 2012). It was designed for simulating walking in a zero-gravity or reduced-gravity environment.

In therapeutic situations, therapists often apply full or partial support to a paretic limb to help reduce the effect of gravity on the patient's motion. This is extremely difficult to do during walking, where the weight of the leg may create problems for the patient whose muscles are weak or who lacks normal neuromuscular control due to a neurological injury (Agrawal and Fattah 2004 c). Hence, new devices have been developed which can compensate the weight of the lower and upper extremity in all configurations (Agrawal and Fattah 2003, 2004c; Agrawal et al. 2001; Cardoso et al. 2002; Fattah and Agrawal 2004, 2005b; Herder 2005; Banala et al. 2004, 2006; Kramer et al. 2007; Agrawal et al. 2007; Agrawal and Agrawal 2005; Stienen et al. 2007; Rizk et al. 2008; Nakayama et al. 2009; Agrawal et al. 2009a).

Let us examine a few examples.

The design concept given in Figs. 2.35 and 2.36 (Agrawal et al. 2009a), which was used for the gravity balancing of the leg, is based on the following hybrid

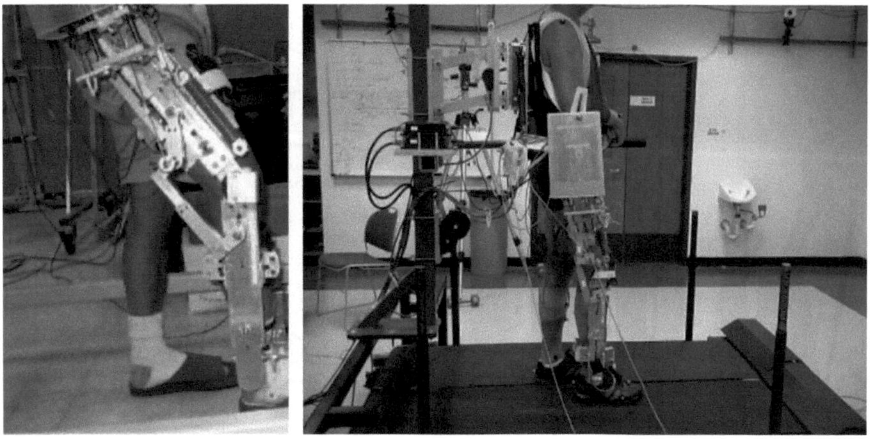

Fig. 2.37 Engineering prototypes. (Banala et al. 2004; Agrawal et al. 2007)

method: first, the centre of mass of the leg is geometrically located using a parallelogram mechanism, then the springs are placed at suitable positions in order to fully compensate the effect of gravity over the range of motion.

Figure 2.37 shows the modified engineering prototypes of the gravity compensation mechanisms, which were used in the experiments (Banala et al. 2004).

A similar problem was also solved in (Arakelian and Ghazaryan 2006, 2008). The gravity balancing mechanism consists of two springs with the same stiffness coefficients: one compression and another extension connected with the shank of the leg and permitting complete gravity compensation of the leg's weight. In order to improve the gravity compensation quality, the spring mass has been included in the balancing condition. It was shown that the mass of the balancing spring increases the unbalanced moment and it must not be neglected. Numerical simulations showed that the error caused by neglect of the spring mass can be up to 8 % (Arakelian and Ghazaryan 2008).

It should be added that the effects of friction on the gravity balanced orthosis was also disclosed (Banala et al. 2009; Agrawal et al. 2007; Rizk et al. 2008; Arakelian et al. 2012). Friction torque can be compensated by a low power motor mounted on the joint of the gravity balanced orthosis. The advantage of such an approach consists in the combination of passive gravity balancing of the leg with the active compensation of friction torque using a low power motor. Numerical simulations have shown that the error caused by neglecting the spring mass and friction torques can be up to 20 %.

Design concepts of passive gravity-balanced assistive devices for sit-to-stand tasks have also been developed (Fattah and Agrawal 2006a; Agrawal et al. 2009b; Ghazaryan et al. 2011).

Several types of upper arm exoskeletons and spring assistive arm supports have also been designed. The gravity compensation problem of these devices was also studied (Cardoso et al. 2002; Herder 2005; Dubey and Agrawal 2011; Lin et al. 2013; Smith et al. 2013).

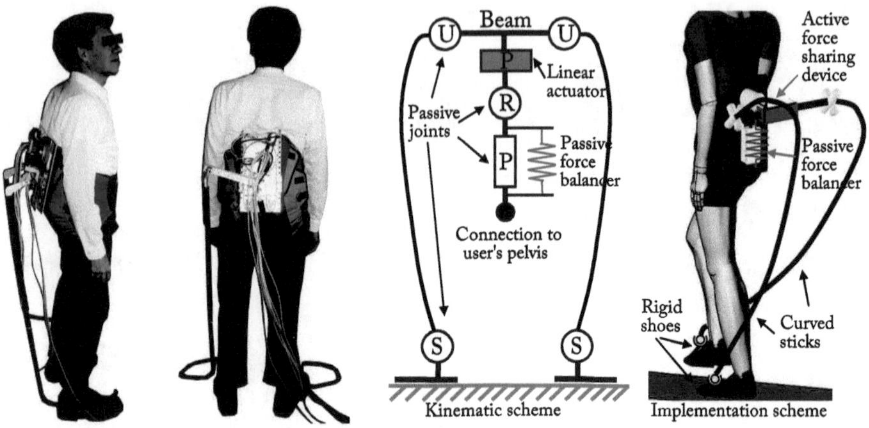

Fig. 2.38 MoonWalker. (Krut et al. 2010)

In (van Dijk et al. 2011) a passive exoskeleton to minimize joint work during walking was developed. The exoskeleton makes use of passive structures, called artificial tendons, acting in parallel with the leg. Artificial tendons are elastic elements that are able to store and redistribute energy over the human leg joints.

The gravity compensation has also been applied on humanoid robots. It should be noted that the performance of a human size humanoid robot is strictly limited by the performance of the motors. In (Shirata et al. 2007), a mechanism has been proposed to reduce the joint torque of the leg required to support the gravitational force of the whole body. Similar studies by using the gas springs have been proposed in (Sugahara et al. 2004).

Another promising field of gravity compensation is the development of walking assist devices with bodyweight support (Krut et al. 2010; Mia et al. 2013; Ashihara 2012a; Tanaka et al. 2011; HONDA 2014).

MoonWalker is a lower limb exoskeleton (Krut et al. 2010,), which is able to sustain part of a user's bodyweight. This orthosis can be used for rehabilitation, to help people who have weak legs, or to help those suffering from a broken leg, to walk (Fig. 2.38).

The main characteristic of MoonWalker is that a passive force balancer provides the force to sustain bodyweight. It is controlled using an actuator that requires very low energy to work on flat terrains, as it is used only to shift that force to the same side as the leg in stance. That motor is also able to provide a part of the energy to climb stairs or slopes. The authors believe that this approach can help improve energetic autonomy of lower limb exoskeletons.

SJTU-EX (Mia et al. 2013) is a powered lower extremity exoskeleton designed to assist and protect soldiers and construction workers. It comprises a rechargeable battery as the power supply, two pseudo-anthropomorphic legs and a backpack-like frame to mount varieties of loads. In this concept, a parallel mechanism with two

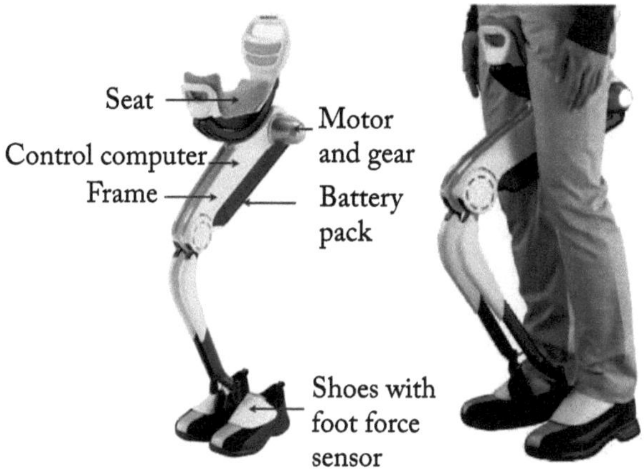

Fig. 2.39 Walking assist device with bodyweight support

degrees of freedom is introduced in place of the hip and knee for a better load-support capability. Springs are mounted on both the active and passive joints in order to eliminate the effect of gravity.

Honda's experimental walking assist device (HONDA 2014; Ikeuchi et al. 2009; Ashihara 2012b; Fig. 2.39) helps support bodyweight to reduce the load on the user's legs while walking. The aim is to lead to reduced fatigue and less physical exertion. Honda's device lightens the load on the user's legs and helps maintain a center of gravity via special mechanisms developed by the company. There are plenty of possible uses for this product helping people afflicted with mobility issues or leg problems. It can also be used for rehabilitation.

It should be noted that the gravity compensation methods continue to be developped and new approaches and solutions are constantly being reported. New physical aspects are introduced into the problems of gravity compensation, such as friction compensation by active driving systems or the improvement of the compensation accuracy by taking into account the mass of the spring. The development of new gravity compensation solutions for exoskeletons, rehabilitation devices and walking assist devices seems promising. The use of active and passive actuations allows a significant reduction of the size and weight of walking assist devices with bodyweight support.

Part II
Balancing of Linkages

Chapter 3
Partial Shaking Force and Shaking Moment Balancing of Linkages

Abstract As was mentioned in Sect. 2.1, the complete shaking force and shaking moment balancing of linkages can only be reached by a considerably complicated design of the initial linkage and by an unavoidable increase of the total mass. This is the reason why in most cases the partial balancing is used in the machinery.

The object of this Chapter is to develop the methods of partial force and moment balancing by means of optimum displacements of counterweighs.

The Sect. 3.1 deals with the shaking moment minimization of fully force balanced planar linkages by transferring the rotation axis of the counterweight mounted on the input crank. Mathematical means for the realization of such a balancing is compact and comprehensible, which allows engineers and technicians to apply the method in short time and without much difficulty.

The similar problems for spatial mechanisms, as well as for balancing by transferring several counterweights are considered in Sects. 3.2 and 3.3.

The balancing method presented in Sect. 3.4 is a symbiosis of two known solutions: approximate shaking force balancing and moment harmonic balancing. To have an identical and simplified mathematical model, the conditions for balancing are formulated by the minimization of the root-mean-square values of the shaking force and shaking moment. Such a solution has two principal advantages: (1) a simple realization, without essential change in the construction of the initial mechanism (i.e., for a shaking force balancing of the linkage, it is enough to add a counterweight mounted on the input crank and for a shaking moment balancing to transfer the axis of rotation of the input link counterweight); (2) a minimization of the shaking moment on the frame of the linkage without increase in the total mass of the counterweights (i.e., only by the use of the counterweight masses designed for the force balancing of the linkage).

V. Arakelian, S. Briot, *Balancing of Linkages and Robot Manipulators*,
Mechanisms and Machine Science 27, DOI 10.1007/978-3-319-12490-2_3

3.1 Shaking Moment Minimization of Fully Force-balanced Planar Linkages by Displacing One Counterweight

3.1.1 Complete Shaking Force and Partial Shaking Moment Balancing of Planar Linkages

3.1.1.1 Balancing of Linkage with Constant Input Angular Velocity

Let us consider an arbitrary n-bar planar linkage with constant input angular velocity $\dot{\varphi}$ (Fig. 3.1). Let $S_{cw}^{st} = m_{cw}r_{cw}$ be the static moment of the counterweight, where m_{cw} is the mass of the counterweight mounted on the input crank, and r_{cw} is the distance of its *com* from the crank axis.

After the balancing of shaking force of this linkage, the shaking moment relative to the center of rotation of the input link can be expressed as (Shchepetilnikov 1982)

$$M_O^{sh} = \sum_{k=1}^{n-1} M_O\left(f_{x_k}^{sh}, f_{y_k}^{sh}\right) - \sum_{k=2}^{n-1} I_{S_k}\ddot{\varphi}_k \tag{3.1}$$

where, for k-th link, $\mathbf{f}_k^{sh^T} = \left[f_{x_k}^{sh} \;\; f_{y_k}^{sh}\right]$ is the shaking force, $M_O\left(f_{x_k}^{sh}, f_{y_k}^{sh}\right)$ is the moment at point O due to the shaking force $\mathbf{f}_k^{sh}$, I_{S_k} is the moment of inertia relative to the *com* $S_k(x_{S_k}, y_{S_k})$ of the link, $\ddot{\varphi}_k$ is the angular acceleration.

If the axis of rotation of the counterweight is moved from center O to the center $O^*(x, y)$ (Fig. 3.1), the balancing of shaking force of the linkage will occur, but a supplementary moment is therewith added:

$$M_O^* = S_{cw}^{st}\dot{\varphi}^2\left(-y\cos\varphi + x\sin\varphi\right) \tag{3.2}$$

In the study (Shchepetilnikov 1982), the shaking moment $M_O^{sh}(\varphi)$ is approximated by the trigonometric series and the first harmonic of this series is balanced by the supplement moment M_O^*. The efficiency of such balancing will be improved by using the approximation method (Arakelian 1995) based on the minimization of the

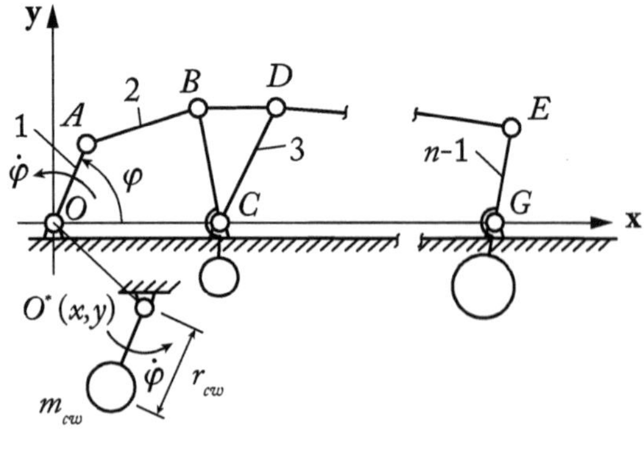

Fig. 3.1 Arbitrary n-bar planar linkage

root-mean-square (*rms*) value:

$$rms = \sqrt{\sum_{i=1}^{N} \left((M_O^{sh})_i + (M_O^*)_i\right)^2 / N} \tag{3.3}$$

where N is the number of calculated positions of the mechanism, $(M_O^*)_i$ is the inertial moment due to the counterweight in the i-th linkage configuration and $(M_O^{sh})_i$ is the shaking moment of the mechanism in the i-th linkage configuration.

For the minimization of the *rms*, it is necessary to minimize the sum:

$$\Delta_M = \sum_{i=1}^{N} \left((M_O^{sh})_i + (M_O^*)_i\right)^2 \rightarrow \min_{x,y} \tag{3.4}$$

For this purpose, we shall achieve the conditions $\partial \Delta_M / \partial x = 0$ and $\partial \Delta_M / \partial y = 0$ from where

$$S_{cw}^{st} \dot{\varphi}^2 \left(y \sum_{i=1}^{N} \cos^2 \varphi_i - x \sum_{i=1}^{N} \sin \varphi_i \cos \varphi_i \right) = - \sum_{i=1}^{N} (M_O^{sh})_i \cos \varphi_i \tag{3.5}$$

$$S_{cw}^{st} \dot{\varphi}^2 \left(x \sum_{i=1}^{N} \sin^2 \varphi_i - y \sum_{i=1}^{N} \sin \varphi_i \cos \varphi_i \right) = - \sum_{i=1}^{N} (M_O^{sh})_i \sin \varphi_i \tag{3.6}$$

From these expressions and taking into account the condition $\sum_{i=1}^{N} \sin \varphi_i \cos \varphi_i = 0$ for $\varphi \in [0 \;\; 2\pi]$, we determine the unknowns:

$$x = \frac{\sum_{i=1}^{N} (M_O^{sh})_i \sin \varphi_i}{S_{cw}^{st} \dot{\varphi}^2 \sum_{i=1}^{N} \sin^2 \varphi_i} \tag{3.7}$$

$$y = -\frac{\sum_{i=1}^{N} (M_O^{sh})_i \cos \varphi_i}{S_{cw}^{st} \dot{\varphi}^2 \sum_{i=1}^{N} \cos^2 \varphi_i} \tag{3.8}$$

3.1.1.2 Balancing of Linkage with Variable Input Angular Velocity

With the variable angular velocity $\dot{\varphi}$ of the input link taken into consideration, the supplementary moment (3.2) that balances the shaking moment will be equal to:

$$M_O^* = S_{cw}^{st} \left(\dot{\varphi}^2 (-y \cos \varphi + x \sin \varphi) - \ddot{\varphi} (r_{cw} - x \cos \varphi - y \sin \varphi) \right) \tag{3.9}$$

In this case, from the conditions $\partial \Delta_M / \partial x = 0$ and $\partial \Delta_M / \partial y = 0$, a new system of linear equations is obtained:

$$\begin{bmatrix} A_1 & B_1 \\ A_2 & B_2 \end{bmatrix} \begin{bmatrix} x \\ y \end{bmatrix} = \begin{bmatrix} C_1 \\ C_2 \end{bmatrix} \tag{3.10}$$

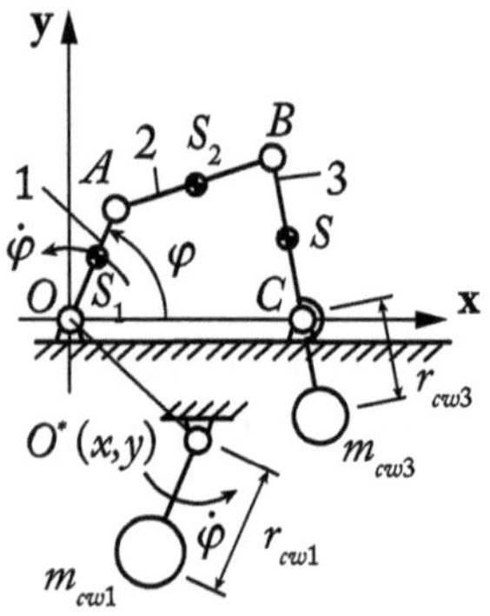

Fig. 3.2 Shaking moment balancing of a four-bar linkage

where

$$A_1 = \sum_{i=1}^{N} \left(\dot{\varphi}_i^2 \sin\varphi_i + \ddot{\varphi}_i \cos\varphi_i\right)^2 \tag{3.11}$$

$$A_2 = \sum_{i=1}^{N} \left(\dot{\varphi}_i^2 \sin\varphi_i + \ddot{\varphi}_i \cos\varphi_i\right)\left(\dot{\varphi}_i^2 \cos\varphi_i - \ddot{\varphi}_i \sin\varphi_i\right) \tag{3.12}$$

$$B_1 = -\sum_{i=1}^{N} \left(\dot{\varphi}_i^2 \sin\varphi_i - \ddot{\varphi}_i \cos\varphi_i\right)\left(\dot{\varphi}_i^2 \cos\varphi_i + \ddot{\varphi}_i \sin\varphi_i\right) \tag{3.13}$$

$$B_2 = -\sum_{i=1}^{N} \left(\dot{\varphi}_i^2 \cos\varphi_i - \ddot{\varphi}_i \sin\varphi_i\right)^2 \tag{3.14}$$

$$C_1 = \sum_{i=1}^{N} \left(\ddot{\varphi}_i r_{cw} - (M_O^{sh})_i / S_{cw}^{st}\right)\left(\dot{\varphi}_i^2 \sin\varphi_i + \ddot{\varphi}_i \cos\varphi_i\right) \tag{3.15}$$

$$C_2 = \sum_{i=1}^{N} \left(\ddot{\varphi}_i r_{cw} - (M_O^{sh})_i / S_{cw}^{st}\right)\left(\dot{\varphi}_i^2 \cos\varphi_i - \ddot{\varphi}_i \sin\varphi_i\right) \tag{3.16}$$

The coordinates of the axis of rotation of the counterweight x and y are determined from the inversion of the system (3.10):

$$\begin{bmatrix} x \\ y \end{bmatrix} = \frac{1}{A_1 B_2 - A_2 B_1} \begin{bmatrix} B_2 & -B_1 \\ -A_2 & A_1 \end{bmatrix} \begin{bmatrix} C_1 \\ C_2 \end{bmatrix} \tag{3.17}$$

3.1.2 Numerical Example and Comparative Analysis

For the four-bar linkage (Fig. 3.2) with parameters:

- $l_{OA} = 0.1$ m, $l_{AB} = 0.3$ m, $l_{BC} = 0.2$ m, $l_{OC} = 0.3$ m, $l_{OS_1} = 0.05$ m, $l_{AS_2} = 0.15$ m, $l_{CS_3} = 0.1$ m,
- $m_1 = 1$ kg, $m_2 = 3$ kg, $m_3 = 2$ kg,
- $I_{S_2} = 0.0225$ kg m^2, $I_{S_3} = 0.01$ kg m^2,
- $\dot{\varphi} = 10$ rad/s,

and taken into account that $r_{cw_1} = 0.05$ m and $r_{cw_3} = 0.1$ m, we achieve the complete shaking force balancing of the mechanism (Shchepetilnikov 1982) and by means of the formulae (3.7) and (3.8) we determine the values $x = 0.189$ m and $y = 0.071$ m.

In Fig. 3.3 are presented the variations of the shaking moments $M_j^{sh}(\varphi)$ for the four-bar linkage under consideration:

1. non-balanced ($j = 1$);
2. only shaking force balanced ($j = 2$);
3. complete shaking force and partial shaking moment balanced by the known method (Shchepetilnikov 1982) ($j = 3$);
4. complete shaking force and partial shaking moment balanced by the proposed method ($j = 4$).

In Fig. 3.3, with the diagrams of shaking moment are presented the coefficients μ_j which characterize the efficiency of balancing methods. These coefficients are calculated by the following expression:

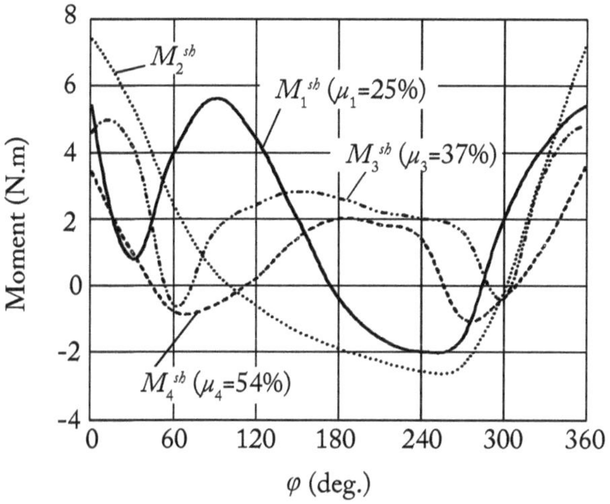

Fig. 3.3 Displacements of the axes of rotation of the counterweights

$$\mu_j = \frac{\max|M_m(\varphi)| - \max|M_j(\varphi)|}{\max|M_m(\varphi)|} \tag{3.18}$$

where

$$|M_m(\varphi)| = \max\{|M_1(\varphi)|, |M_2(\varphi)|, |M_3(\varphi)|, |M_4(\varphi)|\} \tag{3.19}$$

It should be noted that in most of cases: $m = 2$, i.e., the highest value of the shaking moment is obtained for the mechanism which is balanced statically.

3.2 Shaking Moment Minimization of Fully Force-balanced Planar Linkages by Displacing Several Counterweights

3.2.1 *Minimization of the Shaking Moment by Parallel Displacements of Counterweights Mounted on the Frame*

Let us consider now the force-balanced arbitrary n-bar planar linkage depicted in (Fig. 3.4).

If the axis of rotation of the counterweight j is moved from center O_j to the center $O_j^*(x_j, y_j)$ (see Fig. 3.5), the balancing of the shaking force of the linkage can be maintained and the shaking moment relative to the center of rotation of the input link is still equal to (3.1) (Shchepetilnikov 1982) but, in addition, a supplementary moment M_j^* is thereby introduced:

$$M_j^* = S_{cw_j}^{st}\dot{\varphi}_j{}^2\left(-y_j\cos\varphi_j + x_j\sin\varphi_j\right) - S_{cw_j}^{st}\ddot{\varphi}_j\left(r_{cw_j} - x_j\cos\varphi_j - y_j\sin\varphi_j\right) \tag{3.20}$$

where φ_j, $\dot{\varphi}_j$ and $\ddot{\varphi}_j$ are the angle of rotation, the angular velocity and the angular acceleration of the counterweight j, $S_{cw_j}^{st} = m_{cw_j}r_{cw_j}$ is the static moment of the counterweight j relative to its axis of rotation, m_{cw_j} is its mass and r_{cw_j} is the distance between its *com* S_j and its axis of rotation.

For minimization of the root mean square value

$$rms = \sqrt{\sum_{i=1}^{N}\left((M_O^{sh})_i + \sum_{j=1}^{p}(M_j^*)_i\right)^2 / N} \tag{3.21}$$

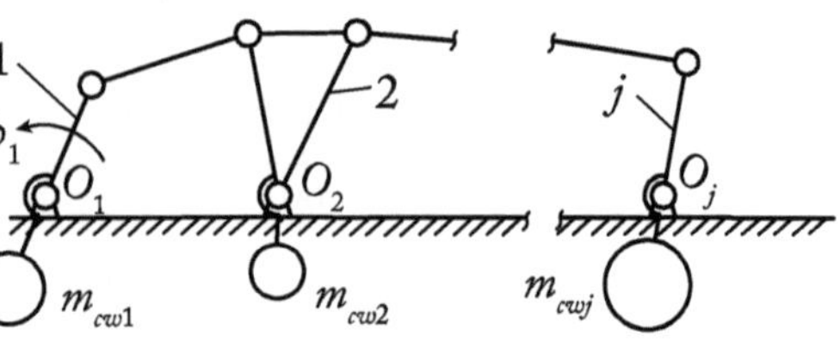

Fig. 3.4 Force balanced n-bar linkage

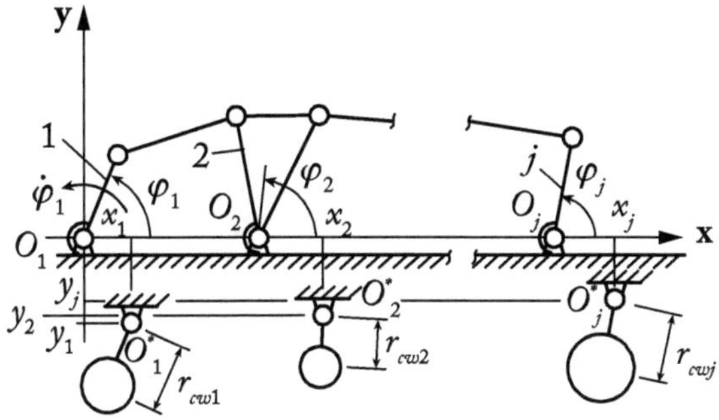

Fig. 3.5 Displacements of the axes of rotation of the counterweights

where $(M_j^*)_i$ is the inertial moment due to the counterweight j in the i-th linkage configuration and $(M_O^{sh})_i$ is the shaking moment of the mechanism in the i-th linkage configuration, it is necessary to minimize the sum:

$$\Delta = \sum_{i=1}^{N} \left((M_O^{sh})_i + \sum_{j=1}^{p} (M_j^*)_i \right)^2 \to \min_{x_j, y_j} \tag{3.22}$$

where N is the number of calculated positions of the linkage and p is the number of counterweights.

For this purpose, the following conditions will be imposed:

$$\frac{\partial \Delta}{\partial x_j} = 0, \quad \frac{\partial \Delta}{\partial y_j} = 0 \tag{3.23}$$

from which we obtain the system of linear equations:

$$\begin{bmatrix} a_{11} & a_{12} & \cdots & a_{1(2p-1)} & a_{1(2p)} \\ a_{21} & a_{22} & \cdots & a_{2(2p-1)} & a_{2(2p)} \\ \vdots & \vdots & \vdots & \vdots & \vdots \\ a_{(2p-1)1} & a_{(2p-1)2} & \cdots & a_{(2p-1)(2p-1)} & a_{(2p-1)(2p)} \\ a_{(2p)1} & a_{(2p)2} & \cdots & a_{(2p)(2p-1)} & a_{(2p)(2p)} \end{bmatrix} \begin{bmatrix} x_1 \\ y_1 \\ \vdots \\ x_p \\ y_p \end{bmatrix} = \begin{bmatrix} b_1 \\ b_2 \\ \vdots \\ b_{2p-1} \\ b_{2p} \end{bmatrix} \tag{3.24}$$

where

$$a_{(2k-1)(2n-1)} = \sum_{i=1}^{N} A_{k\,i} A_{n\,i} \tag{3.25}$$

$$a_{(2k-1)(2n)} = \sum_{i=1}^{N} A_{k\,i} B_{n\,i} \tag{3.26}$$

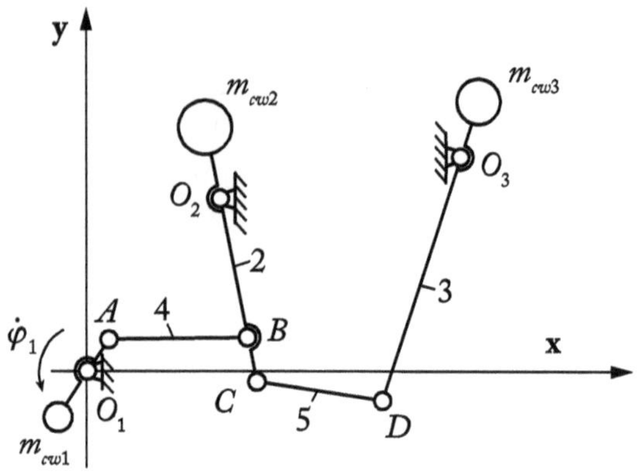

Fig. 3.6 Six-bar force-balanced linkage

$$a_{(2k)(2n-1)} = \sum_{i=1}^{N} B_{k\,i} A_{n\,i} \tag{3.27}$$

$$a_{(2k)(2n)} = \sum_{i=1}^{N} B_{k\,i} B_{n\,i} \tag{3.28}$$

$$b_{2k-1} = \sum_{i=1}^{N} A_{k\,i} (M_O^{sh})_i + \sum_{j=1}^{m} A_{k\,i} C_{j\,i} \tag{3.29}$$

$$b_{2k} = -\sum_{i=1}^{N} B_{k\,i} \left((M_O^{sh})_i + \sum_{j=1}^{m} C_{j\,i} \right) \tag{3.30}$$

with

$$A_{k\,i} = S_{cw_k}^{st} \left((\dot{\varphi}_k)_i^2 \sin(\varphi_k)_i - (\ddot{\varphi}_k)_i \cos(\varphi_k)_i \right) \tag{3.31}$$

$$B_{k\,i} = -S_{cw_k}^{st} \left((\dot{\varphi}_k)_i^2 \cos(\varphi_k)_i + (\ddot{\varphi}_k)_i \sin(\varphi_k)_i \right) \tag{3.32}$$

$$C_{k\,i} = S_{cw_k}^{st} (\ddot{\varphi}_k)_i r_{cw_k} \tag{3.33}$$

and φ_j^i, $(\dot{\varphi}_j)_i$ and $(\ddot{\varphi}_j)_i$ are the values of φ_j, $\dot{\varphi}_j$ and $\ddot{\varphi}_j$ in the i-th linkage configuration.

By inversion of the system of linear equations (3.24), it is possible to determine the unknowns x_j and y_j (Korn and Korn 1968).

3.2.2 *Example: Balancing of a Six-Bar Linkage*

Consider a six-bar planar linkage with revolute joints and constant input angular velocity $\dot{\varphi}_1$ (Fig. 3.6). After balancing the shaking force of this linkage by three counterweights connected to the links O_1A, O_2C and O_3D, the shaking moment relative to the centre O_1 of rotation of input link O_1A can be expressed as (Shchepetilnikov 1982):

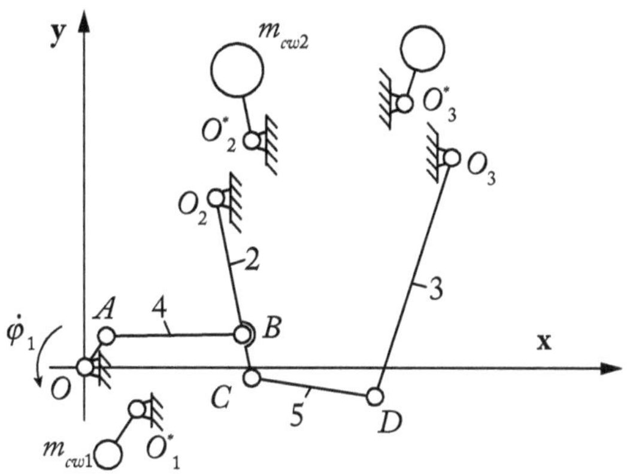

Fig. 3.7 Displacements of the axes of rotation of the counterweights for the six-bar force-balanced linkage

$$M_{O_1}^{sh} = \sum_{k=2}^{5} M_{O_1}\left(f_{x_k}^{sh}, f_{y_k}^{sh}\right) - \sum_{k=2}^{5} I_{S_k}\ddot{\varphi}_k \tag{3.34}$$

After displacements of the axes of rotation of the counterweights (Fig. 3.7) from centres O_j to the centres O_j^* $(j = 1, 2, 3)$, the shaking force balance is maintained but, in addition, supplementary moments M_j^* are thereby introduced:

$$M_1^* = S_{cw_1}^{st}\dot{\varphi}_1{}^2\left(-y_1 \cos\varphi_j + x_1 \sin\varphi_j\right) \tag{3.35}$$

$$M_2^* = S_{cw_2}^{st}\dot{\varphi}_2{}^2\left(-y_2 \cos\varphi_2 + x_2 \sin\varphi_2\right) - S_{cw_2}^{st}\ddot{\varphi}_2\left(r_{cw_2} - x_2 \cos\varphi_2 - y_2 \sin\varphi_2\right) \tag{3.36}$$

$$M_3^* = S_{cw_3}^{st}\dot{\varphi}_3{}^2\left(-y_3 \cos\varphi_3 + x_3 \sin\varphi_3\right) - S_{cw_3}^{st}\ddot{\varphi}_3\left(r_{cw_3} - x_3 \cos\varphi_3 - y_3 \sin\varphi_3\right) \tag{3.37}$$

The three counterweights are driven (by gears or toothed belts for example) at the same rotational speed as links O_1A, O_2C and O_3D. For clarity, the driving mechanisms are not shown here.

In this case, the minimization of the root mean square value is presented by the expression:

$$\Delta = \sum_{i=1}^{N}\left((M_{O_1}^{sh})_i + (M_1^* + M_2^* + M_3^*)_i\right)^2 \rightarrow \min_{x_1, y_1, x_2, y_2, x_3, y_3} \tag{3.38}$$

For this purpose, we shall impose conditions (3.23):

$$\begin{aligned}
&\frac{\partial \Delta}{\partial x_1} = 0, \quad \frac{\partial \Delta}{\partial y_1} = 0,\\
&\frac{\partial \Delta}{\partial x_2} = 0, \quad \frac{\partial \Delta}{\partial y_2} = 0,\\
&\frac{\partial \Delta}{\partial x_3} = 0, \quad \frac{\partial \Delta}{\partial y_3} = 0
\end{aligned} \tag{3.39}$$

from which we obtain the following system of linear equations:

$$\begin{bmatrix} a_{11} & a_{12} & a_{13} & a_{14} & a_{15} & a_{16}\\ a_{21} & a_{22} & a_{23} & a_{24} & a_{25} & a_{26}\\ a_{31} & a_{32} & a_{33} & a_{34} & a_{35} & a_{36}\\ a_{41} & a_{42} & a_{43} & a_{44} & a_{45} & a_{46}\\ a_{51} & a_{52} & a_{53} & a_{54} & a_{55} & a_{56}\\ a_{61} & a_{62} & a_{63} & a_{64} & a_{65} & a_{66} \end{bmatrix} \begin{bmatrix} x_1\\ y_1\\ x_2\\ y_2\\ x_3\\ y_3 \end{bmatrix} = \begin{bmatrix} b_1\\ b_2\\ b_3\\ b_4\\ b_5\\ b_6 \end{bmatrix} \tag{3.40}$$

where the terms a_{kn} are defined in Sect. 3.2.1 using

$$A_{1i} = S^{st}_{cw_1}(\dot{\varphi}_1)^2_i \sin(\varphi_1)_i \tag{3.41}$$

$$B_{1i} = -S^{st}_{cw_1}(\dot{\varphi}_1)^2_i \cos(\varphi_1)_i \tag{3.42}$$

$$C_{1i} = 0 \tag{3.43}$$

$$A_{2i} = S^{st}_{cw_2}\left((\dot{\varphi}_2)^2_i \sin(\varphi_2)_i - (\ddot{\varphi}_2)_i \cos(\varphi_2)_i\right) \tag{3.44}$$

$$B_{2i} = -S^{st}_{cw_2}\left((\dot{\varphi}_2)^2_i \cos(\varphi_2)_i + (\ddot{\varphi}_2)_i \sin(\varphi_2)_i\right) \tag{3.45}$$

$$C_{2i} = S^{st}_{cw_2}(\ddot{\varphi}_2)_i r_{cw_2} \tag{3.46}$$

$$A_{3i} = S^{st}_{cw_3}\left((\dot{\varphi}_3)^2_i \sin(\varphi_3)_i - (\ddot{\varphi}_3)_i \cos(\varphi_3)_i\right) \tag{3.47}$$

$$B_{3i} = -S^{st}_{cw_3}\left((\dot{\varphi}_3)^2_i \cos(\varphi_3)_i + (\ddot{\varphi}_3)_i \sin(\varphi_3)_i\right) \tag{3.48}$$

$$C_{3i} = S^{st}_{cw_3}(\ddot{\varphi}_3)_i r_{cw_3} \tag{3.49}$$

The coordinates of the axes of rotation of the counterweights are determined by:

$$x_j = D_{x_j}/D, \quad y_j = D_{y_j}/D \tag{3.50}$$

where,

$$D = \begin{vmatrix} a_{11} & a_{12} & a_{13} & a_{14} & a_{15} & a_{16} \\ a_{21} & a_{22} & a_{23} & a_{24} & a_{25} & a_{26} \\ a_{31} & a_{32} & a_{33} & a_{34} & a_{35} & a_{36} \\ a_{41} & a_{42} & a_{43} & a_{44} & a_{45} & a_{46} \\ a_{51} & a_{52} & a_{53} & a_{54} & a_{55} & a_{56} \\ a_{61} & a_{62} & a_{63} & a_{64} & a_{65} & a_{66} \end{vmatrix} \tag{3.51}$$

and

$$D_{x_1} = \begin{vmatrix} b_1 & a_{12} & a_{13} & a_{14} & a_{15} & a_{16} \\ b_2 & a_{22} & a_{23} & a_{24} & a_{25} & a_{26} \\ b_3 & a_{32} & a_{33} & a_{34} & a_{35} & a_{36} \\ b_4 & a_{42} & a_{43} & a_{44} & a_{45} & a_{46} \\ b_5 & a_{52} & a_{53} & a_{54} & a_{55} & a_{56} \\ b_6 & a_{62} & a_{63} & a_{64} & a_{65} & a_{66} \end{vmatrix}, \; D_{y_1} = \begin{vmatrix} a_{11} & b_1 & a_{13} & a_{14} & a_{15} & a_{16} \\ a_{21} & b_2 & a_{23} & a_{24} & a_{25} & a_{26} \\ a_{31} & b_3 & a_{33} & a_{34} & a_{35} & a_{36} \\ a_{41} & b_4 & a_{43} & a_{44} & a_{45} & a_{46} \\ a_{51} & b_5 & a_{53} & a_{54} & a_{55} & a_{56} \\ a_{61} & b_6 & a_{63} & a_{64} & a_{65} & a_{66} \end{vmatrix} \tag{3.52}$$

$$D_{x_1} = \begin{vmatrix} a_{11} & a_{12} & b_1 & a_{14} & a_{15} & a_{16} \\ a_{21} & a_{22} & b_2 & a_{24} & a_{25} & a_{26} \\ a_{31} & a_{32} & b_3 & a_{34} & a_{35} & a_{36} \\ a_{41} & a_{42} & b_4 & a_{44} & a_{45} & a_{46} \\ a_{51} & a_{52} & b_5 & a_{54} & a_{55} & a_{56} \\ a_{61} & a_{62} & b_6 & a_{64} & a_{65} & a_{66} \end{vmatrix}, \; D_{y_1} = \begin{vmatrix} a_{11} & a_{12} & a_{13} & b_1 & a_{15} & a_{16} \\ a_{21} & a_{22} & a_{23} & b_2 & a_{25} & a_{26} \\ a_{31} & a_{32} & a_{33} & b_3 & a_{35} & a_{36} \\ a_{41} & a_{42} & a_{43} & b_4 & a_{45} & a_{46} \\ a_{51} & a_{52} & a_{53} & b_5 & a_{55} & a_{56} \\ a_{61} & a_{62} & a_{63} & b_6 & a_{65} & a_{66} \end{vmatrix} \tag{3.53}$$

$$D_{x_1} = \begin{vmatrix} a_{11} & a_{12} & a_{13} & a_{14} & b_1 & a_{16} \\ a_{21} & a_{22} & a_{23} & a_{24} & b_2 & a_{26} \\ a_{31} & a_{32} & a_{33} & a_{34} & b_3 & a_{36} \\ a_{41} & a_{42} & a_{43} & a_{44} & b_4 & a_{46} \\ a_{51} & a_{52} & a_{53} & a_{54} & b_5 & a_{56} \\ a_{61} & a_{62} & a_{63} & a_{64} & b_6 & a_{66} \end{vmatrix}, \; D_{y_1} = \begin{vmatrix} a_{11} & a_{12} & a_{13} & a_{14} & a_{15} & b_1 \\ a_{21} & a_{22} & a_{23} & a_{24} & a_{25} & b_2 \\ a_{31} & a_{32} & a_{33} & a_{34} & a_{35} & b_3 \\ a_{41} & a_{42} & a_{43} & a_{44} & a_{45} & b_4 \\ a_{51} & a_{52} & a_{53} & a_{54} & a_{55} & b_5 \\ a_{61} & a_{62} & a_{63} & a_{64} & a_{65} & b_6 \end{vmatrix} \tag{3.54}$$

3.2.3 Numerical Example

The parameters of the unbalanced six-bar linkage (Fig. 3.6) are the followings:

- $l_{O_1A}=0.6\,\text{m}$, $l_{O_2C}=0.2713\,\text{m}$, $l_{O_3D}=0.364\,\text{m}$, $l_{AB}=0.1965\,\text{m}$, $l_{BC}=0.065\,\text{m}$, $l_{CD}=0.187\,\text{m}$,
- $l_{O_1S_1}=0.03\,\text{m}$, $l_{O_2S_2}=0.136\,\text{m}$, $l_{O_3S_3}=0.364\,\text{m}$, $l_{AS_4}=0.0983\,\text{m}$, $l_{BS_5}=0.033\,\text{m}$,
- $x_{O_2}=0.2\,\text{m}$, $y_{O_2}=0.25\,\text{m}$, $x_{O_3}=0.55\,\text{m}$, $y_{O_3}=0.3\,\text{m}$,
- $m_1=1\,\text{kg}$, $m_2=0.6\,\text{kg}$, $m_3=1.2\,\text{kg}$, $m_4=0.5\,\text{kg}$, $m_5=1\,\text{kg}$,
- $I_{S_1}=0.02\,\text{kg m}^2$, $I_{S_2}=0.004\,\text{kg m}^2$, $I_{S_3}=0.015\,\text{kg m}^2$, $I_{S_4}=0.002\,\text{kg m}^2$, $I_{S_5}=0.003\,\text{kg m}^2$,
- $\dot{\varphi}_1=10\,\text{rad/s}$.

After the complete force balancing of the linkage, we determine by the proposed method the coordinates of the axes of rotation of the counterweights (Fig. 3.7): $x_1=-0.0142\,\text{m}$, $y_1=0.0441\,\text{m}$, $x_2=0.1031\,\text{m}$, $y_2=0.7940\,\text{m}$, $x_3=0.2113\,\text{m}$, $y_3=-0.3658\,\text{m}$.

Figure 3.8 presents the variations in the shaking moment for the following states of the six-bar linkage:

- (a) : shaking force balanced;
- (b) : minimization by the proposed method.

The obtained results show that the suggested balancing method is efficient and that it reduces the maximum values of the shaking moment. The extreme values for the fully force-balanced linkage (curve "a") are −8.74 and 6.47 N m and for the minimized version (curve "b") −1.25 and 0.64 N m. The amplitude of the shaking moment for the fully force-balanced linkage is therefore 15.21 N m compared with 1.89 N m for the optimized linkage. Thus, the numerical example illustrates that a 80 % reduction in the shaking moment is achieved. It should be noted that such a

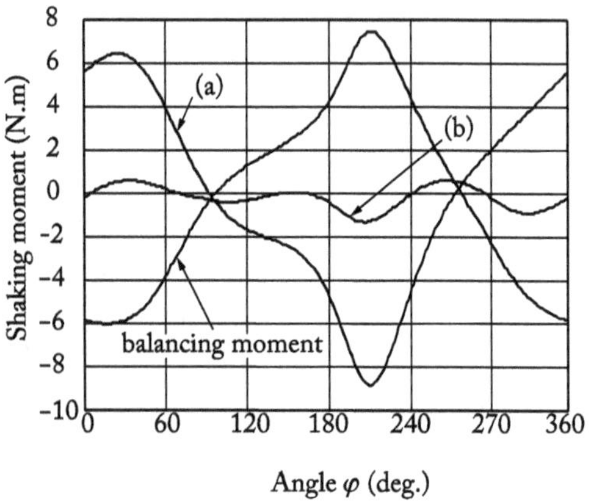

Fig. 3.8 Variations in the shaking moment for fully force-balanced (**a**) and optimized (**b**) linkages

reduction is achieved by only the displacement of the counterweights designed for the full force-balancing of the linkage.

3.3 Shaking Moment Minimization of Fully Force-balanced Spatial Linkages

3.3.1 Complete Shaking Force and Partial Shaking Moment Balancing of Spatial Linkages

3.3.1.1 Balancing of Linkage with Constant Input Angular Velocity

In Fig. 3.9 is presented a spatial *n*-bar linkage with constant input angular velocity $\dot{\varphi}$.

It is considered that this linkage is balanced statically (the shaking force is cancelled). The shaking moment of the linkage can be represented by the expression:

$$\mathbf{m}^{sh} = \mathbf{m}_x^{sh} + \mathbf{m}_y^{sh} + \mathbf{m}_z^{sh} = \sum_{i=1}^{n-1} \mathbf{m}_{x_i}^{sh} + \sum_{i=1}^{n-1} \mathbf{m}_{y_i}^{sh} + \sum_{i=1}^{n-1} \mathbf{m}_{z_i}^{sh} \tag{3.55}$$

where $\mathbf{m}_x^{sh}$, $\mathbf{m}_y^{sh}$ and $\mathbf{m}_z^{sh}$ are the components of the shaking moment of the linkage along **x**, **y** and **z** axes, respectively; $\mathbf{m}_{x_i}^{sh}$, $\mathbf{m}_{y_i}^{sh}$ and $\mathbf{m}_{z_i}^{sh}$ are the components of the shaking moment of the k-th link of the linkage along **x**, **y** and **z** axes, respectively; $n-1$ is the number of moving links.

If the axis of rotation of the counterweight (Fig. 3.9) is moved from center O to the center $O^*(x, y, z)$, the balancing of shaking force of the linkage will occur, but a supplementary moment $\mathbf{m}^*$ is therewith added. In this case (Fig. 3.9), the values of

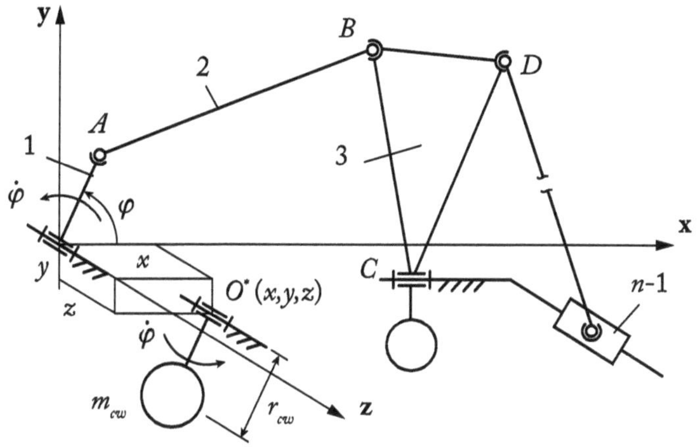

Fig. 3.9 Arbitrary *n*-bar spatial linkage

the components of this additional moment are the followings:

$$M_x^* = S_{cw}^{st}\dot{\varphi}^2 z \sin\varphi \tag{3.56}$$

$$M_y^* = -S_{cw}^{st}\dot{\varphi}^2 z \cos\varphi \tag{3.57}$$

$$M_z^* = S_{cw}^{st}\dot{\varphi}^2 \left(-y\cos\varphi + x\sin\varphi\right) \tag{3.58}$$

where $S_{cw}^{st} = m_{cw} r_{cw}$ is the static moment of the counterweight relative to the axis of rotation.

The problem is the following: to find such coordinates x, y, z of the counterweight centre of rotation which will allow the minimization of the shaking moment of the linkage. For this purpose, it is necessary to minimize the *rms* value of the shaking moment:

$$rms = \sqrt{\sum_{i=1}^{N} \left(\mathbf{m}_i^{sh} + \mathbf{m}_i^*\right)^T \left(\mathbf{m}_i^{sh} + \mathbf{m}_i^*\right)/N} \tag{3.59}$$

where i and N are, respectively, the index and the total number of calculated positions of the linkage.

To obtain a minimum of the *rms* it is necessary to minimize the sum:

$$\Delta = \sum_{i=1}^{N} \left(\mathbf{m}_i^{sh} + \mathbf{m}_i^*\right)^T \left(\mathbf{m}_i^{sh} + \mathbf{m}_i^*\right) \to \min_{x,y,z} \tag{3.60}$$

or

$$\Delta = \sum_{i=1}^{N} \left(M_{xi}^{sh} + M_{xi}^*\right)^2 + \left(M_{yi}^{sh} + M_{yi}^*\right)^2 + \left(M_{zi}^{sh} + M_{zi}^*\right)^2 \to \min_{x,y,z} \tag{3.61}$$

where M_x^{sh}, M_y^{sh} and M_z^{sh} are the norms of vectors $\mathbf{m}_x^{sh}$, $\mathbf{m}_y^{sh}$ and $\mathbf{m}_z^{sh}$, respectively.

For this purpose, it is necessary to ensure the conditions $\partial\Delta/\partial x = 0$, $\partial\Delta/\partial y = 0$ and $\partial\Delta/\partial z = 0$.

From these expressions and taking into account the condition $\sum_{i=1}^{N} \sin\varphi_i \cos\varphi_i = 0$ for $\varphi \in [0 \;\; 2\pi]$, we determine the unknowns:

$$x = \frac{\sum_{i=1}^{N} M_{zi}^{sh} \sin\varphi_i}{S_{cw}^{st}\dot{\varphi}^2 \sum_{i=1}^{N} \sin^2\varphi_i} \tag{3.62}$$

$$y = -\frac{\sum_{i=1}^{N} M_{zi}^{sh} \cos\varphi_i}{S_{cw}^{st}\dot{\varphi}^2 \sum_{i=1}^{N} \cos^2\varphi_i} \tag{3.63}$$

$$z = \frac{\sum_{i=1}^{N} M_{yi}^{sh} \cos\varphi_i - \sum_{i=1}^{N} M_{xi}^{sh} \sin\varphi_i}{S_{cw}^{st}\dot{\varphi}^2 \left(\sum_{i=1}^{N} \cos^2\varphi_i + \sum_{i=1}^{N} \sin^2\varphi_i\right)} \tag{3.64}$$

3.3.1.2 Balancing of Linkage with Variable Input Angular Velocity

If for the studied mechanism, the angular velocity of the input crank 1 (Fig. 3.9) is regarded as variable: $\dot{\varphi} = \dot{\varphi}(t)$, the additional moment which balances the shaking moment can be represented by the expressions:

$$M_x^* = S_{cw}^{st} z \left(\dot{\varphi}^2 \sin\varphi - \ddot{\varphi}\cos\varphi\right) \tag{3.65}$$

$$M_y^* = -S_{cw}^{st} z \left(\dot{\varphi}^2 \cos\varphi + \ddot{\varphi}\sin\varphi\right) \tag{3.66}$$

$$M_z^* = S_{cw}^{st} \left(\dot{\varphi}^2 \left(-y\cos\varphi + x\sin\varphi\right) - \ddot{\varphi}\left(r_{cw} - x\cos\varphi - y\sin\varphi\right)\right) \tag{3.67}$$

In this case, from the conditions $\partial\Delta/\partial x = 0$, $\partial\Delta/\partial y = 0$ and $\partial\Delta/\partial z = 0$, a new system of linear equations is obtained:

$$\begin{bmatrix} A_1 & B_1 & 0 \\ A_2 & B_2 & 0 \\ 0 & 0 & C_3 \end{bmatrix} \begin{bmatrix} x \\ y \\ z \end{bmatrix} = \begin{bmatrix} E_1 \\ E_2 \\ E_3 \end{bmatrix} \tag{3.68}$$

where

$$A_1 = \sum_{i=1}^{N} \left(\dot{\varphi}_i^2 \sin\varphi_i + \ddot{\varphi}_i \cos\varphi_i\right)^2 \tag{3.69}$$

$$A_2 = \sum_{i=1}^{N} \left(\dot{\varphi}_i^2 \sin\varphi_i + \ddot{\varphi}_i \cos\varphi_i\right)\left(\dot{\varphi}_i^2 \cos\varphi_i - \ddot{\varphi}_i \sin\varphi_i\right) \tag{3.70}$$

$$B_1 = -A_2 \tag{3.71}$$

$$B_2 = -\sum_{i=1}^{N} \left(\dot{\varphi}_i^2 \cos\varphi_i - \ddot{\varphi}_i \sin\varphi_i\right)^2 \tag{3.72}$$

$$C_3 = \sum_{i=1}^{N} \left(\dot{\varphi}_i^2 \sin\varphi_i - \ddot{\varphi}_i \cos\varphi_i\right)^2 + \sum_{i=1}^{N} \left(\dot{\varphi}_i^2 \cos\varphi_i + \ddot{\varphi}_i \sin\varphi_i\right)^2 \tag{3.73}$$

$$E_1 = \sum_{i=1}^{N} \left(\ddot{\varphi}_i r_{cw} - M_{zi}^{sh}/S_{cw}^{st}\right)\left(\dot{\varphi}_i^2 \sin\varphi_i + \ddot{\varphi}_i \cos\varphi_i\right) \tag{3.74}$$

$$E_2 = \sum_{i=1}^{N} \left(\ddot{\varphi}_i r_{cw} - M_{zi}^{sh}/S_{cw}^{st}\right)\left(\dot{\varphi}_i^2 \cos\varphi_i - \ddot{\varphi}_i \sin\varphi_i\right) \tag{3.75}$$

$$E_3 = \frac{\sum_{i=1}^{N} M_{yi}^{sh}\left(\dot{\varphi}_i^2 \cos\varphi_i + \ddot{\varphi}_i \sin\varphi_i\right) - \sum_{i=1}^{N} M_{xi}^{sh}\left(\dot{\varphi}_i^2 \sin\varphi_i - \ddot{\varphi}_i \cos\varphi_i\right)}{S_{cw}^{st}} \tag{3.76}$$

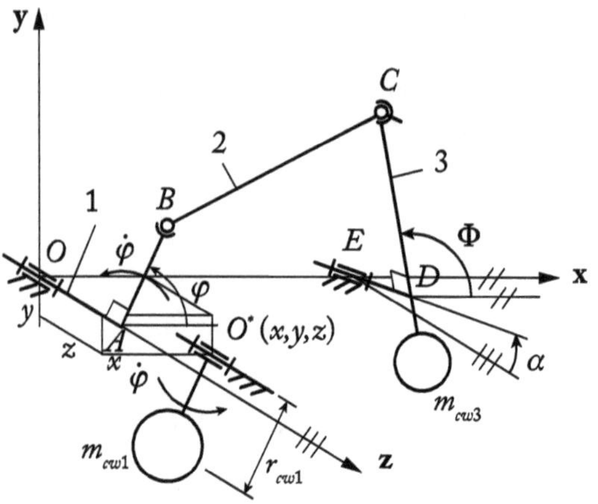

Fig. 3.10 Complete shaking force and partial shaking moment balancing of the *RSS'R* spatial linkage

x, y and z are determined by the inversion of the system (3.68):

$$\begin{bmatrix} x \\ y \\ z \end{bmatrix} = \frac{1}{C_3\left(A_1B_2 - A_2B_1\right)} \begin{bmatrix} B_2C_3 & -B_1C_3 & 0 \\ -A_2C_3 & A_1C_3 & 0 \\ 0 & 0 & A_1B_2 - A_2B_1 \end{bmatrix} \begin{bmatrix} E_1 \\ E_2 \\ E_3 \end{bmatrix} \tag{3.77}$$

3.3.2 *Numerical Example and Comparative Analysis*

Let us examine the shaking moment balancing of the *RSS'R* spatial linkage (Fig. 3.10) with the following parameters:

- $l_{OA} = 0.05$ m, $l_{AB} = 0.2$ m, $l_{BC} = 0.6$ m, $l_{CD} = 0.4$ m, $l_{ED} = 0.1$ m, $l_{OE} = 0.5$ m, $\alpha = 15°$,
- $OA \perp AB$, $ED \perp CD$,
- $l_{AS_1} = l_{AB}/4$, $l_{BS_2} = l_{BC}/2$, $l_{DS_3} = l_{DC}/2.5$
- $m_1 = 3$ kg, $m_2 = 1.8$ kg, $m_3 = 2$ kg,
- $I_{S_2} = 0.0135$ kg m^2, $I_{S_3} = 0.025$ kg m^2,
- $\dot{\varphi} = 10$ rad/s.

The rotation of link 2 about the axis BC is cancelled by a pin and it has a symmetry relative to the *com* S_2.

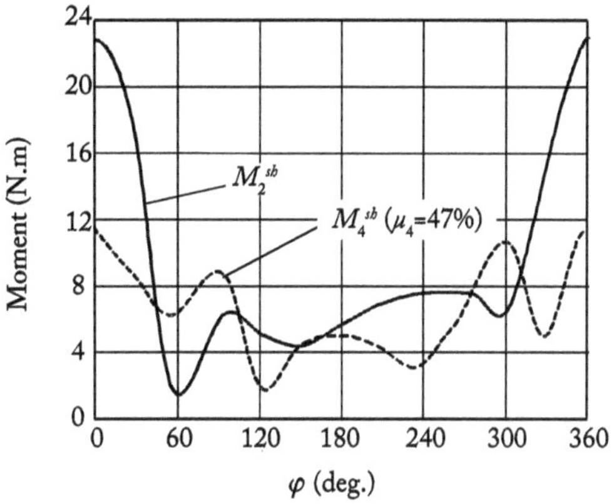

Fig. 3.11 Variations of the shaking moments for statically balanced linkage ($j = 2$) and for the linkage balanced by the suggested method ($j = 4$)

For simplicity we establish the fixed coordinate system with **x**-axis along OE, **z**-axis along OA and **y**-axis determined by the right-hand rule. We consider that this linkage is balanced statically and that the static moments of the counterweights are the followings: $S^{st}_{cw_1} = 0.33$ kg m and $S^{st}_{cw_3} = 0.5$ kg m.

The parameters of the displacement of the counterweight, calculated by the expressions (3.62–3.64), are the following: $x = 0.056$ m, $y = -0.350$ m, $z = 0.047$ m.

In Fig. 3.11 are presented the variations of the shaking moments $M^{sh}_j(\varphi)$ for the *RSS'R* four-bar spatial linkage under consideration:

1. only shaking force balanced ($j = 2$);
2. complete shaking force and partial shaking moment balanced by the proposed method ($j = 4$).

The coefficients μ_j, which characterize the efficiency of balancing methods, are calculated by the following expression:

$$\mu_j = \frac{\max|M_m(\varphi)| - \max\left|M_j(\varphi)\right|}{\max|M_m(\varphi)|} \tag{3.78}$$

where

$$|M_m(\varphi)| = \max\{|M_2(\varphi)|, |M_4(\varphi)|\} \tag{3.79}$$

In this case, the coefficient μ_j is equal to 47 %.

3.4 An Approximate Method of Calculating a Counterweight for the Optimum Shaking Force and Shaking Moment Balancing of Linkages

3.4.1 Shaking Force Balancing

Let us consider an arbitrary n-bar planar linkage (Fig. 3.12). Let the input link of the linkage (the crank) rotate with constant input angular velocity $\dot{\varphi}$. Let $S_{cw}^{st} = m_{cw} r_{cw}$ be the static moment of the counterweight, where m_{cw} is the mass of the counterweight mounted on the input crank, and r_{cw} is the distance of its *com* from the crank axis. The angular position of attachment of the counterweight with respect to crank is given by the angle α_{cw} (see Fig. 3.12). Thus the resultant force (one can also say unbalanced force) of the system can be expressed as:

$$\mathbf{f} = \mathbf{f}_{cw} + \mathbf{f}^{sh} \tag{3.80}$$

where $\mathbf{f}_{cw}$ is the inertial force due to the counterweight,

$$\mathbf{f}_{cw} = S_{cw}^{st}\left(\ddot{\varphi}\begin{bmatrix}-\sin(\varphi+\alpha_{cw})\\ \cos(\varphi+\alpha_{cw})\end{bmatrix} - \dot{\varphi}^2\begin{bmatrix}\cos(\varphi+\alpha_{cw})\\ \sin(\varphi+\alpha_{cw})\end{bmatrix}\right) \tag{3.81}$$

and $\mathbf{f}^{sh^T} = \left[f_x^{sh}\ \ f_y^{sh}\right]$ is the shaking force of the mechanism.

Statement of Problem To find such a selection of counterweights' parameters S_{cw}^{st} and α_{cw} for which the root-mean-square (*rms*) values of the unbalanced force is the least, i.e.

$$\sqrt{\sum_{i=1}^{N}\left(\mathbf{f}_{cw_i} + \mathbf{f}_i^{sh}\right)^2 / N} \rightarrow \min_{S_{cw}^{st},\alpha_{cw}} \tag{3.82}$$

where N is the number of calculated positions of the linkage.

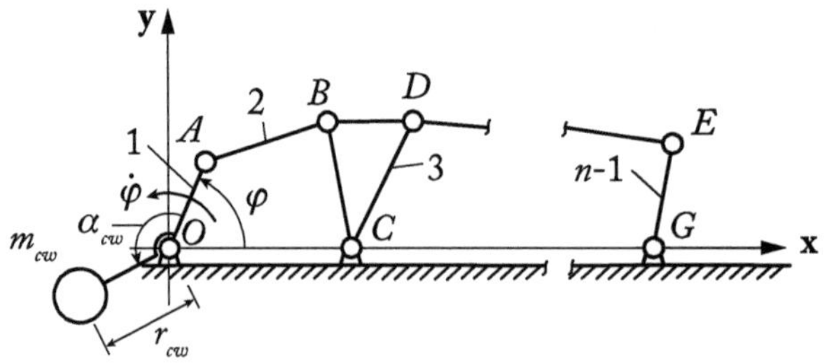

Fig. 3.12 Arbitrary n-bar planar linkage

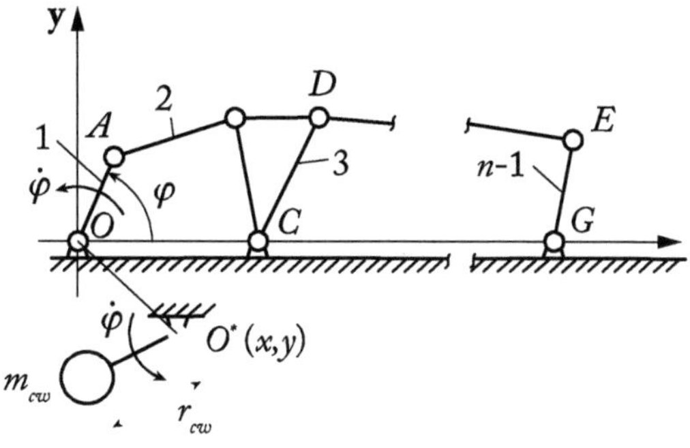

Fig. 3.13 Shaking moment balancing by the parallel displacement of the axis of rotation of the counterweight

For the minimization of the *rms*, it is necessary to minimize the sum:

$$\Delta = \sum_{i=1}^{N} \left(\mathbf{f}_{cw_i} + \mathbf{f}_i^{sh}\right)^2 \rightarrow \min_{S_{cw}^{st},\alpha_{cw}} \tag{3.83}$$

For this purpose, we shall achieve the conditions $\partial\Delta/\partial S_{cw}^{st} = 0$ and $\partial\Delta/\partial\alpha_{cw} = 0$. From these expressions and taking into account that $\sum_{i=1}^{N}\sin\varphi_i\cos\varphi_i{=}0$ (φ_i is the value of φ in the *i*-th linkage configuration) for $\varphi \in [0\ \ 2\pi]$, we determine the unknowns:

$$\alpha_{cw} = \tan^{-1}\left(\frac{\sum_{i=1}^{N} f_{yi}^{sh}\cos\varphi_i - \sum_{i=1}^{N} f_{xi}^{sh}\sin\varphi_i}{\sum_{i=1}^{N} f_{yi}^{sh}\cos\varphi_i + \sum_{i=1}^{N} f_{xi}^{sh}\sin\varphi_i}\right) \tag{3.84}$$

$$S_{cw}^{st} = -\frac{\sum_{i=1}^{N} f_{xi}^{sh}\cos(\varphi_i + \alpha_{cw}) + \sum_{i=1}^{N} f_{yi}^{sh}\sin(\varphi_i + \alpha_{cw})}{\dot{\varphi}^2\left(\sum_{i=1}^{N}\cos^2\varphi_i + \sum_{i=1}^{N}\sin^2\varphi_i\right)} \tag{3.85}$$

3.4.2 Shaking Moment Balancing

As was mentioned above, if the axis of rotation of the counterweight is moved from center O to the center $O^*(x, y)$ (Fig. 3.13), the balancing of shaking force of the linkage will occur, but a supplementary moment is therewith added:

$$M_O^* = S_{cw}^{st}\dot{\varphi}^2\left(-y\cos(\varphi + \alpha_{cw}) + x\sin(\varphi + \alpha_{cw})\right) \tag{3.86}$$

Statement of Problem To find the coordinates x and y of the axis of rotation O^* (see Fig. 3.13) of the counterweight for which the *rms* values of the unbalanced moment is the least, i.e.

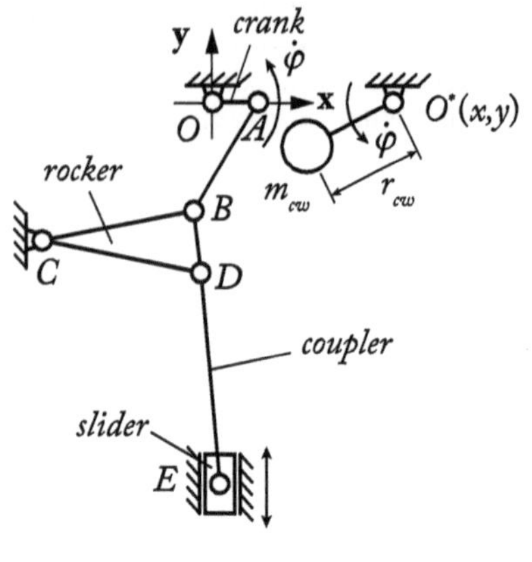

Fig. 3.14 Balancing of the six-bar linkage of the mechanical press

$$\sqrt{\sum_{i=1}^{N}\left((M_O^{sh})_i + (M_O^{*})_i\right)^2 / N} \rightarrow \min_{x,y} \tag{3.87}$$

where $(M_O^{*})_i$ is the inertial moment due to the counterweight in the *i*-th linkage configuration and $((M_O^{sh})_i$ is the shaking moment of the mechanism in the *i*-th linkage configuration.

For the minimization of the *rms*, it is necessary to minimize the sum:

$$\Delta_M = \sum_{i=1}^{N}\left((M_O^{sh})_i + (M_O^{*})_i\right)^2 \rightarrow \min_{x,y} \tag{3.88}$$

For this purpose, we shall achieve the conditions: $\partial\Delta_M/\partial x = 0$ and $\partial\Delta_M/\partial y = 0$ from where we determine the unknowns:

$$x = \frac{\sum_{i=1}^{N}(M_O^{sh})_i \sin(\varphi_i + \alpha_{cw})}{S_{cw}^{st}\dot{\varphi}^2\sum_{i=1}^{N}\sin^2(\varphi_i + \alpha_{cw})} \tag{3.89}$$

$$y = -\frac{\sum_{i=1}^{N}(M_O^{sh})_i \cos(\varphi_i + \alpha_{cw})}{S_{cw}^{st}\dot{\varphi}^2\sum_{i=1}^{N}\cos^2(\varphi_i + \alpha_{cw})} \tag{3.90}$$

3.4.3 Numerical Example

For the six-bar linkage (Fig. 3.14) of the mechanical press with parameters:

- $l_{OA} = 0.1$ m, $l_{AB} = 0.26$ m, $l_{BC} = 0.32$ m, $l_{BD} = 0.34$ m, $l_{CD} = 0.13$ m, $l_{DE} = 0.44$ m, $l_{OS_1} = 0.025$ m, $l_{AS_2} = 0.13$ m, $l_{CS_4} = 0.17$ m,

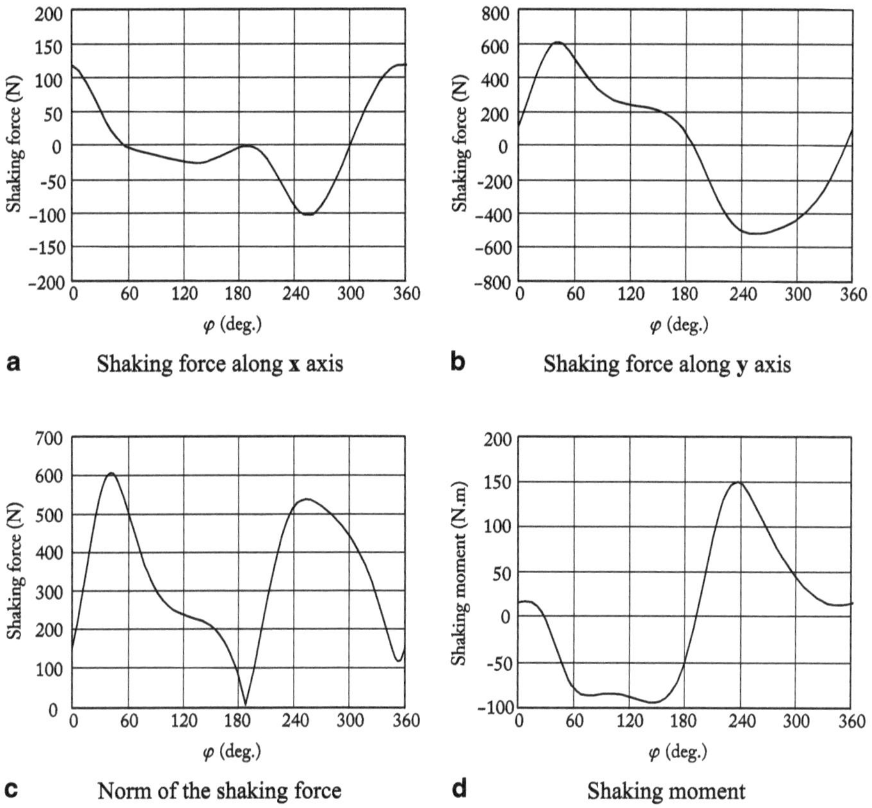

Fig. 3.15 Shaking force and shaking moment for non balanced linkage

- $m_1 = 2\,\text{kg}$, $m_2 = 3\,\text{kg}$, $m_3 = 4\,\text{kg}$, $m_4 = 3\,\text{kg}$, $m_5 = 10\,\text{kg}$,
- $I_{S_2} = 0.0023\,\text{kg}\,\text{m}^2$, $I_{S_3} = 0.04\,\text{kg}\,\text{m}^2$, $I_{S_4} = 0.003\,\text{kg}\,\text{m}^2$,
- $\dot{\varphi} = 5\pi$ rad/s,

we carry out the shaking force balancing and by means of the formulae (3.84) and (3.85) determine $\alpha_{cw} = 186°$ and $S_{cw}^{st} = 1.09\,\text{kg}\,\text{m}$.

Then, for shaking moment balancing we determine the coordinates $x = 0.36\,\text{m}$ and $y = 0.06\,\text{m}$ of the axis of rotation O^* of the counterweight.

Figure 3.15 presents the variations of the shaking force and the shaking moment for non balanced linkage.

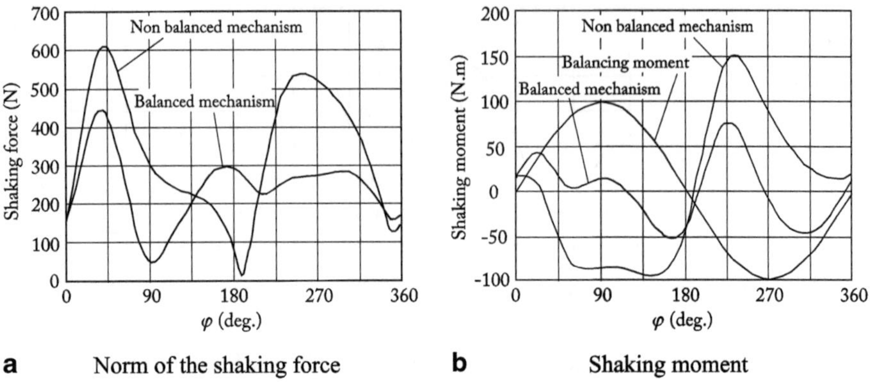

Fig. 3.16 Shaking force and the shaking moment for non balanced and balanced linkages

In Fig. 3.16 the variations of the shaking force and shaking moment for balanced linkage are represented. Thus, by suggested balancing method, the reduction of the *rms* value of the shaking force is about 33.4 % and the shaking moment is about 54.4 %. It should be noted that the minimization of the maximum values of these functions is also achieved. The reduction of the maximum value of the shaking force is about 26.7 % and of the shaking moment is about 48.9 %.

Chapter 4
Complete Shaking Force and Shaking Moment Balancing of Linkages

Abstract In this Chapter, new methods for the full shaking force and shaking moment balancing of linkages are considered. In Sect. 4.1, a new solution for the full shaking force and shaking moment balancing of four-bar linkages is discussed, which allows the complete shaking force and shaking moment balancing of in-line four-bar linkages with constant input speed by adding a class-two Assur group, i.e. a group which does not add any supplementary degree of freedom into the mechanism. It should be noted that the balancing of the shaking moment without counter-rotations of three particular classes of four-bar linkages is known and it was mentioned in the overview (see Sect. 2.1). However, such a method cannot be extended to general four-bar linkages. In the mentioned Section, it is proposed to dynamically balance the in-line four-bar linkages by adding articulated dyads.

The object of the approach presented in Sect. 4.2 is to provide the conditions for a complete shaking force and shaking moment balancing of linkages with a relatively small increase of the total mass of movable links by mounting the gear inertia counterweights on the base of the mechanism. The method involves connecting to the mechanism to be balanced a two-link group forming a pantograph with the crank and coupler. Three versions of sub-linkages are considered: (1) the articulation dyad; (2) the asymmetric link with three rotational pairs; (3) the crank-slider mechanism. The method is illustrated by new balancing schemes for the Stephenson and Watt linkages. An advantage of the schemes outlined here is the fact that all the gear inertia counterweights needed for balancing the shaking moment are mounted on the mechanism frame, which is constructively more efficient.

The complete shaking force and shaking moment balancing of spatial linkages is very complicated and the majority of the works have been concentrated on complete shaking force balancing or partial shaking force and shaking moment balancing. The solution proposed in the Sect. 4.3 shows that it is possible to achieve a complete shaking force and shaking moment balancing of the *RSS'R* spatial linkage by using a coupler with a special shape. In such an approach, the mass of the connecting coupler is substituted dynamically by concentrated masses located at joints. This allows for the transformation of the problem of linkage dynamic balancing into a problem of balancing rotating links carrying concentrated masses.

At the end of this Chapter (Sect. 4.4), we deal with new design methods of self-balanced mechanical systems via duplication of off-set slider-crank mechanisms executing opposite movements.

V. Arakelian, S. Briot, *Balancing of Linkages and Robot Manipulators*,
Mechanisms and Machine Science 27, DOI 10.1007/978-3-319-12490-2_4

4.1 Complete Shaking Force and Shaking Moment Balancing of In-Line Four-Bar Linkages by Adding a Class-Two *RRR* or *RRP* Assur Group

4.1.1 Complete Shaking Force and Shaking Moment Balancing by Adding a Class-Two RRR Assur Group

4.1.1.1 Theoretical Background Related to the Balanced Four-Bar Linkages with Prescribed Geometrical Parameters

Before considering the suggested balancing concept, let us recall basic notions concerning the balanced four-bar linkages with prescribed geometrical parameters.

In the paper of Berkof and Lowen (Lowen and Berkof 1970), the angular momentum H and the shaking moment M^{sh}, expressed at point O, of a force balanced in-line four-bar linkage (Fig. 4.1) were expressed as:

$$H = \sum_{i=1}^{3} I_i \dot{\theta}_i, \quad M^{sh} = \frac{\mathrm{d}H}{\mathrm{d}t} = \sum_{i=1}^{3} I_i \ddot{\theta}_i \tag{4.1}$$

with

$$I_i = m_i \left(k_i^2 + r_i^2 + r_i l_i\right), \quad (i = 1, 3) \tag{4.2}$$

$$I_2 = m_2 \left(k_2^2 + r_2^2 - r_2 l_2\right) \tag{4.3}$$

where r_i is the length of vector $\mathbf{r}_i$ which connects the pivot P_i to the *com* of link i, l_i is the length of vector $\mathbf{l}_i$ which connects the proximal revolute joint P_i to the distal joint on the same link, and k_i is the radius of gyration with respect to the *com* of link i, m_i is the mass of link i. Moreover, θ_i is the angular position of link i with respect to the $\mathbf{x}$-axis.

With regard to the shaking force balancing, the following expressions were obtained:

$$m_1 r_1 = m_2 l_1 \frac{l_2 - r_2}{l_2} \tag{4.4}$$

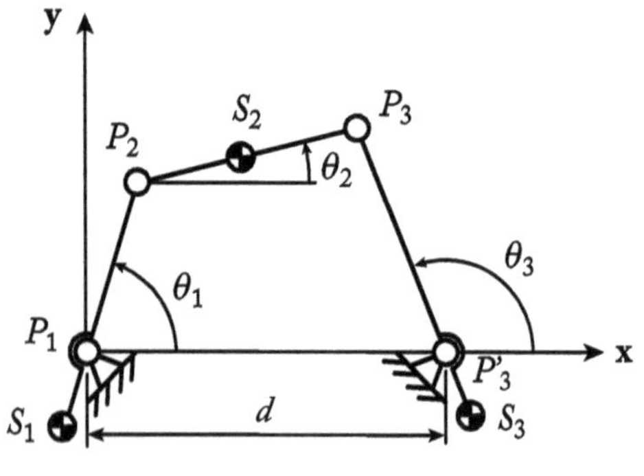

Fig. 4.1 A general in-line four-bar mechanism

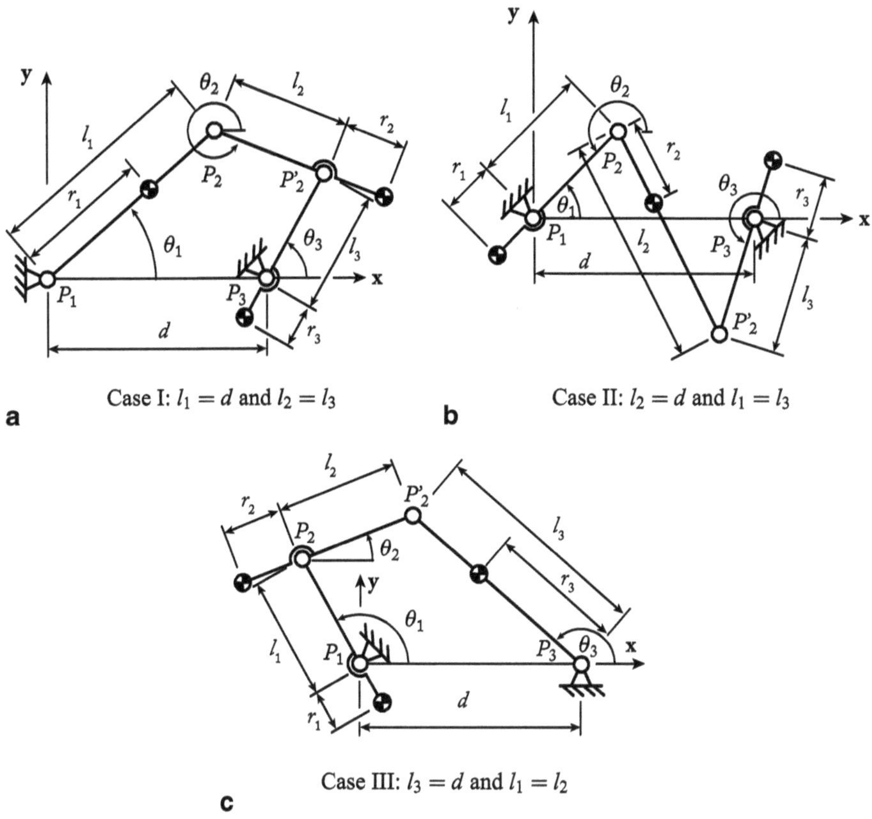

Fig. 4.2 The three kinds of shaking force and shaking moment balanced four-bar mechanisms

$$m_3 r_3 = m_2 r_2 \frac{l_3}{l_2} \tag{4.5}$$

In the works (Wu and Gosselin 2007; Gosselin et al. 2004), it has been shown that it is possible to cancel the shaking moment of a four-bar mechanism by associating mentioned geometric constraints with an optimal distribution of masses. Three kinds of shaking force and shaking moment balanced four-bar mechanisms were found, which are shown in Fig. 4.2.

In order to illustrate the shaking moment balancing, let us consider the mechanism shown in Fig. 4.2b. The geometrical constraints of this mechanism are the following:

$$l_1 = l_3 \tag{4.6}$$

$$d = l_2 \tag{4.7}$$

where d is the length of the base which is the distance between the two fixed joints on the base.

This leads to the following kinematic relationships:

$$\dot{\theta}_1 - \dot{\theta}_2 + \dot{\theta}_3 = 0 \tag{4.8}$$

Thus, from expressions (4.1) and (4.8), it is easy to see that the shaking moment will be cancelled if $I_1 = I_2 = I_3$ (see Eqs. (4.2) and (4.3)). For this purpose the following relationships must be established:

$$k_2^2 = \frac{m_2\left(l_2 r_2 - r_2^2\right) - I_1}{m_2} \tag{4.9}$$

$$k_3^2 = \frac{-m_3\left(l_3 r_3 + r_3^2\right) + I_1}{m_3} \tag{4.10}$$

where

$$I_1 = m_1\left(k_1^2 + r_1^2 + r_1 l_1\right) \tag{4.11}$$

It should be noted that similar results have been obtained for the mechanisms shown in Fig. 4.2a and c.

Statement of the Problem The aim of the suggested balancing approach consists of adding a two-link kinematic chain with prescribed geometrical parameters to an in-line four-bar linkage with arbitrary geometrical parameters. It is important to note that the added structure must be an Assur group, i.e. a group which does not add any supplementary degree of freedom into the mechanism (Leinonen 1991). This allows for the modification of the mass redistribution of the obtained six-bar mechanism without perturbation of the kinematic properties of the initial four-bar linkage. We would like to state that this technique allows for the complete shaking force and shaking moment balancing without counter-rotating masses.

Now let us consider the shaking force and shaking moment balancing of an in-line four-bar mechanism using class-two Assur groups with *RRR* kinematic chain.

4.1.1.2 Shaking Force Balancing

Figure 4.3 shows an in-line four-bar linkage with the added class-two *RRR* Assur group. Let us denote the following vectors as: $\mathbf{l}_1 = \mathbf{d}_{OA}$, $\mathbf{l}_2 = \mathbf{d}_{AB}$, $\mathbf{l}_3 = \mathbf{d}_{CB}$, $\mathbf{l}'_3 = \mathbf{d}_{CP_2}$, $\mathbf{l}_4 = \mathbf{d}_{P_2P'_2}$, $\mathbf{l}_5 = \mathbf{d}_{P_3P'_2}$, $\mathbf{r}_1 = \mathbf{d}_{OS_1}$, $\mathbf{r}_2 = \mathbf{d}_{AS_2}$, $\mathbf{r}_3 = \mathbf{d}_{CS_3}$, $\mathbf{r}_4 = \mathbf{d}_{P_2S_4}$, $\mathbf{r}_5 = \mathbf{d}_{P_3S_5}$, $\mathbf{r}_{cw_1} = \mathbf{d}_{OS_{cw_1}}$, $\mathbf{r}_{cw2} = \mathbf{d}_{CS_{cw_2}}$, $\mathbf{r}_{cw3} = \mathbf{d}_{P_3S_{cw_3}}$.

The added class-two *RRR* Assur group has the above mentioned properties, i.e. it is designed such as:

$$l'_3 = l_5 \tag{4.12}$$

$$e = l_4 \tag{4.13}$$

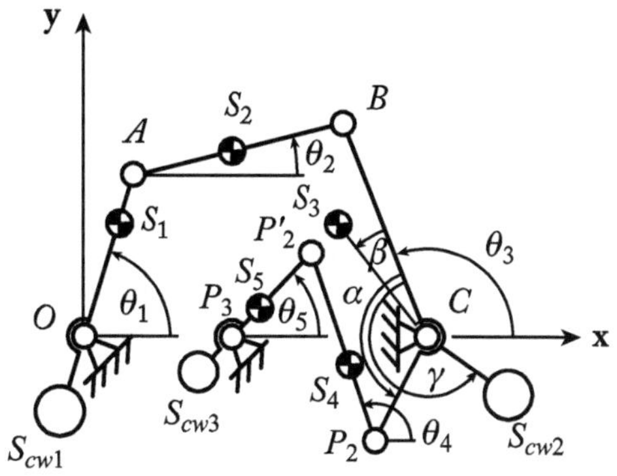

Fig. 4.3 The balanced mechanism with the class-two *RRR* Assur group

where l_i $(i = 3, 4, 5)$ is the norm of vector $\mathbf{l}_i$ and e the distance between C and P_3.

As is shown in Fig. 4.3, the Assur group $P_2P_2'P_3$ is attached to the initial linkage $OABC$ in such a way that it forms a four-bar linkage with link BC, as discussed in Sect. 4.1.1.1.

Let us now derive the expression of the shaking force $\mathbf{f}^{sh}$ of the obtained mechanism:

$$\mathbf{f}^{sh} = \sum_{i=1}^{5} m_i \ddot{\mathbf{d}}_{S_i} \tag{4.14}$$

where $\ddot{\mathbf{d}}_{S_i}$ is the translational acceleration of the *com* S_i. Developing and simplifying, one obtains:

$$\begin{aligned} \mathbf{f}^{sh} = & \left(\tfrac{m_1 r_1}{l_1} + m_2 \tfrac{l_2 - r_2}{l_2}\right) \ddot{\mathbf{d}}_A + m_2 \tfrac{r_2}{l_2} \ddot{\mathbf{d}}_B + m_3 \ddot{\mathbf{d}}_{S_3} \\ & + m_4 \tfrac{l_4 - r_4}{l_4} \ddot{\mathbf{d}}_{P_2} + \left(\tfrac{m_4 r_4}{l_4} + \tfrac{m_5 r_5}{l_5}\right) \ddot{\mathbf{d}}_{P_2'} \end{aligned} \tag{4.15}$$

where r_i is the algebraic values of the norm of vectors $\mathbf{r}_i$, and $\ddot{\mathbf{d}}_A$, $\ddot{\mathbf{d}}_B$, $\ddot{\mathbf{d}}_{S_3}$, $\ddot{\mathbf{d}}_{P_2}$ and $\ddot{\mathbf{d}}_{P_2'}$ represent the acceleration of points A, B, S_3, P_2 and P_2' respectively. Their expressions are:

$$\ddot{\mathbf{d}}_A = l_1 \left(\ddot{\theta}_1 \begin{bmatrix} -\sin\theta_1 \\ \cos\theta_1 \end{bmatrix} - \dot{\theta}_1^2 \begin{bmatrix} \cos\theta_1 \\ \sin\theta_1 \end{bmatrix} \right) \tag{4.16}$$

$$\ddot{\mathbf{d}}_B = l_3 \left(\ddot{\theta}_3 \begin{bmatrix} -\sin\theta_3 \\ \cos\theta_3 \end{bmatrix} - \dot{\theta}_3^2 \begin{bmatrix} \cos\theta_3 \\ \sin\theta_3 \end{bmatrix} \right) \tag{4.17}$$

$$\ddot{\mathbf{d}}_{S_3} = r_3 \left(\ddot{\theta}_3 \begin{bmatrix} -\sin(\theta_3 + \beta) \\ \cos(\theta_3 + \beta) \end{bmatrix} - \dot{\theta}_3^2 \begin{bmatrix} \cos(\theta_3 + \beta) \\ \sin(\theta_3 + \beta) \end{bmatrix} \right) \tag{4.18}$$

$$\ddot{\mathbf{d}}_{P_2} = l_3' \left(\ddot{\theta}_3 \begin{bmatrix} -\sin(\theta_3 + \alpha) \\ \cos(\theta_3 + \alpha) \end{bmatrix} - \dot{\theta}_3^2 \begin{bmatrix} \cos(\theta_3 + \alpha) \\ \sin(\theta_3 + \alpha) \end{bmatrix} \right) \tag{4.19}$$

$$\ddot{\mathbf{d}}_{P_2'} = l_5 \left(\ddot{\theta}_5 \begin{bmatrix} -\sin\theta_5 \\ \cos\theta_5 \end{bmatrix} - \dot{\theta}_5^2 \begin{bmatrix} \cos\theta_5 \\ \sin\theta_5 \end{bmatrix} \right) \tag{4.20}$$

The shaking force $\mathbf{f}^{sh}$ may be cancelled through the addition of three counterweights positioned at points S_{cw_i} (Fig. 4.3), with masses m_{cw_i} $(i = 1, 2, 3)$. With such counterweights, the expression of the shaking force becomes:

$$\mathbf{f}^{sh^*} = \mathbf{f}^{sh} + m_{cw_1} \frac{r_{cw_1}}{l_1} \ddot{\mathbf{d}}_A + m_{cw_2} \ddot{\mathbf{d}}_{S_{cw_2}} + m_{cw_3} \frac{r_{cw_3}}{l_5} \ddot{\mathbf{d}}_{P_2'} \tag{4.21}$$

where r_{cw_i} is the algebraic values of the norm of vectors $\mathbf{r}_{cw_i}$ and

$$\ddot{\mathbf{d}}_{S_{cw_2}} = r_{cw_2} \left(\ddot{\theta}_3 \begin{bmatrix} -\sin(\theta_3 + \gamma) \\ \cos(\theta_3 + \gamma) \end{bmatrix} - \dot{\theta}_3^2 \begin{bmatrix} \cos(\theta_3 + \gamma) \\ \sin(\theta_3 + \gamma) \end{bmatrix} \right) \tag{4.22}$$

Thus the shaking force is cancelled if the distribution of the masses is as follows:

$$m_{cw_1} = -\frac{l_1}{r_{cw_1}} \left(m_1 \frac{r_1}{l_1} + m_2 \frac{l_2 - r_2}{l_2} \right) \tag{4.23}$$

$$\tan\gamma = \frac{m_3 r_3 \sin\beta + m_4 (1 - r_4/l_4)\, l_3' \sin\alpha}{m_2 l_3 r_2/l_2 + m_3 r_3 \cos\beta + m_4 (1 - r_4/l_4)\, l_3' \sin\alpha} \tag{4.24}$$

$$\begin{aligned} m_{cw_2}^2 = {} & \frac{\left(m_3 r_3 \sin\beta + m_4 (1 - r_4/l_4)\, l_3' \sin\alpha\right)^2}{r_{cw_2}^2} \\ & + \frac{\left(m_2 l_3 r_2/l_2 + m_3 r_3 \cos\beta + m_4 (1 - r_4/l_4)\, l_3' \cos\alpha\right)^2}{r_{cw_2}^2} \end{aligned} \tag{4.25}$$

$$m_{cw_3} = -\frac{l_5}{r_{cw_3}} \left(m_4 \frac{r_4}{l_4} + m_5 \frac{r_5}{l_5} \right) \tag{4.26}$$

4.1.1.3 Shaking Moment Balancing

Let us now derive the expression of the shaking moment M^{sh}, expressed at point O, of such a mechanism:

$$M^{sh} = \sum_{i=1}^{5} m_i \left(x_{S_i} \ddot{y}_{S_i} - y_{S_i} \ddot{x}_{S_i} + k_i^2 \ddot{\theta}_i \right) + \sum_{i=1}^{3} m_{cw_i} \left(x_{S_{cw_i}} \ddot{y}_{S_{cw_i}} - y_{S_{cw_i}} \ddot{x}_{S_{cw_i}} \right) \tag{4.27}$$

where x_{S_i}, y_{S_i}, $\ddot{x}_{S_i}$ and $\ddot{y}_{S_i}$ are the position and accelerations along **x** and **y** axes of points S_i, respectively, and $x_{S_{cw_i}}$, $y_{S_{cw_i}}$, $\ddot{x}_{S_{cw_i}}$ and $\ddot{y}_{S_{cw_i}}$ are the position and accelerations along **x** and **y** axes of points S_{cw_i}, respectively, k_i is the radius of gyration of link i.

Now, let us consider that link 2 is a physical pendulum[1] (Berkof 1973), i.e. it can be replaced dynamically by two point masses located at joint centres A and B. This implies that:

$$k_2^2 = r_2 \left(l_2 - r_2 \right) \tag{4.28}$$

Then, considering that the input speed is constant, i.e. $\ddot{\theta}_1 = 0$ and taking into account (4.28), one can simplify (4.27) as:

$$M^{sh} = \sum_{i=3}^{5} I_i \ddot{\theta}_i \tag{4.29}$$

where

$$I_3 = m_3 \left(k_3^2 + r_3^2 \right) + m_2 \frac{r_2}{l_2} l_3^2 + m_{cw2} r_{cw2}^2 + m_4 \frac{l_4 - r_4}{l_4} l_3'^2 \tag{4.30}$$

$$I_4 = m_4 \left(k_4^2 + r_4^2 - r_4 l_4 \right) \tag{4.31}$$

$$I_5 = m_5 \left(k_5^2 + r_5^2 - r_{cw_3} r_5 \right) + m_4 \frac{r_4}{l_4} l_5 \left(l_5 - r_{cw_3} \right) \tag{4.32}$$

Thus, this new six-bar mechanism has the same shaking moment as the four-bar mechanism composed of links $P_1 P_2$, $P_2 P_2'$ and $P_2' P_3$. Therefore, the initial four-bar linkage balancing problem is transformed in the balancing of the four-bar linkage formed by the added Assur group. Note that the latter has specific geometry and its balancing conditions have been examined in Sect. 4.1.1.1.

[1] A "physical pendulum" is a link which has such a distribution of masses that it allows the dynamic substitution of link's mass and inertia by two concentrated masses.

Applying these results to the considered mechanism, we obtain:

$$k_4^2 = \frac{-I_3 - r_4^2 + r_4 l_4}{m_4} \tag{4.33}$$

$$k_5^2 = \frac{m_5\left(-r_5^2 + r_{cw_3} r_5\right) + m_4 r_4 / l_4 l_5 \left(l_5 - r_{cw_3}\right) + I_3}{m_5} \tag{4.34}$$

The substitution of (4.33) and (4.34) into (4.29) leads to:

$$M^{sh} = I_3\left(\ddot{\theta}_3 - \ddot{\theta}_4 + \ddot{\theta}_5\right) \tag{4.35}$$

Taking into account relations (4.6) and (4.6), we have

$$\ddot{\theta}_3 - \ddot{\theta}_4 + \ddot{\theta}_5 = 0 \tag{4.36}$$

and consequently

$$M^{sh} = 0 \tag{4.37}$$

The proposed balancing technique has been illustrated using the four-bar linkage shown in Fig. 4.2b. However, it can also be achieved via the mechanism of Fig. 4.2a or 4.2c.

4.1.2 Complete Shaking Force and Shaking Moment Balancing by Adding a Class-Two RRP Assur Group

4.1.2.1 Shaking Force Balancing

The second solution, which is proposed for the cancellation of the shaking moment of a four-bar linkage is carried out by adding a class-two *RRP* Assur group (Fig. 4.4). Let us denote the following vectors as: $\mathbf{l}_1 = \mathbf{d}_{OA}$, $\mathbf{l}_2 = \mathbf{d}_{AB}$, $\mathbf{l}_3 = \mathbf{d}_{CB}$, $\mathbf{l}'_3 = \mathbf{d}_{CD}$, $\mathbf{l}_4 = \mathbf{d}_{DE}$, $\mathbf{r}_1 = \mathbf{d}_{OS_1}$, $\mathbf{r}_2 = \mathbf{d}_{AS_2}$, $\mathbf{r}_3 = \mathbf{d}_{CS_3}$, $\mathbf{r}_4 = \mathbf{d}_{DS_4}$, $\mathbf{r}_5 = \mathbf{d}_{ES_5}$, $\mathbf{r}_{cw_1} = \mathbf{d}_{OS_{cw_1}}$, $\mathbf{r}_{cw2} = \mathbf{d}_{CS_{cw_2}}$, $\mathbf{r}_{cw3} = \mathbf{d}_{DS_{cw_3}}$.

In this case, the lengths of added links are the following:

$$l'_3 = l_4 \tag{4.38}$$

Thus, the new part created by added links is a Scott-Russell mechanism (Martin 2002), which is attached to the initial linkage at an angle of α.

The relations between the link accelerations are the following for the Scott-Russell mechanism :

$$\dot{\theta}_3 = -\dot{\theta}_4 \tag{4.39}$$

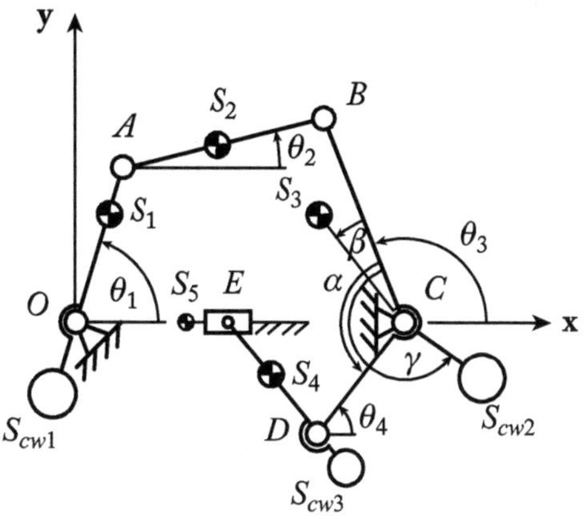

Fig. 4.4 The balanced mechanism with the class-two *RRP* Assur group

Let us now derive the expression of the shaking force $\mathbf{f}^{sh}$ of such a mechanism:

$$\mathbf{f}^{sh} = \sum_{i=1}^{5} m_i \ddot{\mathbf{d}}_{S_i} \tag{4.40}$$

Developing and simplifying, one obtains:

$$\begin{aligned} \mathbf{f}^{sh} = & \left(\tfrac{m_1 r_1}{l_1} + m_2 \tfrac{l_2 - r_2}{l_2}\right) \ddot{\mathbf{d}}_A + m_2 \tfrac{r_2}{l_2} \ddot{\mathbf{d}}_B + m_3 \ddot{\mathbf{d}}_{S_3} \\ & + (m_4 + m_5)\, \ddot{\mathbf{d}}_D + \left(m_4 \tfrac{r_4}{l_4} + m_5\right) \ddot{\mathbf{d}} \end{aligned} \tag{4.41}$$

where r_i and l_i are the algebraic values of the norm of vectors $\mathbf{r}_i$, and $\mathbf{l}_i$, respectively,

$$\ddot{\mathbf{d}}_{S_3} = r_3 \left(\ddot{\theta}_3 \begin{bmatrix} -\sin(\theta_3 + \beta) \\ \cos(\theta_3 + \beta) \end{bmatrix} - \dot{\theta}_3^2 \begin{bmatrix} \cos(\theta_3 + \beta) \\ \sin(\theta_3 + \beta) \end{bmatrix} \right) \tag{4.42}$$

$$\ddot{\mathbf{d}} = l_5 \left(\ddot{\theta}_5 \begin{bmatrix} -\sin\theta_5 \\ \cos\theta_5 \end{bmatrix} - \dot{\theta}_5^2 \begin{bmatrix} \cos\theta_5 \\ \sin\theta_5 \end{bmatrix} \right) \tag{4.43}$$

$$\ddot{\mathbf{d}}_D = l'_3 \left(\ddot{\theta}_3 \begin{bmatrix} -\sin(\theta_3 + \alpha) \\ \cos(\theta_3 + \alpha) \end{bmatrix} - \dot{\theta}_3^2 \begin{bmatrix} \cos(\theta_3 + \alpha) \\ \sin(\theta_3 + \alpha) \end{bmatrix} \right) \tag{4.44}$$

The shaking force $\mathbf{f}^{sh}$ may be cancelled through the addition of three counterweights positioned at points S_{cw_i} (Fig. 4.4), with masses m_{cw_i} $(i = 1, 2, 3)$. With such counterweights, the expression of the shaking force becomes:

$$\mathbf{f}^{sh*} = \mathbf{f}^{sh} + m_{cw_1} \frac{r_{cw_1}}{l_1} \ddot{\mathbf{d}}_A + m_{cw_2} \ddot{\mathbf{d}}_{S_{cw_2}} + m_{cw_3} \ddot{\mathbf{d}}_D + m_{cw_3} \frac{r_{cw_3}}{l_4} \ddot{\mathbf{d}} \tag{4.45}$$

where r_{cw_i} is the algebraic values of the norm of vectors $\mathbf{r}_{cw_i}$ and

$$\ddot{\mathbf{d}}_{S_{cw_2}} = r_{cw_2}\left(\ddot{\theta}_3\begin{bmatrix}-\sin(\theta_3+\gamma)\\ \cos(\theta_3+\gamma)\end{bmatrix} - \dot{\theta}_3^2\begin{bmatrix}\cos(\theta_3+\gamma)\\ \sin(\theta_3+\gamma)\end{bmatrix}\right) \tag{4.46}$$

Thus the shaking force is cancelled if the distribution of the masses is as follows:

$$m_{cw_1} = -\frac{l_1}{r_{cw_1}}\left(m_1\frac{r_1}{l_1} + m_2\frac{l_2 - r_2}{l_2}\right) \tag{4.47}$$

$$\tan\gamma = \frac{m_3 r_3 \sin\beta + (m_4 + m_5)\, l_3' \sin\alpha}{m_2 l_3 r_2/l_2 + m_3 r_3\cos\beta + (m_4 + m_5)\, l_3' \sin\alpha} \tag{4.48}$$

$$m_{cw_2}^2 = \frac{\left(m_3 r_3 \sin\beta + (m_4 + m_5)\, l_3' \sin\alpha\right)^2}{r_{cw_2}^2} + \frac{\left(m_2 l_3 r_2/l_2 + m_3 r_3 \cos\beta + (m_4 + m_5)\, l_3' \cos\alpha\right)^2}{r_{cw_2}^2} \tag{4.49}$$

$$m_{cw_3} = -\frac{l_4}{r_{cw_3}}\left(m_4\frac{r_4}{l_4} + m_5\right) \tag{4.50}$$

4.1.2.2 Shaking Moment Balancing

Let us now derive the expression of the shaking moment M^{sh}, expressed at point O, of such a mechanism:

$$M^{sh} = \sum_{i=1}^{5} m_i\left(x_{S_i}\ddot{y}_{S_i} - y_{S_i}\ddot{x}_{S_i} + k_i^2\ddot{\theta}_i\right) + \sum_{i=1}^{3} m_{cw_i}\left(x_{S_{cw_i}}\ddot{y}_{S_{cw_i}} - y_{S_{cw_i}}\ddot{x}_{S_{cw_i}}\right) \tag{4.51}$$

where x_{S_i}, y_{S_i}, $\ddot{x}_{S_i}$ and $\ddot{y}_{S_i}$ are the position and accelerations along **x** and **y** axes of points S_i, respectively, and $x_{S_{cw_i}}$, $y_{S_{cw_i}}$, $\ddot{x}_{S_{cw_i}}$ and $\ddot{y}_{S_{cw_i}}$ are the position and accelerations along **x** and **y** axes of points S_{cw_i}, respectively, k_i is the radius of gyration of link i.

Now, let us consider as in the previous case, that link 2 is a physical pendulum and that the input speed is constant.

Taking into account (4.28), one can simplify (4.51) as:

$$M^{sh} = I_3\ddot{\theta}_3 + I_4\ddot{\theta}_4 \tag{4.52}$$

where

$$I_3 = m_3\left(k_3^2 + r_3^2\right) + m_2\frac{r_2}{l_2}l_3^2 + m_{cw_2}r_{cw_2}^2 + \left(m_4 + m_5 + m_{cw_3}\right)l_3'^2 \tag{4.53}$$

$$I_4 = m_4\left(k_4^2 + r_4^2\right) + m_5 l_4^2 + m_{cw_3} r_{cw_3}^2 \tag{4.54}$$

Introducing (4.39) into (4.52), we find:

$$M^{sh} = (I_3 - I_4)\,\ddot{\theta}_3 \tag{4.55}$$

Thus, the mechanism will be moment balanced if:

$$I_3 = I_4 \tag{4.56}$$

which can be obtained using a design of link 4 for which the radius of gyration should be equal to

$$k_4 = \sqrt{\frac{I_3 - m_4 r_4^2 - m_5 l_4^2 - m_{cw_3} r_{cw_3}^2}{m_4}} \tag{4.57}$$

It should be mentioned that, in order to avoid the singular configurations of the added structure, the value of angle α should be chosen carefully during the design process.

Let us consider two illustrative examples of the proposed balancing technique.

4.1.3 Illustrative Examples and Numerical Simulations

4.1.3.1 Balancing by Adding a Class-Two *RRR* Assur Group

Let us carry out the complete shaking force and shaking moment balancing of a four-bar linkage with parameters:

- $l_1 = 0.2$ m, $l_2 = 0.27$ m, $l_3 = 0.25$ m, $\beta = 0°$,
- $r_1 = 0.1$ m, $r_2 = 0.135$ m, $r_3 = 0.125$ m,
- $k_1 = 0.056$ m, $k_2 = 0.135$ m, $k_3 = 0.086$ m,
- $m_1 = 1$ kg, $m_2 = 1$ kg, $m_3 = 1$ kg.

The simulations of the proposed mechanism have been carried out using ADAMS software and the obtained results are shown in Fig. 4.5 (full line).

Now we add the *RRR* Assur group with prescribed distribution of the *com* and inertia. Its geometric and mass properties are:

- $l_4 = 0.8$ m, $l_5 = 0.25$ m,
- $r_4 = 0.4$ m, $r_5 = 0.125$ m,
- $m_4 = 1.5$ kg, $m_5 = 1$ kg.

and the locations and masses of the added counterweights are:

- $r_{cw_1} = -0.1$ m, $r_{cw_2} = 0.25$ m, $r_{cw_3} = -0.125$ m,
- $m_{cw_1} = 2$ kg, $m_{cw_3} = 2.5$ kg.

The radii of gyration of elements 4 and 5 and the mass of the counterweight m_{cw_2} are not given as they depend on the value of angle α which is not yet fixed (Fig. 4.3).

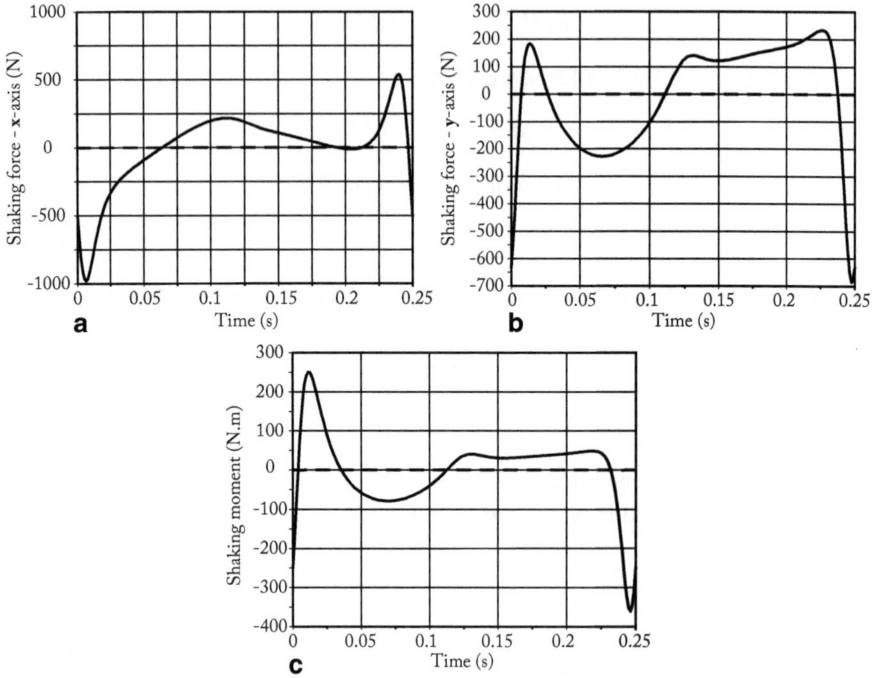

Fig. 4.5 Shaking force and shaking moment of the four-bar mechanism before (*full line*) and after (*dotted line*) balancing

Their variations as a function of α are shown in Fig. 4.6. In these figures, the values of α are bounded between 60 and 180° in order to avoid the *RRR* Assur group to cross a singularity during the motion. In Fig. 4.5 (dotted line), it is shown that after the addition of the Assur group, the shaking force and shaking moment are cancelled.

As angle α is a free parameter that has only influence on the values k_4, k_5 and m_{cw_2}, it can be chosen so that it minimizes one supplementary criterion. In the remainder of the paper, this criterion is chosen to be the linkage input torque (Berkof 1979b; Demeulenaere and Berkof 2008). It should however be mentioned that angle α could be used to minimize another criterion such as the power consumption, the energy, etc.

It should also be noted that the input torque τ, i.e. the torque requested by the actuator to move the mechanism, is computed in the two illustrative examples by using the Lagrange equations (Berkof 1979b):

$$\tau = \frac{\mathrm{d}}{\mathrm{d}t}\left(\frac{\partial L}{\partial \dot{\theta}_1}\right) - \frac{\partial L}{\partial \theta_1} \tag{4.58}$$

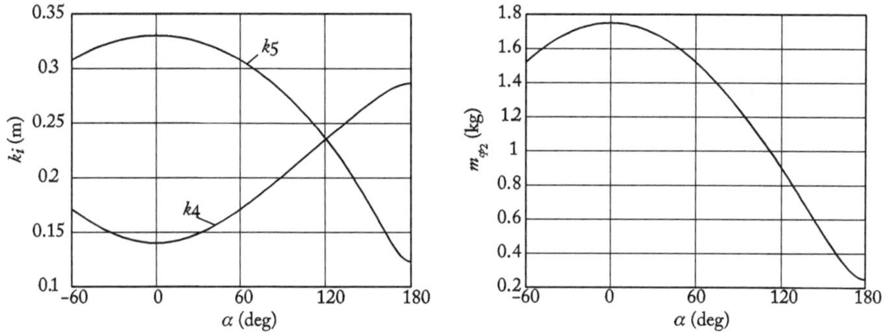

a Variation of k_4 and k_5

b Variation of m_{cw_2}

Fig. 4.6 Variation of the values of parameters k_4, k_5 and m_{cw_2} as a function of angle α

Fig. 4.7 Variation of the maximal input torque absolute value as a function of angle α

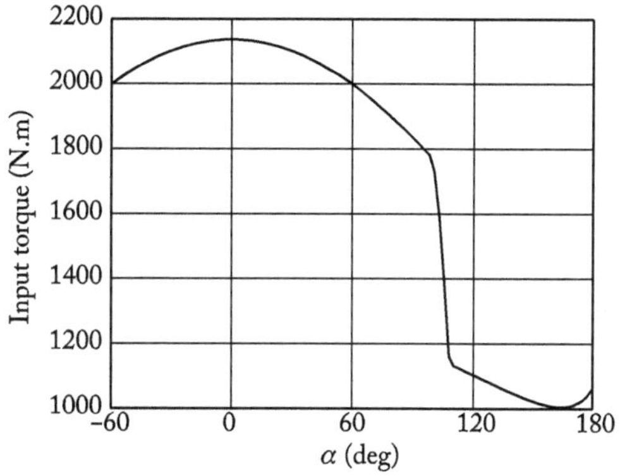

where $L = T - V$ is the Lagrangian of the system, V is the potential energy (equal to 0 in absence of gravity) and T is the kinetic energy:

$$T = \frac{1}{2} \sum_i m_i \left(\dot{x}_{S_i}^2 + \dot{y}_{S_i}^2 \right) + \frac{1}{2} \sum_j I_j \dot{\theta}_j^2 \tag{4.59}$$

$\dot{x}_{S_i}$ and $\dot{y}_{S_i}$ being the velocities along **x** and **y** axes of any *com* (for links and as well as for counterweights).

In Fig. 4.7, the maximum of the input torque absolute value of as a function of angle α is shown. Thus, it is possible to see that if the value of α is chosen arbitrarily, the input torques can grow up to 2140 N m (for $\alpha = 0°$). It also appears that the input torque will be minimal if $\alpha = 164°$. In this case, the value of the input torque is 1010 N m, i.e. about two times less than in the first case.

4.1.3.2 Balancing by Adding a Class-Two *RRP* Assur Group

We now propose obtaining the complete shaking force and shaking moment balancing of the same mechanism by adding a class-two *RRP* Assur group. Its geometric and mass properties are

- $l_4 = 0.25$ m, $l_5 = 0.25$ m, $\alpha = -90°$,
- $r_4 = 0.125$ m,
- $m_4 = 0.35$ kg, $m_5 = 0.1$ kg.

and the locations and masses of the added counterweights are:

- $r_{cw_1} = -0.1$ m, $r_{cw_2} = 0.25$ m, $r_{cw_3} = -0.125$ m,
- $m_{cw_1} = 2$ kg, $m_{cw_3} = 0.55$ kg.

the radius of gyration of element 4 and the mass of the counterweight m_{cw_2} are not expressed in these tables as they depend on the value of angle α (Fig. 4.4). Their variations as a function of α are shown in Fig. 4.8. In these figures, the values of α are bounded between 25 and 100° or 205 and 280° in order to avoid the *RRP* Assur group to cross a singularity during the motion. The simulations of the proposed mechanism have been carried out using ADAMS software and the results are similar to the previous case shown in Fig. 4.5.

As was mentioned above, the angle α is a not fixed design parameter and it can be found from minimization of the input torque of the mechanism. In Fig. 4.9, maximum of the input torque absolute value as a function of angle α is shown. It is possible to see that if the value of α is chosen arbitrarily, the input torques can grow up to 2300 N m (for $\alpha = 25°$). It also appears that the input torque will be minimal if $\alpha = 205°$. In this case, the value of the input torque is 1380 N m, i.e. about 1.7 times less than in the first case.

4.2 Complete Shaking Force and Shaking Moment Balancing of Planar Linkages by Adding the Articulated Dyads

In this Section, we consider the complete shaking force and shaking moment balancing of planar linkages by adding the articulated dyads.

4.2.1 Complete Shaking Force and Shaking Moment Balancing of Sub-linkages

A planar linkage can be described as an assembly of several sub-linkages. Let us consider here three versions of sub-linkages.

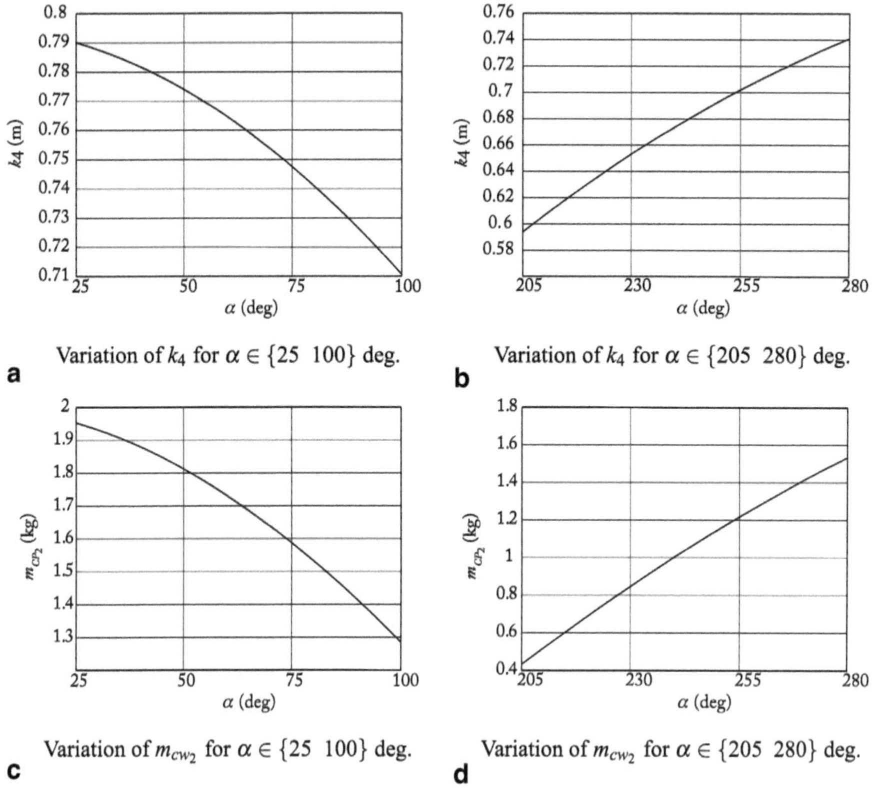

a Variation of k_4 for $\alpha \in \{25\ \ 100\}$ deg.

b Variation of k_4 for $\alpha \in \{205\ \ 280\}$ deg.

c Variation of m_{cw_2} for $\alpha \in \{25\ \ 100\}$ deg.

d Variation of m_{cw_2} for $\alpha \in \{205\ \ 280\}$ deg.

Fig. 4.8 Variation of the values of parameters k_4 and m_{cw_2} as a function of angle α

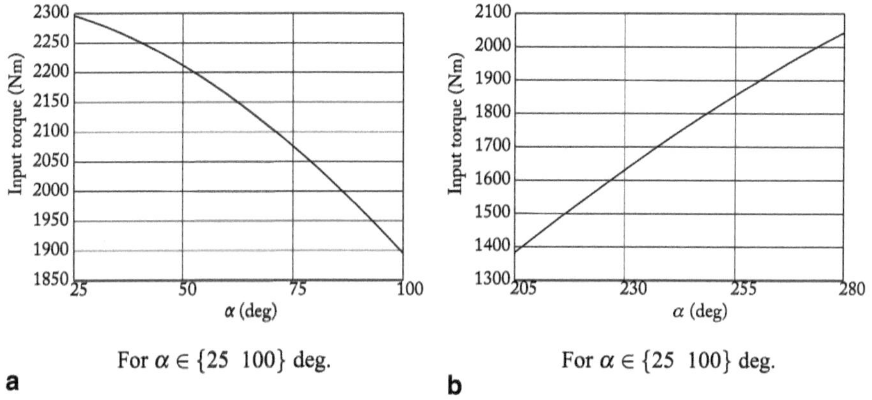

a For $\alpha \in \{25\ \ 100\}$ deg.

b For $\alpha \in \{25\ \ 100\}$ deg.

Fig. 4.9 Variation of the maximal input torque absolute value as a function of angle α

4.2.1.1 Articulation Dyad

The well-known scheme of complete shaking force and shaking moment balancing of an articulation dyad (Gao 1990) is shown in Fig. 4.10a.

The principle of such an approach is as follows. To link 2 is added a counterweight which permits the displacement of the centre of mass of link 2 to joint *A*. Then, by means of a counterweight with mass m_{cw_1} (see Fig. 4.10a), complete balancing of shaking force is achieved. A complete shaking moment balance is realised through four gear inertia counterweights 3–6, one of them being of the planetary type and mounted on link 2 (Gao 1990).

The scheme suggested here (Fig. 4.10b) is distinguished from the earlier scheme by the fact that gear 3 is mounted on the base and is linked kinematically with link 2 through link 1'.

Fig. 4.10 Complete shaking force and shaking moment balancing of articulation dyad

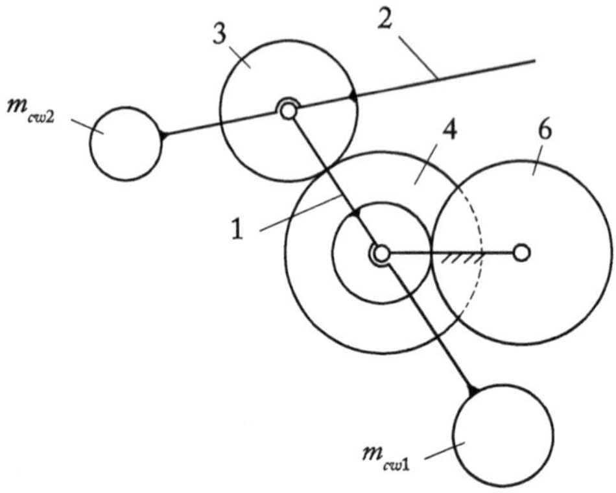

Usual method

a

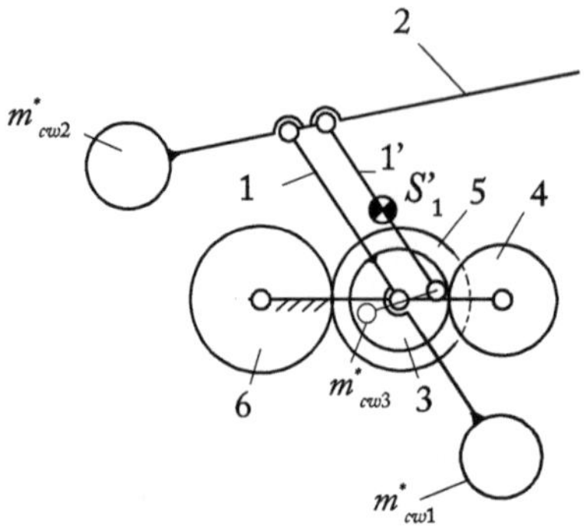

Complete shaking force and shaking moment balancing of articulation dyad by gear inertia counterweights mounted on the base.

b

To have a more illustrative representation of the advantages of such a balancing, let us consider application of the new system with the mass of link 1' not taken into account.

In this case (compared to the usual method Fig. 4.10a), the mass of the counterweight of link 1 will be reduced by an amount:

$$\Delta m_{cw_1} = m_3 l_{OA} / r_{cw_1} \tag{4.60}$$

where m_3 is the mass of the gear 3, l_{OA} is the distance between the centres of hinges O and A, r_{cw_1} is the rotation radius of the *com* of the counterweight.

It is obvious that the moment of inertia of the links is correspondingly reduced. If the gear inertias are made in the form of heavy rims in order to obtain a large moment of inertia, the moments of inertia of the gear inertia counterweights may be presented (Artobolevsky 1977) as $I_i = m_i D_i^2/4$ $(i = 3, \ldots, 6)$. Consequently, the mass of gear 6 will be reduced by an amount

$$\Delta m_6 = 4\left(m_3 l_{OA}^2 + \Delta m_{cw_1} r_{cw_1}^2\right) Z_6 / \left(D_6^2 Z_5\right) \tag{4.61}$$

where Z_5 and Z_6 are the numbers of teeth of the corresponding gears. Thus, the total mass of the system will be reduced by an amount

$$\Delta m = \Delta m_6 + \Delta m_{cw_1} \tag{4.62}$$

Let us now consider the complete shaking force and shaking moment balancing of the articulation dyad with the mass and inertia of link 1' taken into account. For this purpose, initially, we shall statically replace mass m_1' of link 1' by two point masses m_B and m_C at the centres of the hinges B and C:

$$m_B = m_1' l_{CS_1'} / l_{BC} \tag{4.63}$$

$$m_C = m_1' l_{BS_1'} / l_{BC} \tag{4.64}$$

where l_{BC} is the length of link 1, $l_{CS_1'}$ and $l_{BS_1'}$ are the distances between the centres of the joints C and B and the *com* S_1' of link 1', respectively.

After such an arrangement of masses the moment of inertia of link 1' will be equal to[2]

$$I_{S_1'}^* = I_{S_1'} - m_1' l_{BS_1'} l_{CS_1'} \tag{4.65}$$

where $I_{S_1'}$ is the moment of inertia of link 1' about the *com* S_1' of the link.

Thus, we obtain a new dynamic model of the system where the link 1' is represented by two point masses m_B, m_C and has a moment of inertia $I_{S_1'}^*$.

---

[2] After the static substitution of the masses of the link by point masses, it is necessary to take into account the change of the moment of inertia of the link.

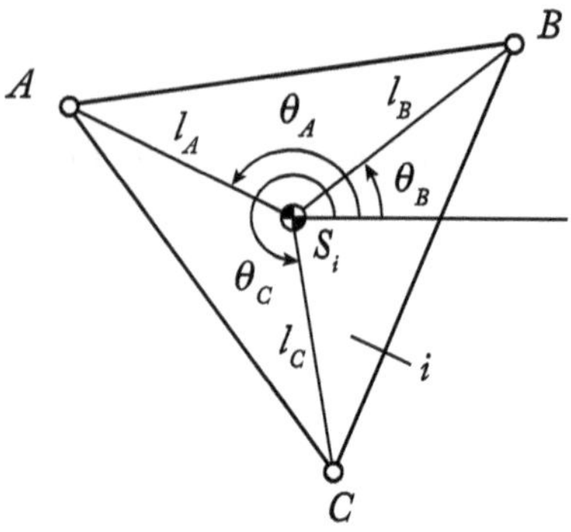

Fig. 4.11 Dynamic substitution of the masses of the link by three rotational pairs

This fact allows for an easy determination of the parameters of the balancing elements as follows:

$$m^*_{cw_2} = \left(m_2 l_{AS_2} + m_B l_{AB}\right) / r_{cw_2} \tag{4.66}$$

where m_2 is the mass of link 2, l_{AB} is the distance between the centres of hinges A and B, l_{AS_2} is the distance of the centre of hinge A from the *com* S_2 of link 2, r_{cw_2} is the rotation radius of the centre of mass of the counterweight with respect to A, and

$$m^*_{cw_1} = \left(\left(m_2 + m^*_{cw_2} + m_B\right) l_{OA} + m_1 l_{OS_1}\right) / r_{cw_1} \tag{4.67}$$

where m_1 is the mass of link 1, l_{OS_1} is the distance of the joint centre O from the centre of mass S_1 of link 1.

Also,

$$m^*_{cw_3} = m_C l_{OC} / r_{cw_3} \tag{4.68}$$

where $l_{OC} = l_{AB}$, r_{cw_3} is the rotation radius of the *com* of the counterweight.

Taking into account the mass of link 1' brings about the correction in (4.62). In this case, $\Delta m = \Delta m_6 + \Delta m_{cw_1} - \Delta m'_1$, where $\Delta m'_1$ is the value characterising the change in the distribution of the masses of the system links resulting from the addition of link 1'.

4.2.1.2 Asymmetric Link with Three Rotational Pairs

In previous work (Gao 1989, 1991) relating to balancing of linkages with a dynamic substitution of the masses of the link by three rotational pairs (see Fig. 4.11), two replacement points A and B are considered. This results in the need to increase the mass of the counterweight. However, such a solution may be avoided by considering the problem of dynamic substitution of link masses by three points. Usually, the *com* of such an asymmetric link is located inside a triangle formed by these points. The conditions for dynamic substitution of masses are the following:

$$\begin{bmatrix} 1 & 1 & 1 \\ l_A e^{i\theta_A} & l_B e^{i\theta_B} & l_C e^{i\theta_C} \\ l_A^2 & l_B^2 & l_C^2 \end{bmatrix} \begin{bmatrix} m_A \\ m_B \\ m_C \end{bmatrix} = \begin{bmatrix} m_i \\ 0 \\ I_{S_i} \end{bmatrix} \tag{4.69}$$

where m_A, m_B and m_C are point masses, l_A, l_B and l_C are the moduli of radius-vectors of corresponding points; θ_A, θ_B and θ_C are angular positions of radius-vectors, m_i is the mass of link, I_{S_i} is the moment of inertia of the link about an axis through S_i (axial moment of inertia of link).

4.2.1.3 Crank-Slider Mechanism

The usual complete shaking force and shaking moment balancing of the offset crank-slider mechanism is shown in Fig. 4.12a. The principle of such an approach resides in the following. On coupler 2 is added a counterweight which transfers the *com* of coupler 2 and slider 3 into the centre of joint A. Then, by means of a counterweight of mass m_{cw_1}, the general centre of mechanism mass is brought to the centre of pivot O. Complete balancing of the shaking moment is achieved by means of the four gear inertia counterweights 4–7.

Three solutions are examined below.

Complete shaking force and shaking moment balancing of the mechanism by mounting the gear inertia counterweights on the links connected directly to the frame.

The balancing scheme illustrated in Fig. 4.12b differs from the traditional scheme by the fact that gear 4 is mounted on the base and linked kinematically with link 2 through an additional link 1' and gear 5.

The complete balancing conditions for the shaking force and shaking moment of the mechanism are similar to those in the previous case when the scheme shown in Fig. 4.10b was considered.

Complete shaking force and shaking moment balancing of the mechanism based on the copying properties of the pantograph.

Figure 4.12c shows an off-set crank-slider mechanism with an articulation dyad *CDE* connected to it which forms a pantograph, with the initial mechanism *OAB*. By selecting, for constructional reasons, the similarity factor of the formed pantograph, $k = l_{OE}/l_{OA}$, we determine the length of the articulation dyad: $l_{CD} = l_{OA} + l_{OE}, l_{DF} = kl_{AB} + l_{AC}$.

By substituting dynamically the mass m_2 of the coupler by point masses at the centres A, B and C and using the following condition:

$$\begin{bmatrix} 1 & 1 & 1 \\ l_{AS_2} & -l_{BS_2} & l_{CS_2} \\ l_{AS_2}^2 & l_{BS_2}^2 & l_{CS_2}^2 \end{bmatrix} \begin{bmatrix} m_A \\ m_B \\ m_C \end{bmatrix} = \begin{bmatrix} m_2 \\ 0 \\ I_{S_2} \end{bmatrix} \tag{4.70}$$

where l_{AS_2}, l_{BS_2} and l_{CS_2} are the distances of joint centres A, B and C from the *com* S_2 of the coupler, I_{S_2} is the axial moment of inertia of the coupler.

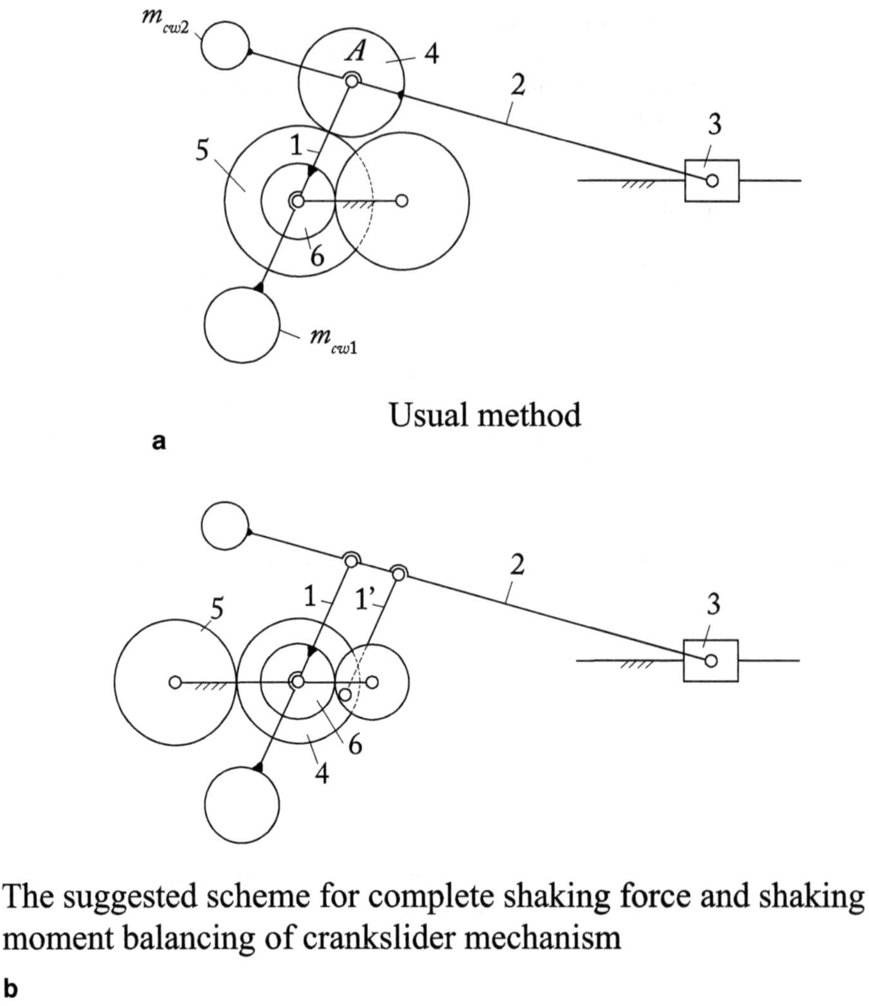

Fig. 4.12 Complete shaking force and shaking moment balancing of crank-slider mechanism

We now require link 1 to be balanced about point O, link 4 (masses m_C, m_D and m_4) about point G and finally masses m_B, m_3, m_G and m_F about point O. The necessary conditions are as follows:

$$\begin{cases} m_A l_{OA} + m_1 l_{OS_1} - m_E l_{OE} = 0 \\ m_D l_{DG} - m_4 \left(l_{CG} - l_{CS_4} \right) - m_C l_{CG} = 0 \\ m_F = \left(m_B + m_3 + \left(m_C + m_D + m_4 \right) l_{AC} / l_{AB} \right) / k \end{cases} \tag{4.71}$$

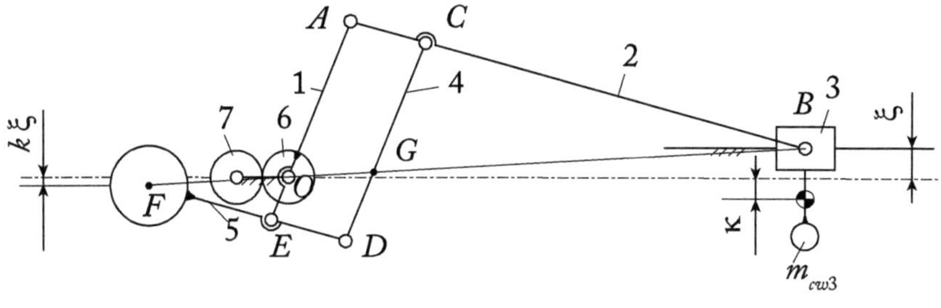

Complete shaking force and shaking moment balancing of crankslider mechanism based on the copying properties of the pantograph

c

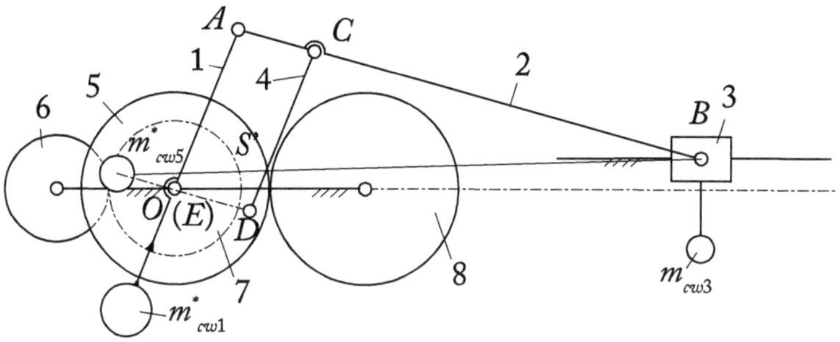

Optimum balancing of the crank-slider mechanism based on the properties of the parallelogram

d

Fig. 4.12 (Continued)

$$\begin{cases} m_D + m_E + m_F = m_5 \\ m_D l_{DS_5} + m_E \left(l_{DS_5} - l_{DE}\right) - m_F \left(l_{DF} - l_{DS_5}\right) = 0 \\ m_D l_{DS_5}^2 + m_E \left(l_{DS_5} - l_{DE}\right)^2 - m_F \left(l_{DF} - l_{DS_5}\right)^2 = I_{S_5} \end{cases} \tag{4.72}$$

where l_{OA}, l_{OE} and l_{OS_1} are the distances of joint centres A, E and of the *com* S_1 of the crank from the pivot centre O, l_{DG}, l_{CG} are the distances of the centres of the joints D, C from the working point G of the pantograph, l_{CS_4} is the distance of the centre of the joint C from the *com* S_4 of link 4, l_{AB}, l_{AC} are the distances of the centres of the joints B, C from the centre of joint A, l_{DE}, l_{DF} are the distances of the centres of the joints E, F from the centre of joint D, m_4 is the mass of link 4, m_D, m_E, m_F are point masses obtained after dynamic substitution, m_5 is the mass of link 5, l_{DS_5} is the distance of the centre of the joint D from the *com* S_5 of link 5, I_{S_5} is the axial moment of inertia of link 5.

We now have the desired parameters

$$\begin{aligned} m_5 &= m_D + m_E + m_F \\ l_{DS_5} &= (m_E l_{DE} + m_F l_{DF}) / m_5 \\ I_{S_5} &= m_D l_{DS_5}^2 + m_E \left(l_{DS_5} - l_{DE}\right)^2 - m_F \left(l_{DF} - l_{DS_5}\right)^2 \end{aligned} \tag{4.73}$$

where

$$\begin{aligned} m_D &= \left(m_C l_{CG} + m_4 \left(l_{CG} - l_{CS_4}\right)\right) / l_{DG} \\ m_E &= \left(m_A l_{OA} + m_1 l_{OS_1}\right) / l_{OE} \end{aligned} \tag{4.74}$$

Thus, we obtain a dynamic model of the mechanism (see Fig. 4.12c) fully equivalent to the real mechanism involving the rotating links 1, 4[3] and four point masses $m_3 + m_B$, m_D, m_F and m_G, three of which perform a translational rectilinear motion in the horizontal sense. As may be seen from this equivalent model, a complete shaking force balancing of the movable links of the mechanism has been achieved: $\mathbf{f}_F^{sh} = \mathbf{f}_B^{sh} + \mathbf{f}_3^{sh} + \mathbf{f}_G^{sh}$ ($\mathbf{f}_G^{sh} = \mathbf{f}_C^{sh} + \mathbf{f}_D^{sh} + \mathbf{f}_4^{sh}$, where $\mathbf{f}_i^{sh} = [f_{xi}^{sh} \; f_{yi}^{sh}]^T$ $(i = B, C, D, F, G, 3, 4)$ is the inertia forces of the corresponding masses).

The shaking moment of the mechanism is determined by the sum:

$$M^{sh} = M_1^{sh} + M_4^{sh} + M_O\left(f_{xi}^{sh}, f_{yi}^{sh}\right) \tag{4.75}$$

where M_1^{sh} and M_4^{sh} are the shaking moments of the rotating links 1 and 4 with the inertia of the replaced point masses taken into account:

$$M_1^{sh} = \ddot{\varphi}\left(I_{S_1} + m_1 l_{OS_1}^2 + m_A l_{OA}^2 + m_E l_{OE}^2\right) \tag{4.76}$$

$$M_4^{sh} = \ddot{\varphi}\left(I_{S_4} + m_4 l_{GS_4}^2 + m_C l_{CG}^2 + m_D l_{DG}^2\right) \tag{4.77}$$

where I_{S_1} and I_{S_4} are the axial moments of inertia of links 1 and 4, $\ddot{\varphi} = \ddot{\varphi}_1 = \ddot{\varphi}_4$ is the angular acceleration of links 1 and 4, $M_O\left(f_{xi}^{sh}, f_{yi}^{sh}\right)$ is the moment resulting from the force of inertia of the masses $m_3 + m_B$, m_G and m_F performing a translational rectilinear motion relative to pivot O.

The moments of the rotating links may be balanced by means of the gears (Berestov 1977b; Berkof 1973; Gao 1990; Ye and Smith 1991) mounted on the base of the mechanism. The moment of inertia of such a gear is given by the following equation:

$$I_{gear} = I_{S_1} + I_{S_4} + m_1 l_{OS_1}^2 + m_4 l_{GS_4}^2 + m_A l_{OA}^2 + m_C l_{CG}^2 + m_E l_{OE}^2 + m_D l_{DG}^2 \tag{4.78}$$

---

[3] The parameters of link 5 are selected so that the *com* of link 4, with the point masses m_C and m_D taken into account, coincides with the working point G of the pantograph, due to which the motion of this link is represented as a translational rectilinear motion of its *com* and a rotary motion relative to point G.

Regarding the moment $M_O\left(f_{xi}^{sh}, f_{yi}^{sh}\right)$, it is necessary to redistribute the masses performing a translational motion, using counterweights mounted on slider 3 (m_{cw_3}) and on link 5 (m'_F).

The necessary conditions for balancing this moment are the following:

$$\begin{cases} m_F + m'_F = k\left(m_B + m_3 + m_{cw_3} + (m_C + m_D + m_4)\, l_{AC}/l_{AB}\right) \\ k\xi\left(m_F + m'_F\right) + \xi\,(m_C + m_D + m_4)\, l_{AC}/l_{AB} = \left(m_B + m_3 + m_{cw_3}\right)\chi \end{cases} \tag{4.79}$$

from which we determine m_{cw_3} and m'_F.

It should be noted that in most constructions of such mechanisms, the eccentricity of slider guides is not significant and the moment $M_O\left(f_{xi}^{sh}, f_{yi}^{sh}\right)$ is relatively small, so that in many mechanism balancing problems this moment may be neglected.

Improvement on previous methods (Berestov 1977b; Gao 1990; Ye and Smith 1991) by mounting the gear inertia counterweight on the mechanism frame.

In Fig. 4.12d is illustrated an off-set mechanism *OAB* and an articulation dyad *CDE* connected to it. This dyad forms a parallelogram with the initial mechanism.

The conditions for balancing the system are determined from the following considerations. With the static substitution of mass m_4 of link 4 by the masses m_C and m_D situated in the centres of corresponding hinges and with the substitution of mass m_2 of the coupler (with the point mass m_C taken into account) by masses m_A and m_B, we obtain a system of point masses performing either a rotational or a translational motion. By adding thereafter a counterweight with a mass m_{cw_3} on the slider, we transfer the slider mass centre with the point mass m_B into the line Ox. However, since the slider performs a translational rectilinear motion, the mass $m_B + m_3 + m_{cw_3}$ may be considered as a point mass in the centre of joint B, since the balance of the inertia forces of the movable masses is not altered by this change. Mounting thereafter a counterweight of mass m_5^* on link 5, we transfer the general *com* $m_B + m_3 + m_{cw_3} + m_5$ onto point S' (see Fig. 4.12d) which performs a rotational motion.

We now obtain a complete shaking force balance by the addition of a counterweight with mass $m_{cw_1}^*$ on the input crank, displacing the *com* of the movable links from point S' to the centre of the pivot O.

In this case, the static moments of the counterweights relative to the pivot will be given by

$$m_1^* r_1^* = m_1 l_{OS_1} + l_{OA}\left(m_2 + m_3 + m_4 l_{DS_4}/l_{CD}\right) \tag{4.80}$$

$$m_5^* r_5^* = m_2 l_{AS_2} + m_3 l_{AB} + m_4 l_{AC} \pm m_5 l_{OS_5} \tag{4.81}$$

where m_i is the mass of link i, $l_{OS_1}, l_{AS_2}, l_{CS_4}, l_{OS_5}$ are the distances of the *com* of links $1, 2, 4, 5$ from the centres of joints $O, A, C, l_{OA}, l_{AB}, l_{AC}, l_{CD}$ are the distances between the centres of corresponding joints.

After such a redistribution of masses, the moment from inertia forces will be balanced by gear inertia counterweights 5–8 (Berestov 1977b; Berkof 1973; Gao 1990; Ye and Smith 1991).

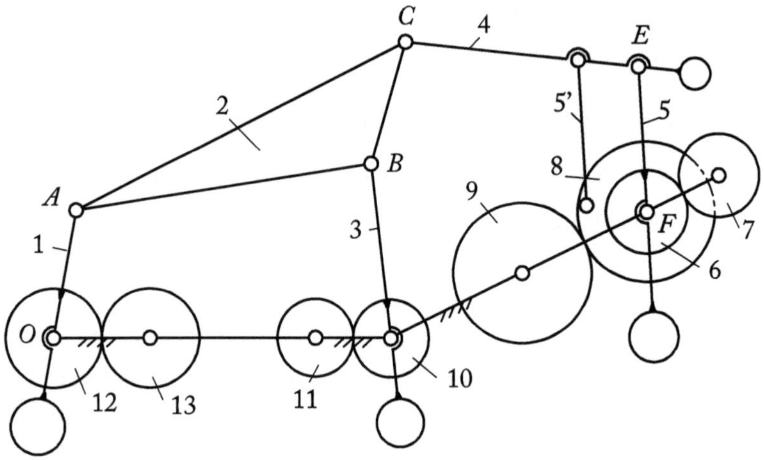

Fig. 4.13 Complete shaking force and shaking moment balancing of Stephenson linkage

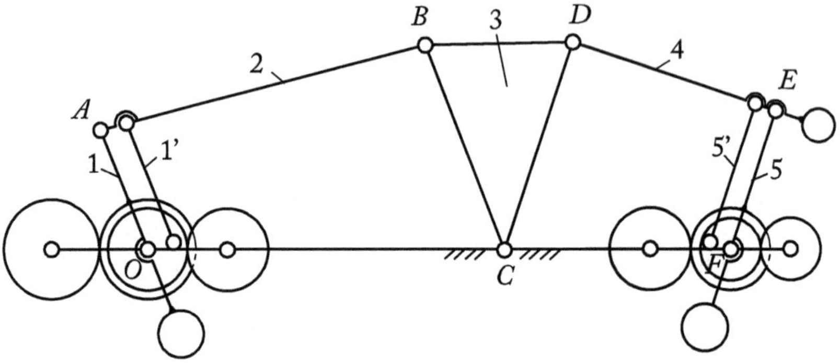

Fig. 4.14 Complete shaking force and shaking moment balancing of Watt linkage-Case 1

4.2.2 Application of the Methods for Complete Shaking Force and Shaking Moment Balancing of Multilink Mechanisms

Let us consider the complete shaking force and shaking moment balancing of Stephenson (Fig. 4.13) and Watt (Figs. 4.14 and 4.15) linkages.

For the complete shaking force and shaking moment balancing of the Stephenson linkage we apply the following approach. First, we replace dynamically the mass of coupler 2 by three point masses located at the centres of the joints A, B and C. That permits us to solve the problem of complete shaking force and shaking moment balancing of the linkage as separate problems of the balancing of sub-linkages (Sect. 4.2.1.1).

For the complete shaking force and shaking moment balancing of the Watt linkage, two methods have been developed. In the first method (Fig. 4.14), we replace

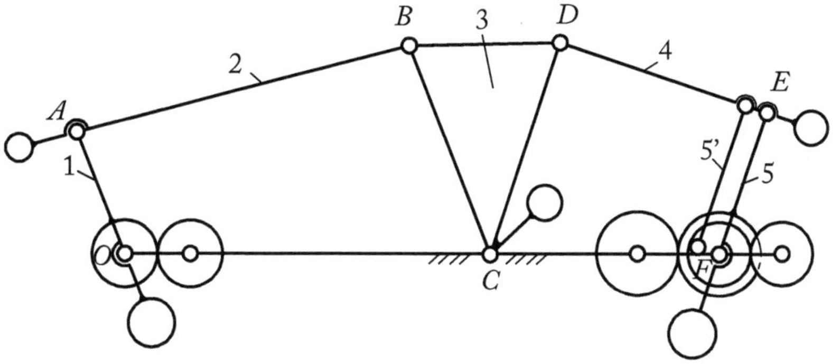

Fig. 4.15 Complete shaking force and shaking moment balancing of Watt linkage-Case 2

dynamically the mass of link 3 by the point masses m_B, m_C and m_D. Then, we consider the problem of sub-linkages *OAB* and *DEF*.

In the second method (Fig. 4.15), we replace dynamically the mass of link 2 by the point masses m_A and m_B. Then, taking into account the point mass m_B we replace dynamically the mass of link 3 by the point masses m_C and m_D. That changes the problem of balancing the linkage into problems of balancing sub-linkages: crank *OA* and articulation dyad *DEF*.

Finally we would like to conclude that the advantage of the schemes outlined here is the fact that all the gear inertia counterweights needed for the balancing of the shaking moments are mounted on the mechanism frame, which is constructively more efficient.

4.3 Complete Shaking Force and Shaking Moment Balancing of *RSS'R* Spatial Linkages

4.3.1 Statement of the Problem

Figure 4.16 shows the *RSS'R* mechanism in which the idle *dof*, i.e. the spinning of the coupler about the line joining the centers of the spherical joints, is canceled by a pin.

The shaking force and the shaking moment of this mechanism are determined by the sum of the forces and moments of the three moving links. Two of these links (2 and 4) are mounted on the frame 1 of the mechanism by revolute joint located at A and D. The third link of the mechanism is the coupler, which carries out spatial motion. The shaking force and shaking moment balancing of the rotating links 2 and 4 does not present any difficulty and the techniques are well-known (Berkof 1973).

Thus, with the dynamic substitution of the mass m_3 of the coupler 3 by two concentrated masses m_{3B} and m_{3C} located at points B and C of the spherical joints, we obtain a dynamic model of the coupler that represents a weightless link with two

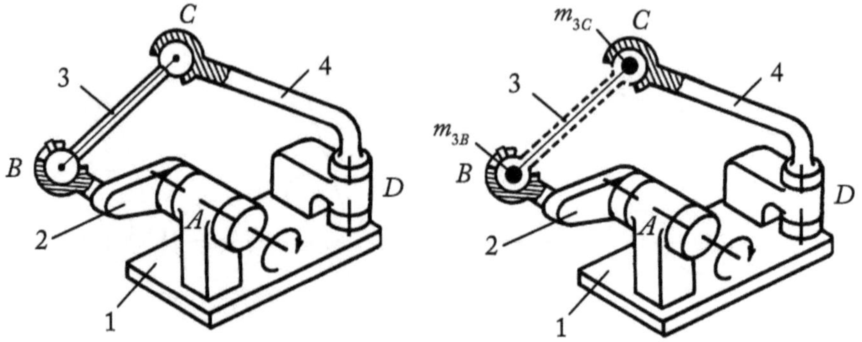

Fig. 4.16 *RSS'R* spatial linkage

point masses (Fig. 4.16). For this it is necessary to find a coupler shape, which allows the dynamic substitution of the coupler mass by the concentrated masses.

4.3.2 *Coupler Shape Design*

Standard coupler of the *RSS'R* linkage is made of two spheres connected with a cylinder. Such a shape of the coupler is not optimal for the realization of the dynamic modeling by masses substitution.

Figure 4.17 shows a special coupler shape composed by two elements: the segments of a sphere and a cylinder.

To obtain the conditions of dynamic substitution, let us consider the masses and the mass moments of inertia for these elements.

Segments of Sphere (Fig. 4.18a)[4]

- Mass:

$$m'_3 = \pi \frac{h^2}{2} (3R - h)\, \rho \tag{4.82}$$

  where, ρ is the coefficient of the material density, which is same for all elements of the coupler; h and R are geometric parameters shown in Fig. 4.18a.
- The location of the *com* S' of the segment of a sphere can by found by the expression:

$$x'_S = \frac{h}{4} \cdot \frac{8R - 3h}{3R - h} \tag{4.83}$$

- Mass moments of inertia:

$$I_{x'x'} = m'_3 \frac{h}{10} \cdot \frac{5R(4R - 3h) + 3h^2}{3R - h} \tag{4.84}$$

[4] The expressions concerning the determination of masses and mass moments of inertia have been taken from (Favorin 1970).

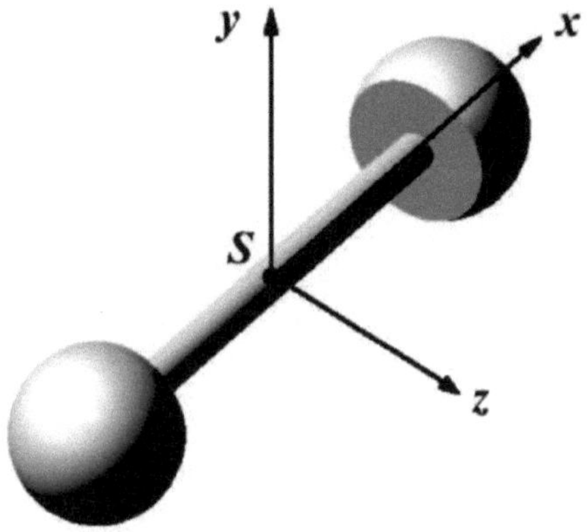

a CAD model in 3D

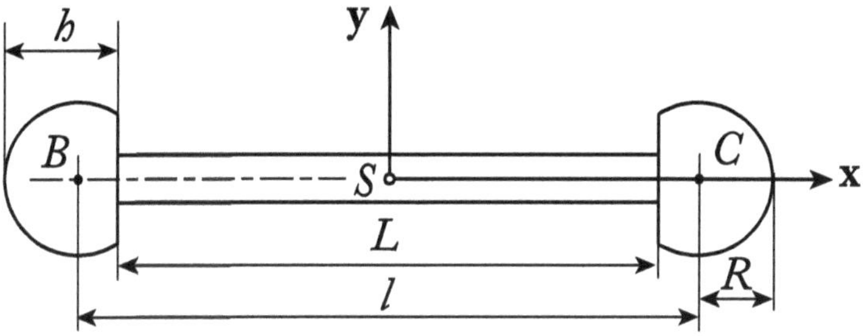

b Drawing in 2D

Fig. 4.17 A special coupler shape

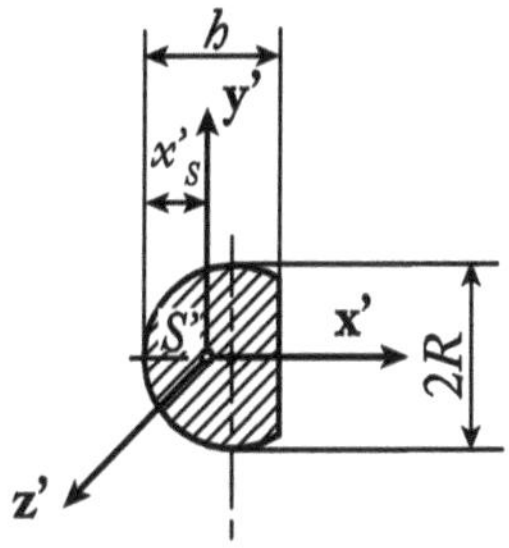

a Segment of a sphere

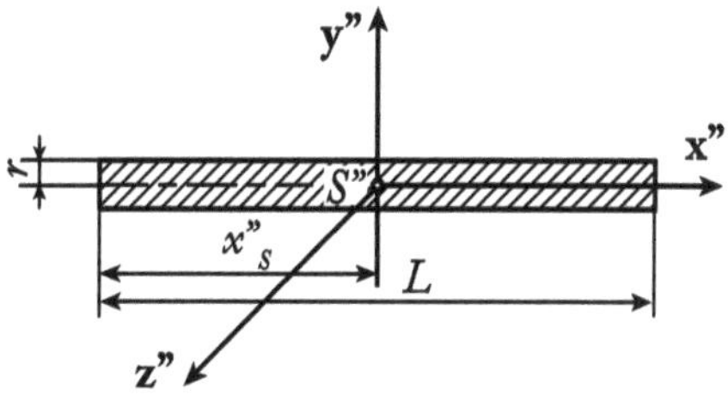

b Circular cylinder

Fig. 4.18 Drawings of the coupler elements

$$I_{y'y'} = I_{z'z'} = \frac{I_{x'x'}}{2} + m'_3 \frac{h^2}{80} \cdot \frac{8R(5R - 3h) + 3h^2}{(3R - h)^2} \tag{4.85}$$

Circular Cylinder (Fig. 4.18b)

- Mass:

$$m_3'' = \pi r^2 L\rho \tag{4.86}$$

where, r is the radius of the circular cylinder and L its length (see Fig. 4.18b).
- The location of the *com* S'' of the segment of a sphere can found by the expression:

$$x_S'' = 0.5L \tag{4.87}$$

- Mass moments of inertia:

$$I_{x''x''} = m_3'' r^2 \tag{4.88}$$

$$I_{y''y''} = I_{z''z''} = \frac{m_3''\left(3r^2 + L^2\right)}{12} \tag{4.89}$$

Taking into account that the suggested coupler shape is symmetric and the rotation around **x** axis is locked by pin, the condition of the dynamic substitution of the coupler mass by the concentrated masses m_{3B} and m_{3C} located at the centers B and D of the spherical joints can be reduced to the expression:

$$2I_{z'z'} + 2m_3'\left(R - x_S' + 0.5l\right)^2 + I_{z''z''} = \left(m_{3B} + m_{3C}\right)(0.5l)^2 \tag{4.90}$$

where $m_{3B} = m_{3C} = m_3' + \frac{m_3''}{2}$.

Introducing the relationships (4.82– 4.89) into (4.90), we obtain the following equation:

$$\begin{aligned}
&\left(\frac{2\pi h^2}{3}(3R-h)\right)\left(\frac{h}{20}\cdot\frac{5R(4R-3h)+3h^2}{3R-h} + \frac{h^2}{80}\cdot\frac{8R(5R-3h)+3h^2}{(3R-h)^2}\right) + \\
&\left(\frac{2\pi h^2}{3}(3R-h)\right)\left(R - x_S' + 0.5l\right)^2 + \pi r^2 L\frac{\left(3r^2+L^2\right)}{12} \\
&= (0.5l)^2\left(2\left(\pi\frac{h^2}{3}(3R-h)\right) + \pi r^2 L\right)
\end{aligned} \tag{4.91}$$

from which we determine the radius of the connecting cylinder:

$$r_{1,2} = \sqrt{\frac{-b \pm \sqrt{b^2 - 4ac}}{2a}} \tag{4.92}$$

where,

$$a = \frac{\pi L}{4} \tag{4.93}$$

$$b = \frac{1}{12}\pi L^3 - \frac{1}{4}\pi L\, l^2 \tag{4.94}$$

$$c = \left(\frac{2\pi h^2}{3}(3R-h)\right)\left(\frac{h}{20}\cdot\frac{5R(4R-3h)+3h^2}{3R-h}+\frac{h^2}{80}\cdot\frac{8R(5R-3h)+3h^2}{(3R-h)^2}\right)$$
$$+\left(\frac{2\pi h^2}{3}(3R-h)\right)\left(R-x'_S+0.5l\right)^2-\frac{l^2}{2}\left(\pi\frac{h^2}{3}(3R-h)\right) \tag{4.95}$$

Coupler Shape Design Example Let us now consider the creation of coupler shape from the given length $l = 0.14$ m. First we must choose the geometric parameters of the segments of the sphere, for example $R = 0.02$ m and $h = 0.03$ m. Then we compute from (4.92) the radius $r_1 = 0.005$ m and $r_2 = 0.122$ m. For design reasons, we choose $r = r_1 = 0.005$ m and obtain a special coupler shape, which allows the substitution of the mass of the coupler by two concentrated masses: $m_{3B} = m_{3C} = 0.25$ kg.

Thus, connecting two segments of the sphere by the circular cylinder with radius r determined from expression (4.92), we obtain such a mass distribution of the coupler, which can be substituted by two concentrated point masses situated at the centers B and C.

It is obvious that one can find various shapes of the coupler which also achieve the substitution of the masses of the coupler by two concentrated masses. But we will be satisfied by the considered example which is sufficient for the demonstration of the shaking force and shaking moment balancing of the linkage.

4.3.3 Numerical Example

Let us examine the complete shaking force and shaking moment balancing of the *RSS'R* spatial linkage (Fig. 4.19) with following parameters:

- $l_{OA_0} = 0.1$ m, $l_{A_0B} = 0.05$ m, $l_{BC} = 0.3$ m, $l_{CD} = 0.15$ m, $l_{D_0D} = 0.05$ m,
- $\overrightarrow{AA_0} \perp \overrightarrow{A_0B}$, $\overrightarrow{DD_0} \perp \overrightarrow{CD_0}$,
- $m_2 = 3$ kg, $m_3 = 4$ kg, $m_4 = 4$ kg,
- mass moment of inertia of the link 3 relatively to the plane which is perpendicular to the $\overrightarrow{BC}$: $I^{(3)}_{S_{z_3z_3}} = I^{(3)}_{S_{y_3y_3}} = 0.09$ kg m^2.

The axis of the revolute joint at A is parallel to the axis $\mathbf{y}$, the axis of the revolute joints at D and E are located in the horizontal plane $O\mathbf{xz}$ and they are parallel to the axis $\mathbf{z}$ ($x_D = 0.25$ m, $z_{D_0} = -0.05$ m and $x_E = 0.35$ m).

The parameters of the *com* of links 1–3 are the followings: $r_2 = l_{AB}/4$, $r_3 = l_{BC}/2$, $r_4 = l_{D_0C}/2$. The input angular velocity is constant: $\dot{\varphi}_2 = 10\pi$ rad/s. The rotation of link 3 about the axis $\overrightarrow{BC}$ is cancelled by a pin.

The complete shaking force and shaking moment balancing of the linkage can be achieved in the following sequence:

1. By selecting the geometrical parameters $R = 0.04$ m and $h = 0.06$ m, we find from (4.92) the radius of the connecting cylinder of the coupler 3: $r = 0.0083$ m.

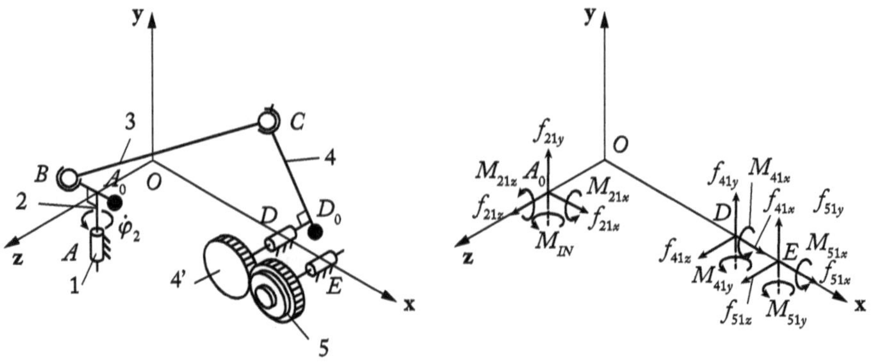

a Scheme of the mechanism **b** Ground bearing forces and moments

Fig. 4.19 Complete shaking force and shaking moment balancing of the *RSS'R* spatial linkage

2. Compute the concentrated masses: $m_{3B} = m_{3C} = 2$ kg.
3. Taking into account that $r_{cw_2} = 0.025$ m and $r_{cw_4} = 0.075$ m, we achieve the complete shaking force balancing of the linkage by adding two counterweights.
4. The complete shaking moment balancing is achieved by the inertia counterweight mounted on gear 5, which is meshed with gear 4' of the output link 4 (see Berkof 1973).

Figure 4.19c describes the ground bearing forces f_{j1x}, f_{j1y} and f_{j1z} $(j = 2, 4, 5)$ along the **x**, **y** and **z** axes, respectively, as well as the ground bearing moments M_{j1x}, M_{j1y} and M_{j1z} around the **x**, **y** and **z** axes, respectively, with respect to the points $A_0 = (0, 0, 0.1)$ m, $D = (0.25, 0, 0)$ m and $E = (0.35, 0, 0)$ m.

In Fig. 4.20 are presented the variations of the shaking forces (sum of the ground bearing forces) after balancing of the linkage. It is seen that the resultant shaking force is equal to zero. The shaking moment (sum of the ground bearing moments) of the balanced linkage is also equal to zero (Fig. 4.21).

It should be noted that the resultant shaking moment with respect to the **y** axis represents the sum of the moments of the ground bearing forces and the reaction of the input torque on the ground (i.e. moment exerted by driving motor on link 1).

To reduce the impact forces in the oscillating gears 4' and 5, it is desirable to use the anti-backlash gears.

Finally, it is necessary to remind that the complete shaking force and shaking moment balancing of linkages can only be reached by the increase of the total mass and input torque. The examined case is not an exception.

4.3.4 Input Torque of the Balanced Linkage

Now we shall exanimate the input torque of the balanced *RSS'R* linkage.

Lagrange's equation for a single *dof* rigid-body linkage can be written as:

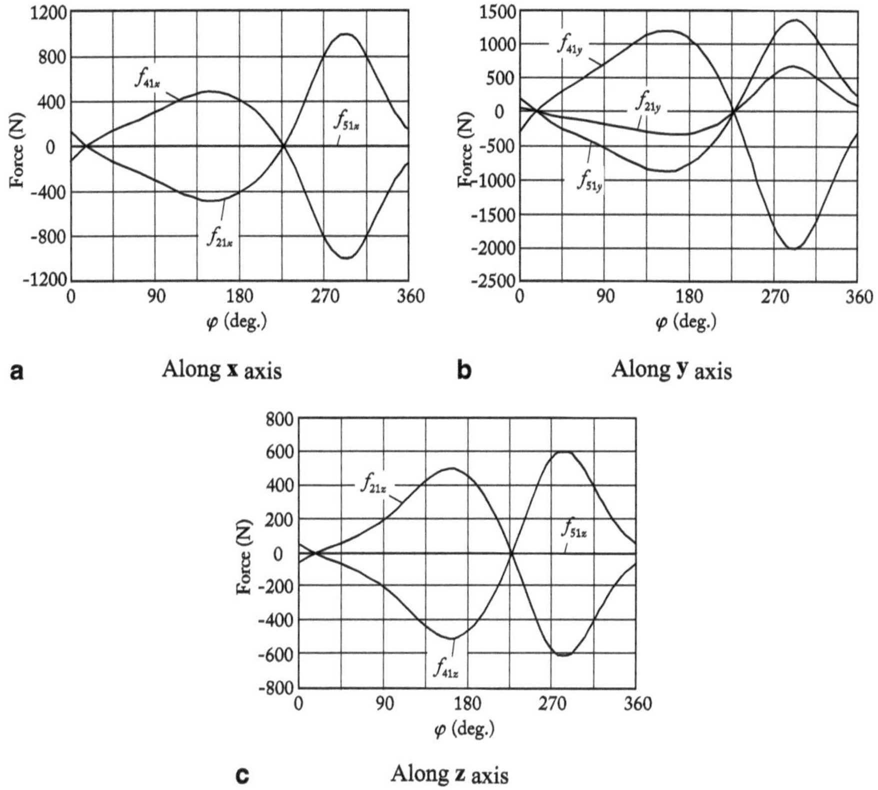

Fig. 4.20 Variation of the shaking forces (ground bearing forces) of the balanced linkage

$$\tau = \frac{\mathrm{d}}{\mathrm{d}t}\left(\frac{\partial T}{\partial \dot{\varphi}_2}\right) - \frac{\partial T}{\partial \varphi_2} \tag{4.96}$$

where, τ is the generalized force acting on the system, $T = \sum_{i=2}^{n} T_i$ is the sum of the kinetic energies of all the n links, and φ_2 is the chosen generalized coordinate.

The input torque of the *RSS'R* spatial linkage is a very complex function but we shall show below that for balanced linkage with constant input angular velocity this function can be reduced to single-term expression.

For this purpose, we shall exanimate the kinetic energy of the connecting coupler and two concentrated masses m_{3B} and m_{3C}. Considering that the connecting coupler 3 is a rigid body with two spherical joints and with arbitrary mass distribution (Fig. 4.22), the kinetic energy can be written as (Wittenburg 1977):

$$T_{coupler} = \frac{1}{2} m_3 \mathbf{v}_{S_3}^T \mathbf{v}_{S_3} + \frac{1}{2} \omega_3^T \mathbf{I}_3 \omega_3 \tag{4.97}$$

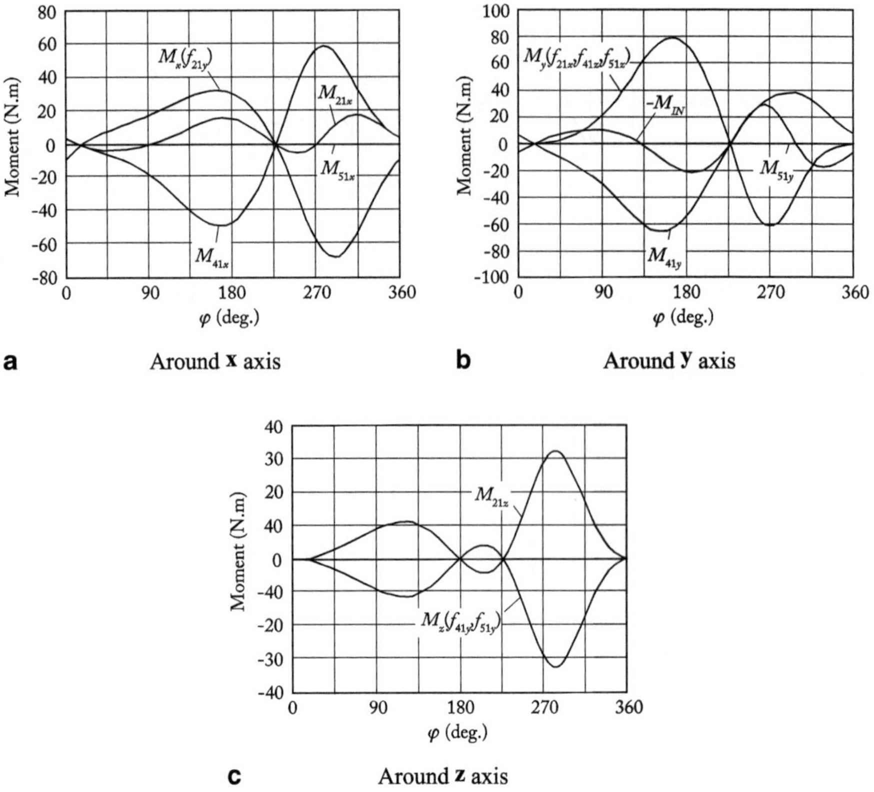

Fig. 4.21 Variation of the shaking forces (ground bearing forces) of the balanced linkage

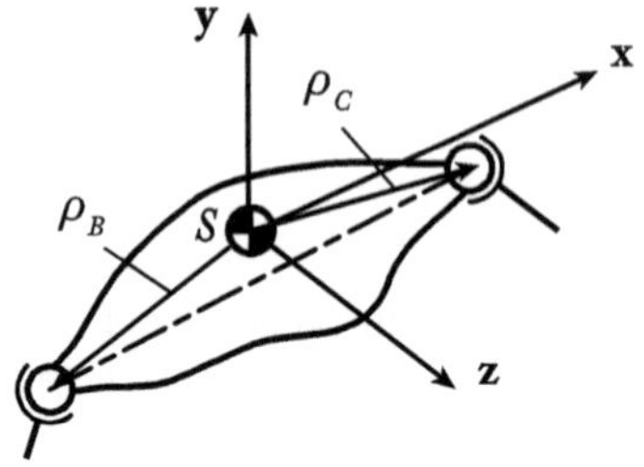

Fig. 4.22 Connecting coupler with arbitrary link mass distribution

or in the scalar form as:

$$T_{coupler} = \frac{1}{2}\left(m_3 \left\|\mathbf{v}_{S_3}\right\|^2 + I_{xx}^{(3)}\omega_{3_x}^2 + I_{yy}^{(3)}\omega_{3_y}^2 + I_{zz}^{(3)}\omega_{3_z}^2\right) - I_{xy}^{(3)}\omega_{3_x}\omega_{3_y} - I_{xz}^{(3)}\omega_{3_x}\omega_{3_z} - I_{yz}^{(3)}\omega_{3_y}\omega_{3_z} \tag{4.98}$$

where, $\mathbf{v}_{S_3}$ is the linear velocity of the *com* S_3 of the connecting coupler, $\omega_3^T = \left[\omega_{3_x}\ \ \omega_{3_y}\ \ \omega_{3_z}\right]$ is the angular velocity of the connecting coupler, and $\mathbf{I}_3$ is the tensor of inertia at point S_3:

$$\mathbf{I}_3 = \begin{bmatrix} I_{xx}^{(3)} & -I_{xy}^{(3)} & -I_{xz}^{(3)} \\ -I_{xy}^{(3)} & I_{yy}^{(3)} & -I_{yz}^{(3)} \\ -I_{xz}^{(3)} & -I_{yz}^{(3)} & I_{zz}^{(3)} \end{bmatrix} \tag{4.99}$$

$I_{xx}^{(3)}$, $I_{yy}^{(3)}$ and $I_{zz}^{(3)}$ are axial moments of inertia, $I_{xy}^{(3)}$, $I_{xz}^{(3)}$ and $I_{yz}^{(3)}$ are the products of inertia.

The kinetic energy of two concentrated masses m_{3B} and m_{3C} located at the centers B and C of the spherical joints can be written as:

$$T_{cm} = \frac{1}{2}\left(m_{3B}\left\|\mathbf{v}_B\right\|^2 + m_{3C}\left\|\mathbf{v}_C\right\|^2\right) \tag{4.100}$$

where, $\mathbf{v}_B$ and $\mathbf{v}_C$ are the linear velocities of centers B and C of the spherical joints.

Taking into account that $\mathbf{v}_B = \mathbf{v}_{S_3} + \omega_3 \times \mathbf{r}_{BS_3}$ and $\mathbf{v}_C = \mathbf{v}_{S_3} + \omega_3 \times \mathbf{r}_{CS_3}$ (where $\mathbf{r}_{BS_3} = [x_B \;\; y_B \;\; z_B]^T$ and $\mathbf{r}_{CS_3} = [x_C \;\; y_C \;\; z_C]^T$ are the vectors between points B and S_3 and C and S_3, respectively (Fig. 4.22)), (4.100) can be written as:

$$\begin{aligned} T_{cm} &= \frac{1}{2}\Big((m_{3B}+m_{3C})\left\|\mathbf{v}_{S_3}\right\|^2 + 2m_{3B}\mathbf{v}_{S_3}^T\left(\omega_3 \times \mathbf{r}_{BS_3}\right) + 2m_{3C}\mathbf{v}_{S_3}^T\left(\omega_3 \times \mathbf{r}_{CS_3}\right) \\ &\quad + m_{3B}\left(\omega_3 \times \mathbf{r}_{BS_3}\right)^T\left(\omega_3 \times \mathbf{r}_{BS_3}\right) + m_{3C}\left(\omega_3 \times \mathbf{r}_{CS_3}\right)^T\left(\omega_3 \times \mathbf{r}_{CS_3}\right)\Big) \\ &= \frac{1}{2}\Big((m_{3B}+m_{3C})\left\|\mathbf{v}_{S_3}\right\|^2 + 2\mathbf{v}_{S_3}^T\left(\omega_3 \times \left(m_{3B}\mathbf{r}_{BS_3} + m_{3C}\mathbf{r}_{CS_3}\right)\right) \\ &\quad + m_{3B}\left(\omega_3 \times \mathbf{r}_{BS_3}\right)^T\left(\omega_3 \times \mathbf{r}_{BS_3}\right) + m_{3C}\left(\omega_3 \times \mathbf{r}_{CS_3}\right)^T\left(\omega_3 \times \mathbf{r}_{CS_3}\right)\Big) \end{aligned} \tag{4.101}$$

The sum $m_{3B}\left(\omega_3 \times \mathbf{r}_{BS_3}\right)^T\left(\omega_3 \times \mathbf{r}_{BS_3}\right) + m_{3C}\left(\omega_3 \times \mathbf{r}_{CS_3}\right)^T\left(\omega_3 \times \mathbf{r}_{CS_3}\right)$ can be written in scalar form as:

$$\begin{aligned} & m_{3B}\left(\omega_3 \times \mathbf{r}_{BS_3}\right)^T\left(\omega_3 \times \mathbf{r}_{BS_3}\right) + m_{3C}\left(\omega_3 \times \mathbf{r}_{CS_3}\right)^T\left(\omega_3 \times \mathbf{r}_{CS_3}\right) = \\ & m_{3B}\left(\left(\omega_{3_y}z_B - \omega_{3_z}y_B\right)^2 + \left(\omega_{3_z}x_B - \omega_{3_x}z_B\right)^2 + \left(\omega_{3_x}y_B - \omega_{3_y}x_B\right)^2\right) + \\ & m_{3C}\left(\left(\omega_{3_y}z_C - \omega_{3_z}y_C\right)^2 + \left(\omega_{3_z}x_C - \omega_{3_x}z_C\right)^2 + \left(\omega_{3_x}y_C - \omega_{3_y}x_C\right)^2\right) = \\ & \omega_{3_x}^2\left(m_{3B}(y_B^2+z_B^2) + m_{3C}(y_C^2+z_C^2)\right) + \omega_{3_y}^2\left(m_{3B}(x_B^2+z_B^2) + m_{3C}(x_C^2+z_C^2)\right) + \\ & \omega_{3_z}^2\left(m_{3B}(x_B^2+y_B^2) + m_{3C}(x_C^2+y_C^2)\right) - 2\omega_{3_x}\omega_{3_y}\left(m_{3B}x_By_B + m_{3C}x_Cy_C\right) - \\ & 2\omega_{3_x}\omega_{3_z}\left(m_{3B}x_Bz_B + m_{3C}x_Cz_C\right) - 2\omega_{3_y}\omega_{3_z}\left(m_{3B}y_Bz_B + m_{3C}y_Cz_C\right) \end{aligned} \tag{4.102}$$

Now, taking into account the following conditions of the dynamic substitution of masses of the connecting coupler by two concentrated masses m_{3B} and m_{3C}:

$$m_{3B} + m_{3C} = m_3 \tag{4.103}$$

$$m_{3B}x_{BS_3} + m_{3C}x_{CS_3} = 0 \tag{4.104}$$

$$m_{3B}y_{BS_3} + m_{3C}y_{CS_3} = 0 \tag{4.105}$$

$$m_{3B}z_{BS_3} + m_{3C}z_{CS_3} = 0 \tag{4.106}$$

$$m_{3B}\left(y_B^2 + z_B^2\right) + m_{3C}\left(y_C^2 + z_C^2\right) = I_{xx}^{(3)} \tag{4.107}$$

$$m_{3B}\left(x_B^2 + z_B^2\right) + m_{3C}\left(x_C^2 + z_C^2\right) = I_{yy}^{(3)} \tag{4.108}$$

$$m_{3B}\left(x_B^2 + y_B^2\right) + m_{3C}\left(x_C^2 + y_C^2\right) = I_{zz}^{(3)} \tag{4.109}$$

$$m_{3B}x_By_B + m_{3C}x_Cy_C = I_{xy}^{(3)} \tag{4.110}$$

$$m_{3B}x_Bz_B + m_{3C}x_Cz_C = I_{xz}^{(3)} \tag{4.111}$$

$$m_{3B}y_Bz_B + m_{3C}y_Cz_C = I_{yz}^{(3)} \tag{4.112}$$

and substituting Eq. (4.102) into (4.101), we obtain:

$$\begin{aligned} T_{cm} = \; & \frac{1}{2}\left(m_3\left\|\mathbf{v}_{S_3}\right\|^2 + I_{xx}^{(3)}\omega_{3_x}^2 + I_{yy}^{(3)}\omega_{3_y}^2 + I_{zz}^{(3)}\omega_{3_z}^2\right) \\ & -I_{xy}^{(3)}\omega_{3_x}\omega_{3_y} - I_{xz}^{(3)}\omega_{3_x}\omega_{3_z} - I_{yz}^{(3)}\omega_{3_y}\omega_{3_z} \end{aligned} \tag{4.113}$$

which is identical to Eq. (4.97).

Thus, the kinetic energy of the connecting coupler of the examined linkage and the kinetic energy of the concentrated masses m_{3B} and m_{3C}, obtained from dynamic substitution of masses of the coupler by two points masses are the same.

Applying this condition, the total kinetic energy of the *RSS'R* balanced linkage can be expressed as:

$$T = T_2 + T_4 + T_5 + T_{cm} \tag{4.114}$$

where

$$T_2 = \frac{1}{2}\left(I_{zz}^{(2)} + m_2 l_{A_0S_2}^2\right)\dot{\varphi}_2^2 \tag{4.115}$$

$$T_4 = \frac{1}{2}\left(I_{zz}^{(4)} + m_4 l_{DS_4}^2\right)\dot{\varphi}_4^2 \tag{4.116}$$

$$T_5 = \frac{1}{2} I_{zz}^{(5)} \dot{\varphi}_5^2 \tag{4.117}$$

$$T_{cm} = \frac{1}{2} m_{3B} l_{A_0B}^2 \dot{\varphi}_2^2 + \frac{1}{2} m_{3C} l_{CD_0}^2 \dot{\varphi}_4^2 \tag{4.118}$$

m_2 and m_4 are the masses of the links 2 and 4, $I_{zz}^{(2)}$, $I_{zz}^{(4)}$ and $I_{zz}^{(5)}$ are the axial inertia moments of links 2, 4 and 5 relative to the axis of rotation, $\dot{\varphi}_2$, $\dot{\varphi}_4$ and $\dot{\varphi}_5$ are the angular velocities of links 2, 4 and 5 (see Fig. 4.19).

Thus, Eq. (4.114) can be expressed as:

$$T = \frac{1}{2} K_2 \dot{\varphi}_2^2 + \frac{1}{2} K_4 \dot{\varphi}_4^2 \tag{4.119}$$

where $K_2 = I_{zz}^{(2)} + m_2 l_{A_0S_2}^2 + m_{3B} l_{A_0B}^2$ and $K_4 = I_{zz}^{(4)} + I_{zz}^{(5)} + m_4 l_{DS_4}^2 + m_{3C} l_{CD_0}^2$.

In (Berkof 1979b), it is shown that Lagrange's equation for a single *dof* rigid-body linkage can be reduced into the form:

$$\tau = \frac{1}{\dot{\varphi}_2} \frac{\mathrm{d}T}{\mathrm{d}t} \tag{4.120}$$

Substituting Eq. (4.119) into (4.120), we obtain:

$$\tau = K_2 \ddot{\varphi}_2 + K_4 \ddot{\varphi}_4 \frac{\dot{\varphi}_4}{\dot{\varphi}_2} \tag{4.121}$$

Moreover, for constant input angular velocity $\dot{\varphi}_2$, (4.121) is reduced to:

$$\tau = K_4 \ddot{\varphi}_4 \frac{\dot{\varphi}_4}{\dot{\varphi}_2} \tag{4.122}$$

The obtained results allow the minimization of the input torque of the balanced *RSS'R* spatial linkage by the technique shown in (Berkof 1979b).

Now let us consider novel approaches for the design of self balanced mechanisms.

4.4 Design of Self-balanced Mechanical Systems

4.4.1 Shaking Force Balancing

Figure 4.23 shows the proposed self-balanced mechanical system, which consist of two identical crank-slider mechanisms OAB and OCD, as well as a pair of gears 6 and 7 connected to the input crank 1. Please note that the guides of sliders are common, i.e. the asymmetric model of duplication is applied.

Let us firstly consider the frequency components of the proposed mechanical system.

The position of the slider 3 can be determined by the following expression:

$$d_B = \sqrt{(r+l)^2 - e^2} - r\cos\varphi - l\cos\psi \tag{4.123}$$

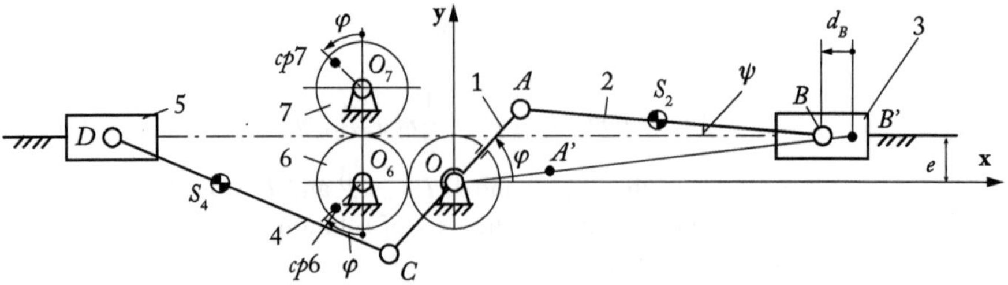

Fig. 4.23 Kinematic diagram of the suggested self-balanced mechanical system consisting of two identical off-set crank-slider mechanisms and a pair of counter-rotating masses

where, $r = l_{OA}$ is the distance between the centers of joints O and A; $l = l_{AB}$ is the distance between the centers of joints A and B; $e = y_B$ is the eccentricity of slider guide; φ is the rotating angle of the input crank and ψ is the acute angle that the rod makes with the sliding axis:

$$\psi = \cos^{-1}\left(1 - \left(\frac{r\sin\varphi - e}{l}\right)^2\right)^{1/2} \tag{4.124}$$

Expression (4.124) can be represented using Newton's binomial series as follows:

$$\left(1 - \left(\frac{r\sin\varphi - e}{l}\right)^2\right)^{1/2} = 1 - \frac{1}{2}\left(\frac{r\sin\varphi - e}{l}\right)^2 - \frac{1}{8}\left(\frac{r\sin\varphi - e}{l}\right)^4 - \frac{1}{16}\left(\frac{r\sin\varphi - e}{l}\right)^6 + \ldots \tag{4.125}$$

Substituting (4.125) into (4.123) and taking into account that it is sufficient to keep the two first terms of this series, we obtain the following expression for the position of the slider 3:

$$d_B = \sqrt{(r+l)^2 - e^2} - r\cos\varphi - l\left(1 - \frac{1}{2}\left(\frac{r\sin\varphi - e}{l}\right)^2\right) \tag{4.126}$$

After differentiating this equation with respect to time, we obtain the following expressions for the velocity and acceleration of the slider 3:

$$\dot{d}_B = \dot{\varphi}(r\sin\varphi + 0.5\lambda r\sin 2\varphi - \lambda e\cos\varphi) \tag{4.127}$$

$$\ddot{d}_B = \dot{\varphi}^2(r\cos\varphi + \lambda r\cos 2\varphi + \lambda e\sin\varphi) \tag{4.128}$$

where $\dot{\varphi}$ is the constant velocity of the input crank, and $\lambda = r/l$.

Now let us consider a second crank-slider mechanism, which is identical to the initial linkage. Please note that these identical mechanisms are mounted on the same guide, i.e. an asymmetric mode of duplication is applied (see Fig. 4.23).

In a similar way the acceleration of slider 5 can be written as:

$$\ddot{d}_D = -\dot{\varphi}^2(r\cos\varphi + \lambda r\cos 2\varphi - \lambda e\sin\varphi) \tag{4.129}$$

where, $r = l_{OC}$ is the distance between the centers of joints O and C; $l = l_{CD}$ is the distance between the centers of joints C and D.

Thus, the shaking force of reciprocating moving mass (rm) is the following:

$$F^{sh} = m_{rm}(\ddot{d}_B + \ddot{d}_D) = 2m_{rm}\lambda e\dot{\varphi}\sin\varphi \tag{4.130}$$

where $m_{rm} = m_3 + m_2 l_{AS_2}/l_{AB} = m_5 + m_4 l_{CS_4}/l_{CD}$.

This force can be balanced by two counterweights ($m_{cw6}r_{cw6} = m_{cw7}r_{cw7} = m_{rm}\lambda e$), which are mounted on gears 6, 7 and connected to the input crank as it is shown in Fig. 4.23.

4.4.2 Shaking Moment Balancing

Now that the inertia force balancing is achieved, we will consider the shaking moment. After shaking force balancing the shaking moment of the suggested system can be expressed as:

$$M^{sh} = (\ddot{\varphi}_2 + \ddot{\varphi}_4)(I_S - m\,s\,s') \tag{4.131}$$

where $\ddot{\varphi}_2$ is the angular acceleration of link 2; $\ddot{\varphi}_4$ is the angular acceleration of link 4; $I_S = I_{S_2} = I_{S_4}$ is the moment of inertia of links 2 and 4; $s = s_2 = s_4 = l_{AS_2} = l_{CS_4}$ and $s' = s_2' = s_4' = l_2 - l_{AS_2} = l_4 - l_{CS_4}$.

The angular acceleration of link 2 can be expressed as:

$$\ddot{\varphi}_2 = -\lambda\dot{\varphi}^2(1-\lambda^2)\frac{\sin\varphi}{\cos^3\psi} \tag{4.132}$$

or taking into account expression (4.125):

$$\ddot{\varphi}_2 = -\lambda\dot{\varphi}^2(1-\lambda^2)\sin\varphi\left(1-\frac{1}{2}\left(\frac{r\sin\varphi - e}{l}\right)^2\right)^{-3} \tag{4.133}$$

The angular acceleration of link 4 can be expressed as:

$$\ddot{\varphi}_4 = \lambda\dot{\varphi}^2(1-\lambda^2)\sin\varphi\left(1-\frac{1}{2}\left(\frac{r\sin\varphi - e}{l}\right)^2\right)^{-3} \tag{4.134}$$

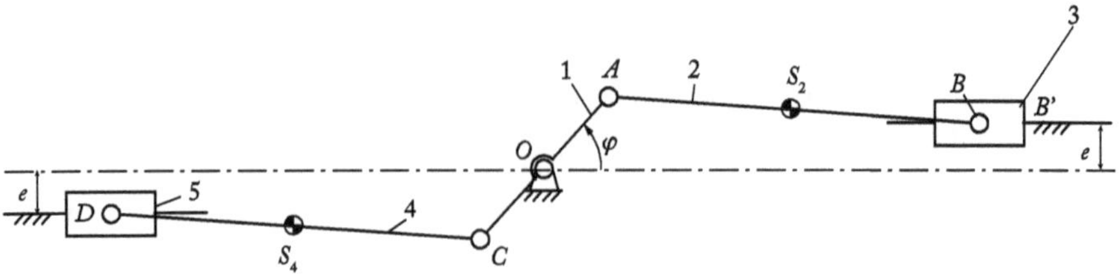

Fig. 4.24 Self-balanced mechanical system via symmetric duplication of two identical off-set crank-slider mechanisms

Thus, the shaking moment of the self-balanced system is the following:

$$M^{sh} = (I_S - m\,s\,s')\lambda\dot{\varphi}^2(1-\lambda^2)\sin\varphi\cdot \left(\left(1-\frac{1}{2}\left(\frac{r\sin\varphi-e}{l}\right)^2\right)^{-3} - \left(1-\frac{1}{2}\left(\frac{r\sin\varphi-e}{l}\right)^2\right)^{-3}\right) \tag{4.135}$$

When $e = 0$, the shaking moment will be fully cancelled. This result corresponds to the self-balanced system with symmetric duplication of two axial crank-slider mechanisms. In the presence of eccentricity e the shaking moment cannot be fully cancelled. However, it will be very small taking into account that $e << l$.

To illustrate the proposed balancing concept let us now consider a numerical example and simulation results carried out by using ADAMS software.

4.4.3 *Numerical Example and Simulation Results*

The off-set crank-slider mechanism OAB shown in Fig. 4.23 has the following parameters: $r = 0.05$ m, $l = 0.2$ m, $e = 0.025$ m, $l_{AS_2} = l_{BS_2}$ (S_2 is the center of mass of the coupler link 2), $I_{S_2} = 0.01$ kg m^2 (I_{S_2} is the axial moment of coupler link 2), $m_2 = 1$ kg (m_2 is the mass of coupler link 2), $m_3 = 3$ kg (m_3 is the mass of slider 3), the centers of the slider's mass is located on the axis B. The simulation of this mechanism, as well as a symmetrically duplicated crank-slider mechanisms (Fig. 4.24) has been carried out using ADAMS software.

The shaking force and shaking moment variations of these structures for input angular velocity are shown in Fig. 4.25.

It should be noted that the mechanical system based on the symmetric duplication of two identical off-set crank-slider mechanisms allows the full cancellation of the shaking force but it increases the shaking moment (see Fig. 4.25, Case b).

Let us now consider a mechanical system with an identical second crank-slider mechanism having the same parameters and a pair of gears with static moments $m_{cw6}r_{cw6} = m_{cw7}r_{cw7} = (0.5m_2 + m_3)\lambda e = 0.022$ kg m (Fig. 4.23). The shaking

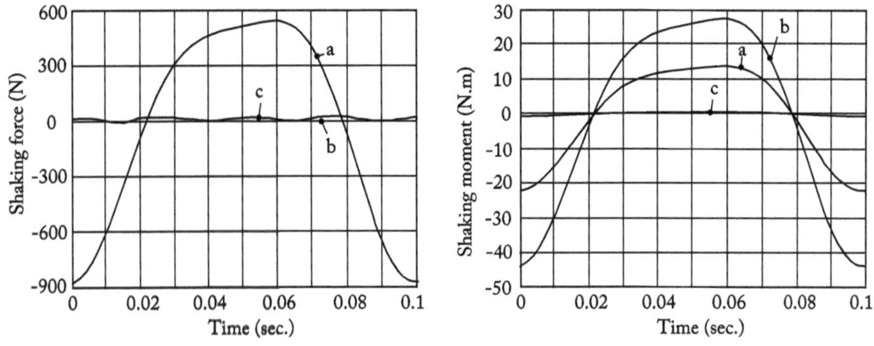

a Shaking force **b** Shaking moment

Fig. 4.25 Shaking force and shaking moment variations: **a** off-set crank-slider mechanism; **b** self-balanced mechanical system based on the symmetric duplication of mechanisms (Fig. 4.24); **c** suggested self-balanced mechanical system (Fig. 4.23)

force and shaking moment of the suggested self-balanced mechanical system is shown in Fig. 4.25.

In order to evaluate the efficiency of the suggested design concept, we compare the maximum values of the shaking force and shaking moment of self-balanced systems with the initial crank-slider mechanism. For the three examined cases we have:

- Case a: $\left|F_{max}^{sh}\right| = 868$ N and $\left|M_{max}^{sh}\right| = 22$ N m,
- Case b: $\left|F_{max}^{sh}\right| = 0$ N and $\left|M_{max}^{sh}\right| = 44$ N m,
- Case c: $\left|F_{max}^{sh}\right| = 32$ N and $\left|M_{max}^{sh}\right| = 0.43$ N m.

Thus, we can note that the known self-balanced mechanical system (Case b) allows the complete cancellation of the shaking force, but the shaking moment doubles. In the case of the suggested solution (Fig. 4.25, Case c), the reduction of the maximum value of the shaking force is 96 % and of the shaking moment is 98 %.

Now let us deal new design methods of self-balanced mechanical systems via duplication of mechanisms executing similar but opposite movements. In the known methods of the duplication of mechanisms, the created systems have symmetry relative to the input crank fixed axis. In the proposed design concept the symmetry is created relating to the vertical plan which decomposes the self-balanced system into two identical subsystems. Such a self-balanced system ensures not only the shaking force but also the shaking moment balancing.

Figure 4.26 shows a self-balanced mechanical system composed of two identical off-set crank-slider mechanisms mounted symmetrically relative to the vertical axis **y**.

The shaking force balancing of the mechanical system based on the following considerations. If the masses of connecting rods 3 and 4 will be statically replaced by concentrated masses located at the centers of their joints, the masses of the sliders 5 and 6 will be fully balanced due to the symmetry of mechanisms ABC and $A'B'C'$

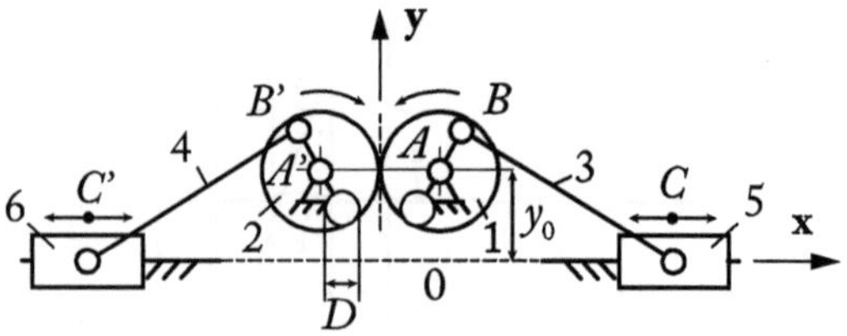

Fig. 4.26 Self-balanced mechanical system

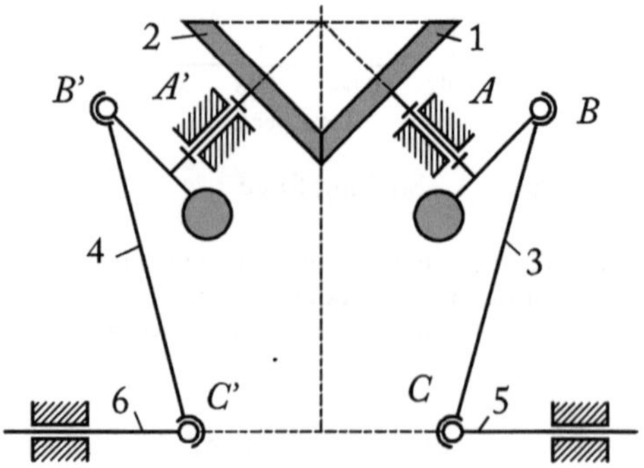

Fig. 4.27 Self-balanced spatial mechanical system

in which the sliders execute similar but opposite movements. The other concentrated masses located at the centers of corresponding joints will be balanced by means of counterweights (see Fig. 4.26).

Finally it should be noted that can be designed in the same way the self-balancing spatial mechanisms. Figure 4.27 shows an example of such a self-balancing spatial mechanism. An additional condition in this case is to use such a shape of the connecting rod that ensures the dynamic substitution of masses of the rod by concentrated masses located at the centers of their spherical elements (see Sect. 4.3).

Chapter 5
Balancing of Slider-Crank Mechanisms

Abstract In this Chapter, the balancing methods of slider-crank mechanisms are presented. In Sect. 5.1, the generalized Lanchester balancer is proposed. It allows the balancing of primary and secondary shaking forces of off-set slider-crank mechanisms. Section 5.2 deals with the solution of the problem of the shaking force and shaking moment balancing of slider-crank mechanisms based on the properties of the Watt gear-slider mechanism. In the examined system, the output gear mounted on the frame meshes with a second identical gear attached to the connecting rod of the slider-crank mechanism. The output gear thus rotates at approximately twice the speed of the input crank and has the same angular acceleration as the connecting rod. These properties are used for the generation of the movements of counterweights intended for balancing. Such a solution allows significant minimization of the shaking force and shaking moment of the mechanism. It also provides the conditions for balancing with only a small increase of the input torque. The efficiency of the suggested method is illustrated by a numerical example.

The Sect. 5.3 presents a solution for improving the balancing of double slider-crank mechanical systems. In these systems the shaking force balancing is achieved by two identical slider-crank mechanisms, which execute similar but opposite movements. Section 5.4 disclose with the simultaneous inertia force/moment balancing and torque compensation of slider-crank mechanisms by using a cam/spring auxiliary mechanism.

At the end of the Chapter (Sect. 5.5), the shaking force and shaking moment balancing of slider-crank mechanisms via optimal generation of the input crank rotation is disclosed.

5.1 Generalized Lanchester Balancer

5.1.1 Shaking Force Balancing of Off-set Crank-Slider mechanism

Figure 5.1 shows an off-set crank-slider mechanism. Let us first consider the kinematic analysis of off-set slider-crank mechanism and the inertia forces of the reciprocating motion.

V. Arakelian, S. Briot, *Balancing of Linkages and Robot Manipulators*,
Mechanisms and Machine Science 27, DOI 10.1007/978-3-319-12490-2_5

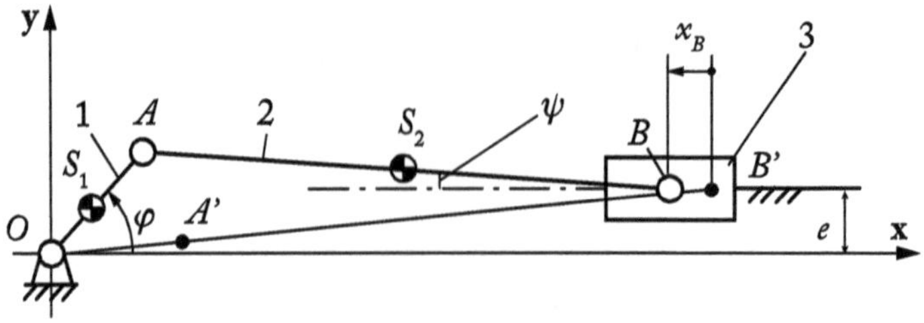

Fig. 5.1 Off-set crank-slider mechanism

The position of the slider can be determined by the following expression:

$$x_B = \sqrt{(l_{OA} + l_{AB})^2 - e^2} - l_{OA}\cos\varphi - l_{AB}\cos\psi \tag{5.1}$$

where $e = y_B$ is the eccentricity of slider guide; φ is the rotating angle of the input crank and ψ is the acute angle that the rod makes with the sliding axis (Fig. 5.1) determined from the following expression:

$$\cos\psi = (1 + p)^q \tag{5.2}$$

where $p = -\left(\frac{l_{OA}\sin\varphi - e}{l_{AB}}\right)$ and $q = 0.5$.

This expression can be represented using Newton's binomial series as follows:

$$(1 + p)^q = 1 + q\,p + \frac{q\,(q-1)}{1\cdot 2}p^2 + \frac{q\,(q-1)\,(q-2)}{1\cdot 2\cdot 3}p^3 + \ldots \tag{5.3}$$

Substituting (5.3) into (5.2), which is then substituted into (5.1), and taking into account that it is sufficient to keep the two first terms of this series, we obtain the following expression for the position of the slider:

$$x_B = \sqrt{(l_{OA} + l_{AB})^2 - e^2} - l_{OA}\cos\varphi - l_{AB}\left(1 - 0.5\left(\frac{l_{OA}\sin\varphi - e}{l_{AB}}\right)^2\right) \tag{5.4}$$

It should be noted that for off-set slider-crank mechanisms $\psi \neq 90$ deg, i.e. $|p| < 1$ (see (5.2)) and for the reasonable parameters of the mechanism's links angle ψ is an acute angle with small value, i.e. $|p| << 1$. Therefore, we can state that:

$$1 + q\,p >> \frac{q\,(q-1)}{1\cdot 2}p^2 + \frac{q\,(q-1)\,(q-2)}{1\cdot 2\cdot 3}p^3 + \ldots \tag{5.5}$$

After differentiating this equation with respect to time, we obtain the following expressions for the velocity and acceleration of the slider:

$$\dot{x}_B = \dot{\varphi}\,(l_{OA}\sin\varphi + 0.5\lambda l_{OA}\sin 2\varphi - \lambda e\cos\varphi) \tag{5.6}$$

$$\ddot{x}_B = \dot{\varphi}^2\,(l_{OA}\sin\varphi + 0.5\lambda l_{OA}\sin 2\varphi + \lambda e\cos\varphi) \tag{5.7}$$

where $\dot{\varphi}$ is the constant velocity of the input crank and $\lambda = l_{OA}/l_{AB}$.

Thus, the shaking force of the reciprocating motion can be expressed as:

$$F_B^{sh} = -m_B\ddot{x}_B = F_{(1)} + F_{(1\prime)} + F_{(2)} \tag{5.8}$$

where,

$$F_{(1)} = -m_B\dot{\varphi}^2 l_{OA}\cos\varphi \tag{5.9}$$

$$F_{(1\prime)} = -m_B\dot{\varphi}^2\lambda\ e\sin\varphi \tag{5.10}$$

$$F_{(2)} = -m_B\dot{\varphi}^2\lambda\ l_{OA}\cos 2\varphi \tag{5.11}$$

where $m_A = m_1 l_{OS_1}/l_{OA} + m_2 l_{BS_2}/l_{AB}$, $m_B = m_2 l_{AS_2}/l_{AB} + m_3$, m_1 is the mass of the crank, m_2 is the coupler link's mass and m_3 is the mass of the slider. The term $m_A\ddot{\mathbf{d}}_A$ can be balanced by an optimal redistribution of the moving masses via the use of a counterweight. Then,

$$F_B^{sh} = k_1\ \dot{\varphi}^2\cos(\varphi+\alpha) - k_2\,(2\dot{\varphi})^2\cos 2\varphi \tag{5.12}$$

where,

$$\alpha = \tan^{-1}(-e/l_{AB}) \tag{5.13}$$

$$k_1 = -\frac{m_B l_{OA}}{\cos\alpha} = \frac{m_B\lambda e}{\sin\alpha} \tag{5.14}$$

$$k_2 = 0.25 m_B\lambda l_{OA} \tag{5.15}$$

In the obtained expression, the first term is the primary shaking force and the second is the secondary shaking force. The primary shaking force can be balanced by counterweights that rotate at the input speed but are out of phase with the input crank by angle α (Fig. 5.2). The secondary shaking force can be balanced by counterweights that rotate at two times the input speed.

The parameters of the added counterweights are the following:

$$2m_{cw_1}r_{cw_1} = k_1 \tag{5.16}$$

$$2m_{cw_2}r_{cw_2} = k_2 \tag{5.17}$$

where, m_{cw_1} and m_{cw_2} are the masses of the counterweights, r_{cw_1} and r_{cw_1} are the distances of the pivot centers from the *com* of the counterweights.

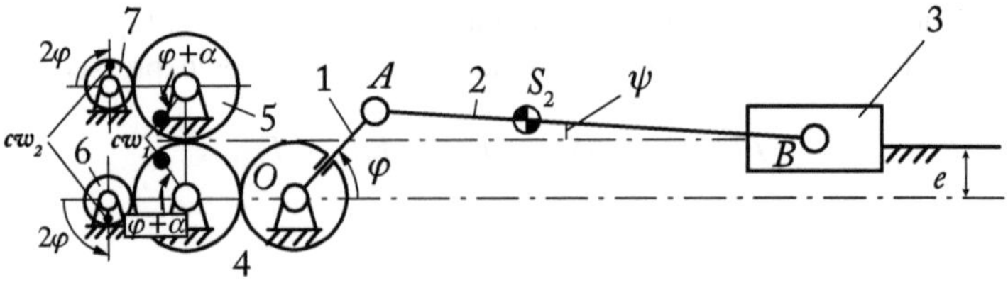

Fig. 5.2 Balanced mechanism

Fig. 5.3 Shaking force variations for unbalanced and balanced mechanisms

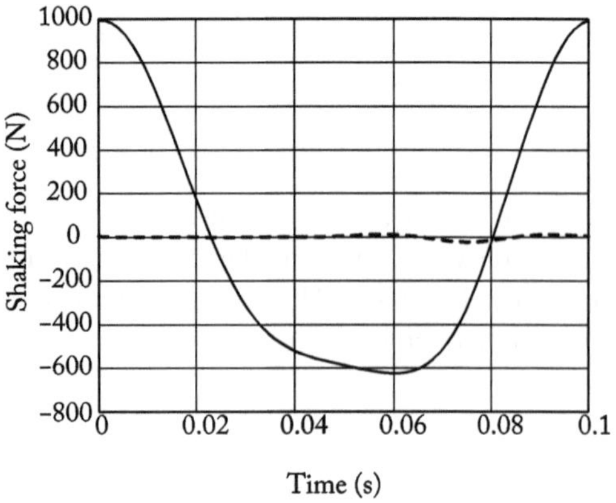

5.1.2 *Numerical Simulations*

In the given numerical example, the off-set crank-slider mechanism OAB shown in Fig. 5.1 has the following parameters:

- $l_{OA} = 0.05$ m, $l_{AB} = 0.2$ m, $l_{OS_1} = 0.025$ m, $l_{OS_1} = 0.1$ m,
- $m_1 = 2$ kg, $m_2 = 2$ kg, $m_3 = 3$ kg.

Thus, for the given parameters of the mechanism we obtain:

- $\alpha = -7.125°$,
- $k_1 = 0.2$ kg m, $k_2 = -0.0125$ kg m,

Then, by selecting $r_{cw_1} = 0.0336$ m and $r_{cw_2} = 0.0125$ m, we determine $m_{cw_1} = 3$ kg and $m_{cw_2} = 0.5$ kg.

The simulation of this mechanism with obtained balancing parameters has been carried out using the software ADAMS. The shaking force variations of unbalanced and balanced mechanisms for input angular velocity are shown in Fig. 5.3.

The suggested balancing technique allows the reduction of the maximum value of the shaking force of the studied off-set slider-crank mechanism by 98 %.

It should be noted that, in general, the complete shaking force balancing of off-set crank-slider mechanisms can only be reached by a considerable increase of link

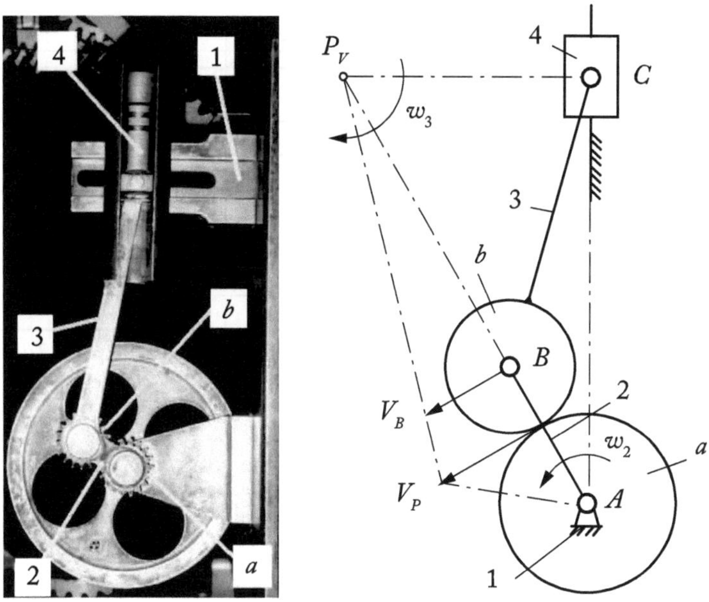

Fig. 5.4 Watt gear-slider mechanism

masses of the mechanism. The harmonic balancing has not been applied to the offset crank-slider mechanisms. As shown in the obtained results, the quasi-complete shaking force balancing has been achieved by a small increase in the total mass of the mechanism.

5.2 Balancing via the Properties of the Watt Gear-Slider Mechanism

5.2.1 Watt Gear-Slider Mechanism

Figure 5.4 shows the Watt gear-slider mechanism. It contains a slider-crank mechanism ABC in which the connecting rod 3 is attached to the gear "b". This meshes with the output gear "a" mounted on the frame 1 concentric with the axis of the crank 2 (see Fig. 5.4). Thus, the rotation of the input crank 2 produces the translation of the slider 4, as well as the secondary output rotation of the gear "a".

The transmission ratio of gears "a" and "b" is represented by the following expression (Arakelian 1997):

$$i_{ab}^{(2)} = \frac{\dot{\varphi}_a - \dot{\varphi}_2}{\dot{\varphi}_b - \dot{\varphi}_2} = -\frac{z_b}{z_a} \tag{5.18}$$

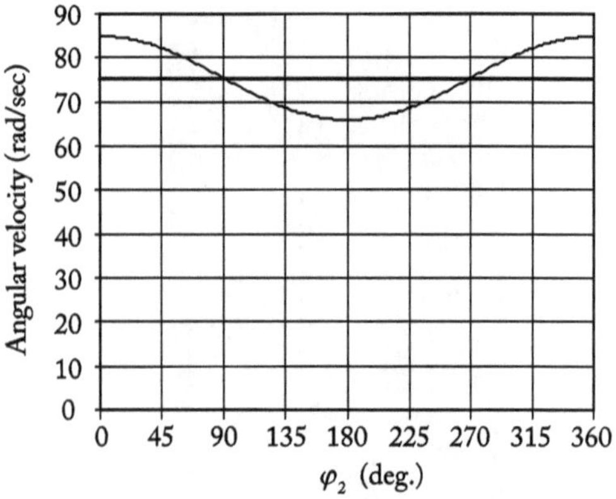

Fig. 5.5 Angular velocity variations of the gear "a"

where $\dot{\varphi}_a$, $\dot{\varphi}_b$ and $\dot{\varphi}_2$ are the angular velocities of the gears "a", "b" and the crank 2; z_a and z_b are the tooth numbers of the gears "a", "b".

Taking into account that $\dot{\varphi}_b = \dot{\varphi}_3$, we determine

$$\dot{\varphi}_a = \dot{\varphi}_2 \left(1 + \frac{z_b}{z_a}\right) - \dot{\varphi}_3 \frac{z_b}{z_a} \tag{5.19}$$

When $z_a = z_b$, this becomes $\dot{\varphi}_a = 2\dot{\varphi}_2 - \dot{\varphi}_3$ and a turn of the crank 2 corresponds to the two turns of the gear "a". Thus, the speed of the gear "a" is almost twice the speed of the input crank 2 and turns in the same direction. It should be noted that in general $\dot{\varphi}_2 > \dot{\varphi}_3$[1] and $\dot{\varphi}_a \approx 2\dot{\varphi}_2$ (see Fig. 5.5).

The angular acceleration of the gear "a" determined by differentiating the expression $\dot{\varphi}_a = 2\dot{\varphi}_2 - \dot{\varphi}_3$ is the following: $\ddot{\varphi}_a = -\ddot{\varphi}_3$ ($\dot{\varphi}_2 = const$). These properties of the Watt mechanism are used for the shaking force and shaking moment balancing of slider-crank linkages.

5.2.2 *Shaking Force and Shaking Moment of the Slider-Crank Mechanism*

The shaking force of the slider-crank mechanism may be expressed as:

$$\mathbf{f}^{sh} = \sum_{i=1}^{3} \mathbf{f}_i^{sh} \tag{5.20}$$

where $\mathbf{f}_i^{sh}$ is the shaking force of the link i.

[1] In the examined numerical example (see Sect. 5.2.4), $l_{BC}/l_{AB} = 4$ and $2\max(\dot{\varphi}_2)/\max(\dot{\varphi}_3) = 16$.

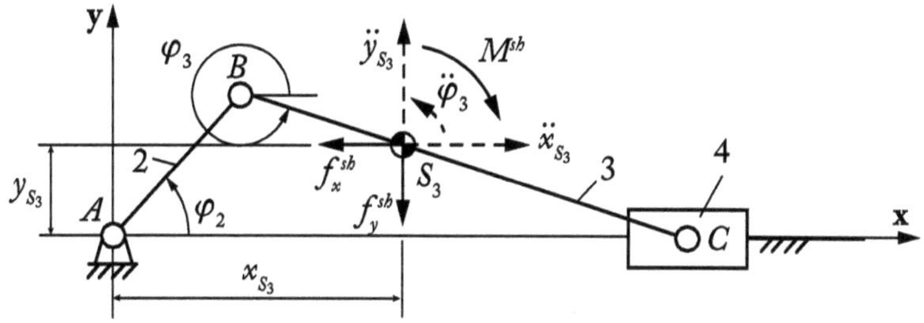

Fig. 5.6 Shaking force and shaking moment of the connecting rod of the mechanism

The shaking moment of the mechanism with constant input angular velocity ($\dot{\varphi}_2 = const$) can be expressed as (see Fig. 5.6):

$$M^{sh} = -I_{S_3}\ddot{\varphi}_3 - m_3\ddot{y}_{S_3}x_{S_3} + m_3\ddot{x}_{S_3}y_{S_3} \tag{5.21}$$

where I_{S_3} is the moment of inertia relative to the *com* S_3 of the connecting rod 3; $\ddot{\varphi}_3$ is the angular acceleration of the link 3; m_3 is the mass of the link 3; $\ddot{x}_{S_3}$ and $\ddot{y}_{S_3}$ are the linear accelerations of the *com* S_3 along the **x** and **y** axes; x_{S_3} and y_{S_3} are the coordinates of the centre of mass S_3 along the **x** and **y** axes.

The linear accelerations and coordinates of the *com* S_3 along the **x** and **y** axes can be written as (Arakelian 1997):

$$x_{S_3} = l_{AB}\cos\varphi_2 + l_{BS_3}\cos\varphi_3 \tag{5.22}$$

$$y_{S_3} = l_{AB}\sin\varphi_2 + l_{BS_3}\sin\varphi_3 \tag{5.23}$$

$$\ddot{x}_{S_3} = -l_{AB}\dot{\varphi}_2^2\cos\varphi_2 - l_{BS_3}\dot{\varphi}_3^2\cos\varphi_3 - \ddot{\varphi}_3\sin\varphi_3 \tag{5.24}$$

$$\ddot{y}_{S_3} = -l_{AB}\dot{\varphi}_2^2\sin\varphi_2 - l_{BS_3}\dot{\varphi}_3^2\sin\varphi_3 + \ddot{\varphi}_3\cos\varphi_3 \tag{5.25}$$

The rotation angles φ_2, φ_3 and the angular velocities $\dot{\varphi}_2, \dot{\varphi}_3$ are given by (Arakelian 1997):

$$\frac{\sin\varphi_3}{\sin\varphi_2} = -\frac{l_{AB}}{l_{BC}}, \text{ and } \frac{\dot{\varphi}_3}{\dot{\varphi}_2} = -\frac{l_{AB}\cos\varphi_2}{l_{BC}\cos\varphi_3} \tag{5.26}$$

The angular velocity $\dot{\varphi}_3$ can also be expressed as (Arakelian 1997):

$$\dot{\varphi}_3^2 = \frac{\ddot{\varphi}_3 l_{BC} \cos\varphi_3 - \dot{\varphi}_2^2 l_{AB} \sin\varphi_2}{l_{BC} \sin\varphi_3} \tag{5.27}$$

By substituting Eqs. (5.22–5.25) into (5.21) and after some algebraic manipulation taking into account Eqs. (5.26) and (5.27), the shaking moment of the mechanism can be expressed as:

$$M^{sh} = -\left(I_{S_3} - m_3 l_{BS_3} r_3^*\right)\ddot{\varphi}_3 \tag{5.28}$$

where $r_3^* = l_{BC} - l_{BS_3}$.

The obtained result can be interpreted just as easily by the method of static and dynamic substitution of distributed masses by concentrated point masses. By substituting statically the mass m_3 of the connecting rod by two concentrated point masses m_{3B} and m_{3C} at the centres B and C, we obtain an identical static model of the mechanism but the dynamic model of the mechanism is transformed.

The deviation between the inertia moment of the link with distributed masses and the link with the concentrated point masses can be expressed as (Shchepetilnikov 1982):

$$\Delta I_{S_3} = m_3\left(\rho_3^2 - m_3 l_{BS_3} r_3^*\right) \tag{5.29}$$

where $\rho_3 = \sqrt{I_{S_3}/m_3}$.

Thus, to have an identical dynamic model, we must to add a complementary shaking moment:

$$\Delta M_3^{sh} = -\Delta I_{S_3}\ddot{\varphi}_3 \tag{5.30}$$

Taking into account that the moment generated by the inertia forces of the concentrated point masses m_{3B} and m_{3C} about the A is equal to zero, the shaking moment of the mechanism can be written as:

$$M^{sh} = \Delta M_3^{sh} = -\Delta I_{S_3}\ddot{\varphi}_3 \tag{5.31}$$

Thus, Eqs. (5.28) and (5.31) are the same: the shaking moment of the mechanism is proportional to the angular acceleration of the connecting rod.

5.2.3 *Shaking Force and Shaking Moment Balancing*

The balanced scheme illustrated in Fig. 5.7 is based on the Watt gear-slider mechanism and it differs from the traditional scheme by the added gears 6′, 6″ and the toothed-belt transmission (2′, 5′ and 2″, 5″) connected to the crank 2. The conditions for balancing of the mechanism are determined from the following considerations.

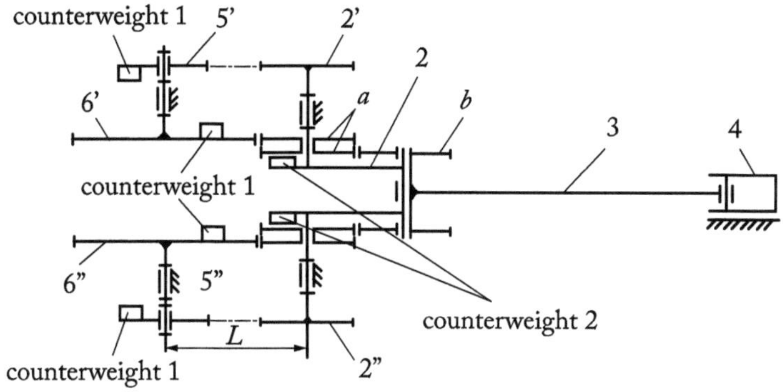

Fig. 5.7 Shaking force and shaking moment balancing of mechanism

5.2.3.1 Shaking Force Balancing

The gears 5′ and 5″ have the same angular velocity as the input crank 2 ($\dot{\varphi}_2 = \dot{\varphi}_5$) and the directions of rotation are the same. The transmission ratio between the gears "a" and 6 is $\dot{\varphi}_a/\dot{\varphi}_5 = 2$. Thus the angular velocity of the gears 6′ and 6″ is given by $\dot{\varphi}_a/2 = \dot{\varphi}_2 - \dot{\varphi}_3/2$ and the direction is opposite to the input crank 2. However the magnitude of $\dot{\varphi}_3/2$ is very small relative to that of $\dot{\varphi}_2$ (see Eq. (5.26)) and it can be neglected. So we can consider that the gears 5′, 5″ and 6′, 6″ turn with the same angular velocity $\dot{\varphi}_2$ but in the opposite direction.

We now balance the first harmonic of the shaking force by the addition of the counterweights with mass m_{cw_1} on the gears 5′, 5″ and 6′, 6″:

$$m_{cw_1} = \frac{l_{AB}\left(m_3 l_{BS_3} + m_4 l_{BC}\right)}{4 r_{cw_1} l_{BC}} \tag{5.32}$$

where r_{cw_1} is the rotation radius of the *com* of the counterweights.

It should be noted that two counterweights with mass

$$m_{cw_2} = \frac{1}{2r_{cw_2}}\left(m_2 l_{AS_2} + \frac{m_3 l_{AB} r_3^*}{l_{BC}}\right) \tag{5.33}$$

must be added on the input crank 2 (see Fig. 5.7).

5.2.3.2 Shaking Moment Balancing

As was shown in the paragraph above: $\ddot{\varphi}_a = -\ddot{\varphi}_3$. This means that the moments of the rotating links may be balanced by gears with prescribed inertia moments. The moment of inertia of each gear "a" is given by the following equation:

$$I_a = I_{S_6} + m_{cw_1} r_{cw_1}^2 + 0.5 I_{S_3} \tag{5.34}$$

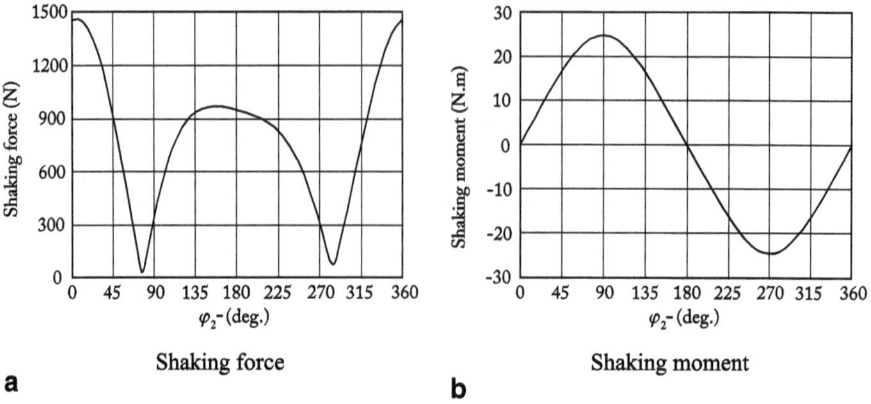

Fig. 5.8 Shaking force and shaking moment of unbalanced mechanism

However there is a moment generated by the difference of the inertia forces of the counterweights mounted on the gears 5′, 5″ and 6′, 6″. This moment represents the shaking moment of the mechanism and can be expressed as:

$$M_O^{sh} = 2m_{cw_1} r_{cw_1} L\left((\dot{\varphi}_2 - \dot{\varphi}_3/2)^2 \sin\left(\varphi_2 - \varphi_3^*/2\right) - \dot{\varphi}_2^2 \sin\varphi_2\right) \tag{5.35}$$

where L is the distance between the fixed rotation axes (see Fig. 5.7), $\varphi_3^* = 2\pi - \varphi_3$.

It should be noted that this moment is small and that the difference between the inertia forces of the counterweights mounted on the gears 5′, 5″ and 6′, 6″ is almost negligible.

5.2.4 *Numerical Example*

Figure 5.8 shows the shaking force and shaking moment variations of a slider-crank mechanism with initial parameters:

- $l_{AB} = 0.1$ m, $l_{BC} = 0.4$ m, $l_{AS_2} = 0$ m, $l_{BS_3} = 0.2$ m,
- $\dot{\varphi}_2 = 37.7$ rad/s (360 rev/min),
- $m_2 = 3$ kg, $m_3 = 4$ kg, $m_4 = 4$ kg,
- $I_{S_3} = 0.0216$ kg m^2.

After balancing, we obtain the following link parameters:

- $l_{BS_3} = 0.14$ m, $r_{cw_1} = r_{cw_2} = 0.1$ m,
- $m_3 = 4$ kg, $m_{cw_1} = 1.35$ kg, $m_{cw_2} = 1.3$ kg,
- $I_{S_3} = 0.08$ kg m^2.

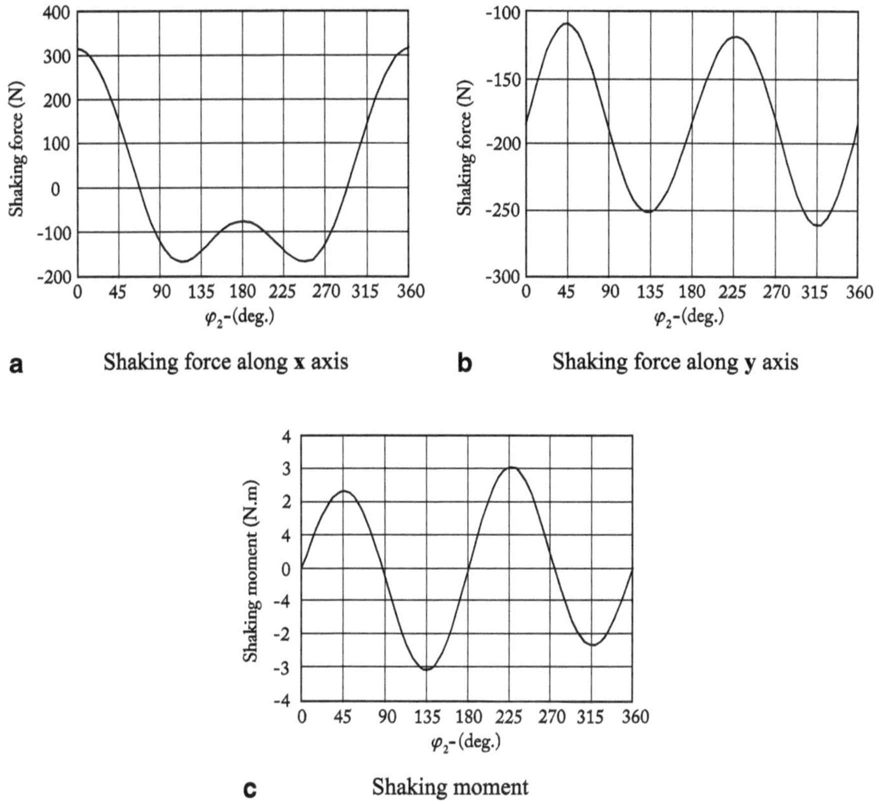

Fig. 5.9 Shaking force and shaking moment of the balanced mechanism

Figure 5.9 represents the variations of the shaking force and shaking moment for the balanced mechanism.

Thus, using the suggested balancing method, a reduction in the *rms* value of the shaking force of 71 % is obtained and an even greater reduction of 88.5 % for the shaking moment (Fig. 5.9). It should be noted that minimization of the maximum values of these functions is also achieved. The reduction of the maximum value of the shaking force is 75 % and of the shaking moment is 87.5 %.

It should also be noted that the suggested solution achieves the conditions for balancing with only a small increase of the input torque. Figure 5.10 shows the torque variations for a slider-crank mechanism: (a) unbalanced; (b) shaking force balanced by two counterweights added to the crank ($r_{cw_2} = 0.1$ m) and the connecting rod ($r_{cw_3} = 0.2$ m); (c) balanced by the suggested method. The results show that while the torques for the cases (a) and (c) are the same, there is a significant increase for the balancing using technique (b).

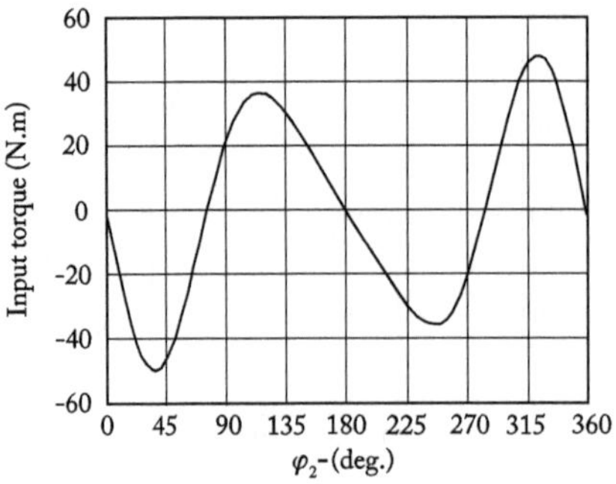

a for unbalanced mechanism

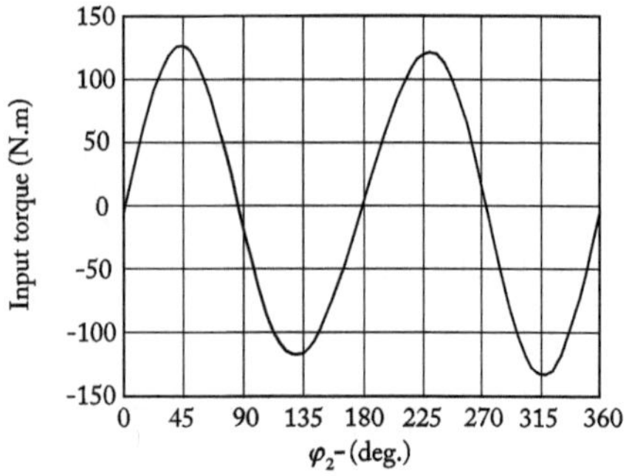

b for shaking force balanced mechanism by two counterweights added to the crank

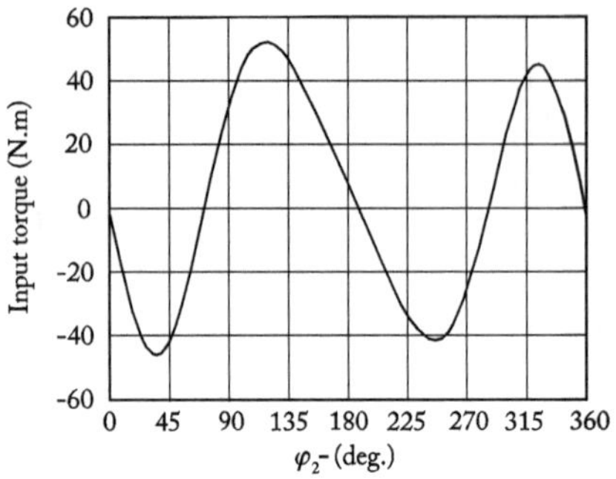

c for shaking force balanced mechanism using the suggested method

Fig. 5.10 Torque variations

Thus, it has been presented a new balancing scheme based on the properties of the Watt gear-slider mechanism. In the examined system the output gear mounted on the frame meshes with a second identical gear attached to the connecting rod of the slider-crank mechanism. The motion of the output gear is transferred to the counterweights which rotate with approximately the same angular velocity as the input crank but in the opposite direction. The rotation of the second pair of counterweights is generated by the input crank. Thus the balancing of the first harmonic of the shaking force is achieved by these counterweights. Shaking moment minimization is achieved by the prescribed inertia moment of the output gear which rotates with the same angular acceleration as the connecting rod. The high efficiency of such a balancing method is demonstrated by a numerical example.

5.3 Shaking Moment Cancellation of Self-balanced Slider-Crank Mechanical Systems by Means of Optimum Mass Redistribution

5.3.1 Shaking Force and Shaking Moment Balancing

5.3.1.1 Shaking Force Balancing Using Two Identical Slider-Crank Mechanisms

Figure 5.11 shows well-known self-balanced mechanical system, in which two identical slider-crank mechanisms execute similar but opposite movements (see Dresig 2001). Such mechanical systems found a successful application in engines, agricultural machines, mills and in various automatic machines.

In this mechanical system the symmetry relative the input crank achieves the complete shaking force balancing. However the shaking moment is not balanced and for the mechanism with constant input angular velocity ($\dot{\varphi}_2 = const$) can be presented by the following expression :

$$M^{sh} = -\sum_{i=3}^{4} I_{S_i}\ddot{\varphi}_i - \sum_{i=3}^{4} m_i \ddot{y}_{S_i} x_{S_i} + \sum_{i=3}^{4} m_i \ddot{x}_{S_i} y_{S_i} \tag{5.36}$$

where I_{S_i} is the moment of inertia relative to the *com* S_i of the connecting coupler (i = 3, 4); $\ddot{\varphi}_i$ is the angular acceleration of the connecting coupler; m_i is the mass of the connecting coupler; $\ddot{x}_{S_i}$ and $\ddot{y}_{S_i}$ are the linear accelerations of the *com* S_i along the **x** and **y** axes; x_{S_i} and y_{S_i} are the coordinates of the *com* S_i along the **x** and **y** axes.

It should be noted that the shaking moment (5.36) is independent of the reference point because the shaking force is fully balanced.

Fig. 5.11 Self-balanced slider-crank system

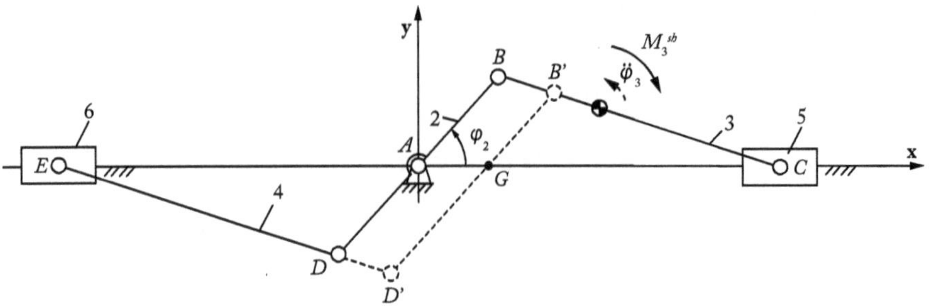

Fig. 5.12 Self-balanced slider-crank system with a imagined articulation dyad $B'D'E$

5.3.1.2 Shaking Moment Cancellation Using the Copying Properties of the Pantograph

Figure 5.12 shows a self-balanced slider-crank system with a imagined articulation dyad $B'D'E$, which forms a pantograph with the initial system. The similarity factor of the formed pantograph is $k = l_{AD}/l_{AB} = 1$ and $l_{BB'} = l_{DD'}, l_{B'D'} = l_{AD} + l_{AB}$.

By substituting dynamically the mass m_3 of the connecting coupler 3 by point masses at the centers B, B' and C and using following condition

$$\begin{bmatrix} 1 & 1 & 1 \\ l_{BS_3} & -l_{CS_3} & l_{B'S_3} \\ l_{BS_3}^2 & l_{CS_3}^2 & l_{B'S_3}^2 \end{bmatrix} \begin{bmatrix} m_B \\ m_C \\ m_{B'} \end{bmatrix} = \begin{bmatrix} m_3 \\ 0 \\ I_{S_3} \end{bmatrix} \tag{5.37}$$

we determine the value of the point masses m_B, m_C and $m_{B'}$.

We now require imagined link $B'D'$ to be balanced about point G formed pantograph, i.e.

$$m_{D'} = m_{B'} \frac{l_{B'G}}{l_{D'G}} \tag{5.38}$$

The concentrated point masses m_G ($m_G = m'_B + m'_D$), m_C and m_E must be balanced about center A, i.e.

$$m_E = \frac{m_G l_{BB'} + m_C l_{BC}}{l_{DE}} \tag{5.39}$$

Finally, the concentrated point masses m_B and m_D must also be balanced about center A, i.e.

$$m_D = m_B \frac{l_{AB}}{l_{AD}} \tag{5.40}$$

Thus, we obtain the values of three concentrated point masses $m_{D'}$, m_D and m_E, which allow the determination of the mass and inertia parameters of the connecting coupler 4:

$$m_4^* = m_{D'} + m_D + m_E \tag{5.41}$$

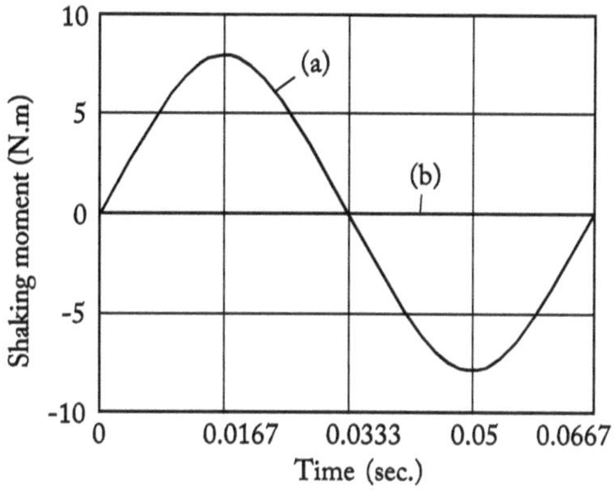

Fig. 5.13 Shaking moment of initial **a** and balanced **b** mechanical systems

$$l^*_{ES_4} = \frac{m_D l_{DE} + m_{D'} l_{D'E}}{m_4} \tag{5.42}$$

$$I^*_{S_4} = m_D l^2_{DS_4} + m_{D'} l^2_{D'S_4} + m_E l^2_{ES_4} \tag{5.43}$$

Finally, on the base of the obtained mass and inertia moment parameters we can design the shape of the connecting coupler 4 of the self-balanced slider-crank mechanism.

5.3.2 *Numerical Example*

The parameters of the self-balanced slider-crank system to be shaking moment balanced are the followings:

- $l_{AB} = l_{AD} = 0.05$ m, $l_{BC} = l_{DE} = 0.2$ m, $l_{CS_3} = l_{ES_4} = 0.1$ m,
- $m_3 = m_4 = 0.35$ kg, $m_5 = m_6 = 2$ kg,
- $I_{S_3} = I_{S_4} = 0.005$ kg m^2,
- $\dot{\varphi}_2 = 30\pi$ rad/s.

Figures 5.13 and 5.14 show the variations of the shaking moment and the input torque of the initial mechanical system (curves "a").

For cancellation of the shaking moment it is necessary to redistribute the masses of the second connecting coupler. By substituting dynamically the mass m_3 of the connecting coupler 3 by point masses at the centers B, B' and C and taking into account the conditions (5.38–5.40) we calculate the mass and inertia parameters of the connecting coupler 4: $m^*_4 = 0.35$ kg; $l^*_{ES_4} = 0.1$ m and $I^*_{S_4} = 0.002$ kg m^2.

Figures 5.13 and 5.14 illustrate the obtained results. So, by optimal redistribution of the masses of the connecting coupler 4, the shaking moment is cancelled and the input torque of the balanced mechanical system is reduced (curves "b").

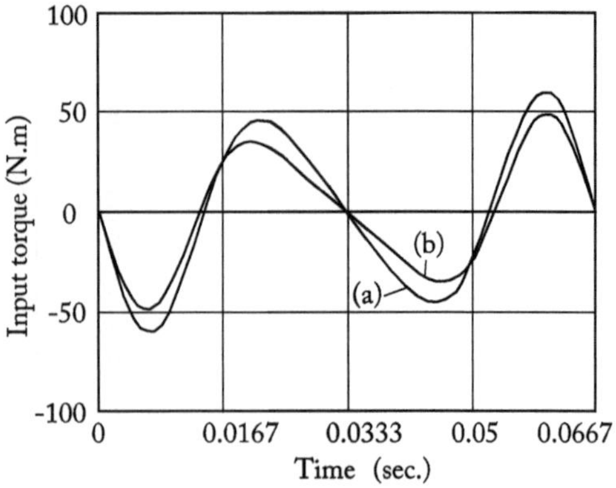

Fig. 5.14 Input torques of initial **a** and balanced **b** mechanical systems

Figure 5.15 shows the shape and the dimensions of the new connecting coupler. It is made of aluminum and corresponds to the calculated values of mass and inertia moment parameters.

Thus, we have presented a solution for improving the balancing of double slider-crank mechanical systems. In the presented system, the shaking force balancing is achieved by two identical slider-crank mechanisms, which execute similar but opposite movements. However the shaking moments are not balanced and can be source for vibrations. By modification of the parameters of the second connecting rod of the system, the complete shaking moment balancing has achieved. The conditions for shaking moment balancing have been formulated by using the copying properties of the pantograph linkage and the method of dynamic substitution of the connecting coupler mass by the concentrated point masses.

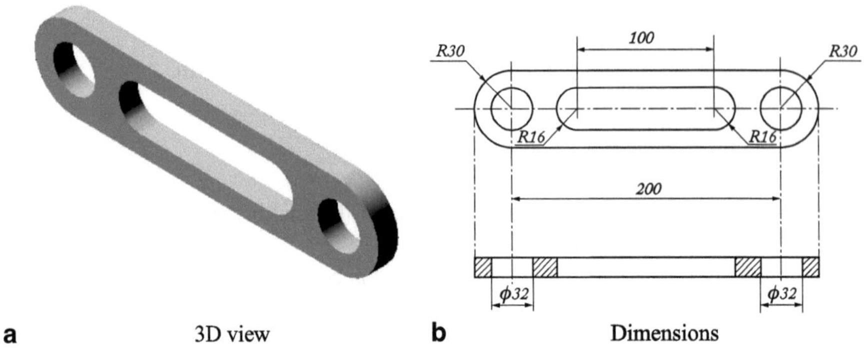

Fig. 5.15 Shape and dimensions of the calculated connecting coupler 4

5.4 Simultaneous Inertia Force/Moment Balancing and Torque Compensation of Slider-Crank Mechanisms

5.4.1 *Design of the Inertia Force/Moment Balanced and Torque Compensated Slider-Crank Mechanism*

Figure 5.16 shows an off-set slider-crank mechanism, which contains an initial slider-mechanism *OAB* with crank 1 mounted on the frame, rod 2 and slider 3, as well as cams 4, 5 with followers 6, 7 and a compression spring 8.

Let us first consider the inertia force and moment balancing of the slider-crank mechanism. For this purpose, we consider that rod 2 is a "physical pendulum" link (Berkof 1973) (see also Arakelian 2007), i.e. its mass distribution allows the dynamic substitution of the rod's mass by two point masses, also,

$$\begin{bmatrix} 1 & 1 & 1 \\ l_{AS_2} & 0 & -l_{BS_2} \\ l_{AS_2}^2 & 0 & l_{BS_2}^2 \end{bmatrix} \begin{bmatrix} m_A \\ 0 \\ m_B \end{bmatrix} = \begin{bmatrix} m_2 \\ 0 \\ I_{S_2} \end{bmatrix} \tag{5.44}$$

where, m_A and m_B are point masses; m_2 is the mass of rod 2; I_{S_2} is the axial moment inertia of the rod about the centre of mass S_2 of the link; l_{AS_2} and l_{BS_2} are the distances between the centres of the joints A and B and the *com* S_2 of link 2, respectively.

Thus, the dynamic model of the rod represents a weightless link with two point masses m_A and m_B situated in the centres of corresponding joints. This dynamic model is fully equivalent to the real rod.

We now require rotating masses to be balanced about point O. Therefore, the displacement of the counterweight mounted on follower 6 is selected in such a manner that the inertia force of the follower 6 with counterweight will be opposite to the inertia force of the masses m_3 and m_8 carried out reciprocating motion:

$$\ddot{x}_{S_6} = -\frac{m_B + m_3}{m_6}\ddot{x}_B \tag{5.45}$$

where m_3 is the mass of the slider 3, m_6 the mass of follower 6 with counterweight, $\ddot{x}_{S_6}$ the acceleration of the follower 6 and $\ddot{x}_B$ the acceleration of the slider.

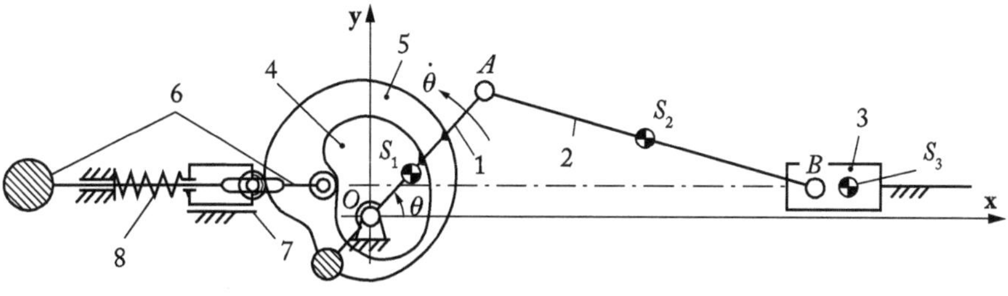

Fig. 5.16 Balanced and torque compensated off-set slider-crank mechanism

Therefore, in order to generate a prescribed reciprocating motion of follower 6 with acceleration $\ddot{x}_{S_6}$, the cam 4 is used. Then, by means of a counterweight, the *com* of crank 1 plus the cam 4 and point mass m_A is brought to the centre of the pivot O.

After such a redistribution of moving masses, the shaking force and shaking moment are cancelled and the slider-crank mechanism transmits no inertia loads to surrounding. However, it is known that the added masses destined to balance the inertia force increase the input torque of the mechanism. For minimisation of the input torque of mechanism, one uses another technique which consists in adding a cam-spring compensation device.

The above-mentioned literature review shows that these two problems, i.e. inertia force balancing and torque compensations, are studied separately and it is considered that they are not coupled. Thus, according with the known design approaches, two devices will be developed and coupled with the slider-crank mechanism.

It will be shown that the design of these two problems can be considered together and the spring used for maintaining the contact between the counterweight and the cam may also be used for balancing the input torque. For this purpose, we use the elastic force of spring 8 designed for maintaining contact between follower 6 and cam 4. In other words, we will use the spring 8 for the generation of a complementary moment on the input crank.

Let us now consider the input torque compensation.

The input torque τ of the dynamically balanced mechanism with the spring can be written under the form:

$$\tau = \frac{\mathrm{d}}{\mathrm{d}t}\left(\frac{\partial L}{\partial \dot{\theta}}\right) - \frac{\partial L}{\partial \theta} \tag{5.46}$$

where $L = T - V$ is the Lagrangian of the system, T is kinetic energy and V its potential energy. Neglecting the mass of the spring and follower 7, T and V can be written as:

$$\begin{aligned} T &= 0.5\left(I_{S_1}\dot{\theta}^2 + m_1\dot{x}_{S_1}^2 + m_{cw}\dot{x}_{cw}^2 + I_{S_2}\dot{\theta}_2^2 + m_2\dot{x}_{S_2}^2 + m_B\dot{x}_B^2 + m_6\dot{x}_{S_6}^2\right) \\ &= 0.5\left(I_{S_1} + l_{OA}^2\left(m_{cw}r_{cw}^2 + m_1 r_1^2 + m_A\right)\right)\omega^2 + 0.5\left(m_B + m_3\right) \\ &\left(1 + \frac{m_B + m_3}{m_6}\right)\dot{x}_B^2 \end{aligned} \tag{5.47}$$

$$V = 0.5k\delta^2 \tag{5.48}$$

where

- $\dot{\theta} = \omega$ is a constant for steady-state conditions,
- $\dot{x}_{S_2}$ is the velocity of the centre of masses of element 2 and its angular velocity,
- k is the spring constant,
- δ is the displacement of the end of the spring from its equilibrium position,
- m_{cw} is the mass of the counterweight mounted on the element 1,

- r_{cw} is the dimensionless position of this counterweight ($r_{cw} = (m_1 l_{OS_1}/l_{OA} + m_A)/m_{cw}$),
- m_1 is the mass of the element 1 plus the cam 4 and I_{S_1} their global axial moment of inertia,
- r_1 is the dimensionless position of the global *com* of these elements ($r_1 = l_{OS_1}/l_{OA}$).

Thus, the first and second terms of the Lagrange equation (5.46) become:

$$\frac{\mathrm{d}}{\mathrm{d}t}\left(\frac{\partial L}{\partial \dot{\theta}}\right) = 2\left(m_B + m_3\right)\left(1 + \frac{m_B + m_3}{m_6}\right)\left(\frac{\partial x_B}{\partial \theta}\right)\left(\frac{\partial^2 x_B}{\partial \theta^2}\right)\omega^2 \tag{5.49}$$

$$\frac{\partial L}{\partial \theta} = \left(m_B + m_3\right)\left(1 + \frac{m_B + m_3}{m_6}\right)\left(\frac{\partial x_B}{\partial \theta}\right)\left(\frac{\partial^2 x_B}{\partial \theta^2}\right)\omega^2 - k\frac{\partial \delta}{\partial \theta}\delta \tag{5.50}$$

taking into account that

$$x_B = l_{OA}\cos\theta \pm \sqrt{l_{AB}^2 - (a - l_{OA}\sin\theta)^2} \tag{5.51}$$

Therefore, the input torque can be deduced:

$$\tau = \left(m_B + m_3\right)\left(1 + \frac{m_B + m_3}{m_6}\right)\left(\frac{\partial x_B}{\partial \theta}\right)\left(\frac{\partial^2 x_B}{\partial \theta^2}\right)\omega^2 + k\frac{\partial \delta}{\partial \theta}\delta \tag{5.52}$$

In order to avoid torque fluctuation, the spring has to create a force that makes the input torque constant. In many cases, when a mean value of the torque moment is equal to zero, as in the case of unloaded slider-crank mechanism, this constant is equal to zero. Thus, in such a case, the compensation of the input torque is equivalent to its cancellation. From the point of view of the energy fluctuation, after such a compensation, the periodic variations of the input torque is cancelled and the required input torque is equal to zero, i.e. in the stationary operating mode, the generation of motion may be accomplished by only a very small input torque, which is needed for overcoming friction.

Thus, under the condition that the input torque is equal to zero with the cam-follower system, Eq. (5.52) admits the integral:

$$\left(m_B + m_3\right)\left(1 + \frac{m_B + m_3}{m_6}\right)\left(\frac{\partial x_B}{\partial \theta}\right)^2 \omega^2 + k\delta^2 = A \tag{5.53}$$

where A is an integration constant. Note that the value of A represents two times the sum of the kinetic energy stored in the mechanism (without rod 1 and the cam 5) plus the potential energy in the spring.

For the continuity of contact between the cam and the follower, the force created by the spring has to be always superior to the inertia force of the follower 6 with the counterweight, i.e.

$$k\delta \geq \left(m_3 + m_B\right)\ddot{x}_B \tag{5.54}$$

or also,

$$\delta \geq \frac{(m_3 + m_B)\ddot{x}_B}{k} \tag{5.55}$$

From expression (5.52), the value of δ may also be computed:

$$\delta^2 = \frac{A - (m_B + m_3)\left(1 + \frac{m_B+m_3}{m_6}\right)\left(\frac{\partial x_B}{\partial \theta}\right)^2 \omega^2}{k} \tag{5.56}$$

Combining Eqs. (5.55) and (5.56) leads to,

$$A \geq (m_3 + m_B)\left(\left(1 + \frac{m_B + m_3}{m_6}\right)\left(\frac{\partial x_B}{\partial \theta}\right)^2 + \frac{m_3 + m_B}{k}\frac{\partial^2 x_B}{\partial \theta^2}\right)\omega^2 \tag{5.57}$$

In order Eq. (5.57) to be valuable for any value of θ, the constant A should be equal to:

$$A = \max_{\theta \in [0\ 2\pi]}\left((m_3 + m_B)\left(\left(1 + \frac{m_B + m_3}{m_6}\right)\left(\frac{\partial x_B}{\partial \theta}\right)^2 + \frac{m_3 + m_B}{k}\frac{\partial^2 x_B}{\partial \theta^2}\right)\omega^2\right) \tag{5.58}$$

To avoid resonance, the spring should be stiff enough so that the lowest natural frequency of the system is considerably higher than the highest significant harmonic of the output motion of the follower (Angeles and Wu 2001). After the appropriate spring constant is selected, the displacement of the spring can be determined from Eqs. (5.56) and (5.58), namely:

$$\delta = \sqrt{\frac{A - (m_B + m_3)\left(1 + \frac{m_B+m_3}{m_6}\right)\left(\frac{\partial x_B}{\partial \theta}\right)^2 \omega^2}{k}} \tag{5.59}$$

Thus here we find the cam profile for torque compensation.

The next part presents an illustrative example of the proposed approach. The simulations have been carried out by using ADAMS software.

5.4.2 *Illustrative Example*

5.4.2.1 Shaking Force and Shaking Moment Balancing

The following parameters of mechanism's links are specified for the simulations:

- $l_{OA} = 0.292$ m; $l_{AB} = 0.427$ m; $r_1 = 0.5$; $y_B = 0.1$ m;
- $m_1 = 2$ kg; $m_2 = 3$ kg; $m_3 = 4$ kg;
- $I_{S_1} = 0.03$ kg m^2, $I_{S_2} = 0.14$ kg m^2.

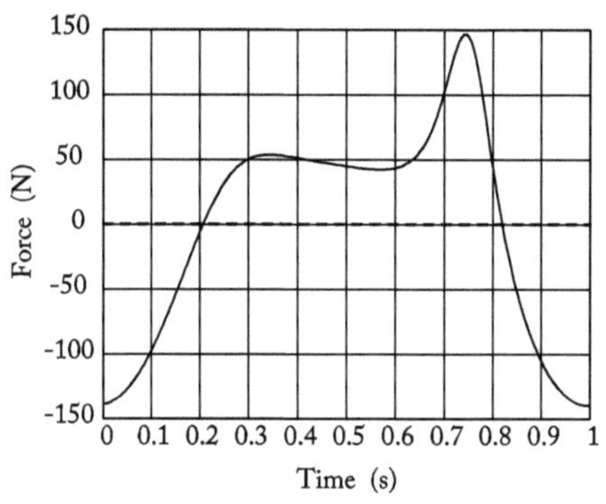

a Shaking force with respect to the **x** axis

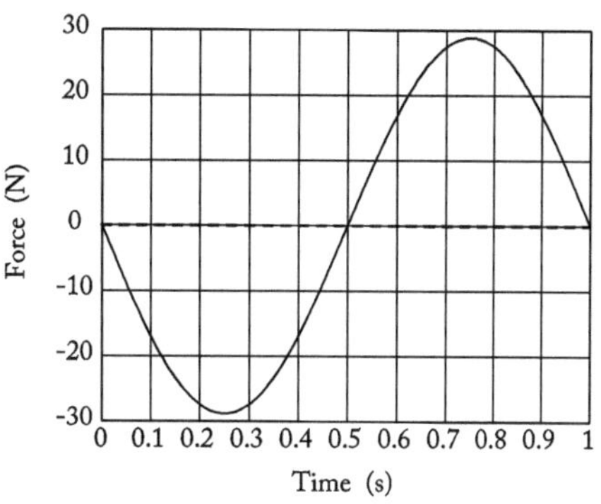

b Shaking force with respect to the **y** axis

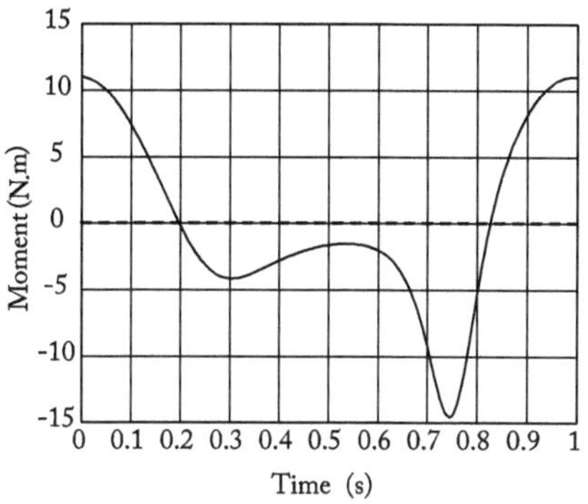

c Shaking moment

Fig. 5.17 Variations of the shaking forces and shaking moment before (*full line*) and after (*dashed line*) balancing

The period of the mechanism is fixed to 1 sec.

The shaking forces and shaking moment of the above-mentioned mechanism are represented in Fig. 5.17 (full line).

By selecting $l_{cw} = r_{cw} l_{OA} = 0.2$ m, we obtain $m_{cw} = 3.25$ kg. This counterweight will be mounted on the input crank. The second counterweight is carried out the reciprocating motion. In order to reduce the size of the cam, the displacement of the *com* S_6 of link 6 is three times smaller than the displacement of point B. Therefore $m_6 = 13.5$ kg.

Figure 5.18 shows the obtained cam profile. The variations of the shaking forces and shaking moment of the mechanisms with redistributed moving masses are given in Fig. 5.17 by dashed lines. The simulation results show that after balancing the shaking forces and moment are cancelled.

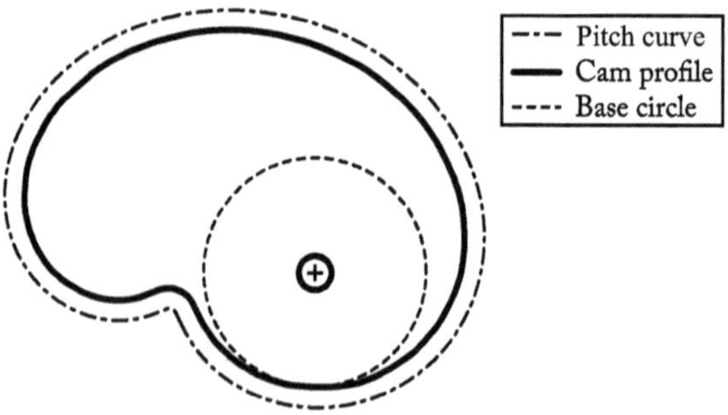

Fig. 5.18 Profile of the cam for the displacement of link 6 assuming the inertia forces balancing

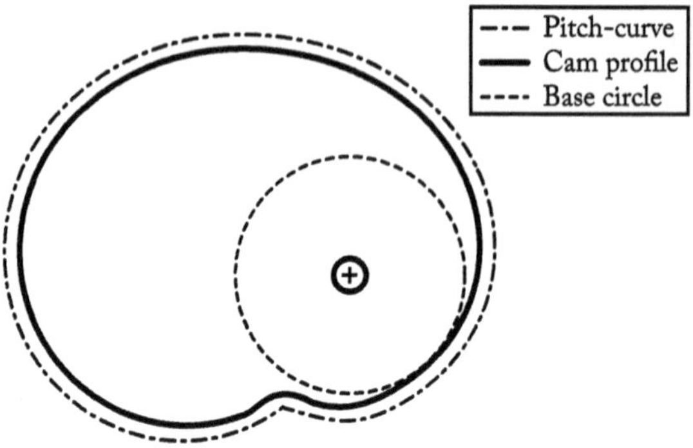

Fig. 5.19 Profile of the cam for the displacement of link 7 assuming the torque compensation

5.4.2.2 Input Torque Compensation

The parameters are identical to those used in the previous part. The constant spring is fixed to $k = 10$ kN/m. The joint between the cam and the follower has been modeled by a contact between two solid bodies. The linear spring 8 is designed in such a manner that it ensures a permanent contact in the cam mechanisms. Thus, the displacements of the follower 7 are determined from Eq. (5.59) and the obtained cam profile is shown in Fig. 5.19.

The input torques before and after compensation are shown in Fig. 5.20.

Thus we can note that the suggested approach allows not only to carry out complete shaking force and shaking moment balancing of the offset slider-crank mechanism but also it assumes its input torque compensation.

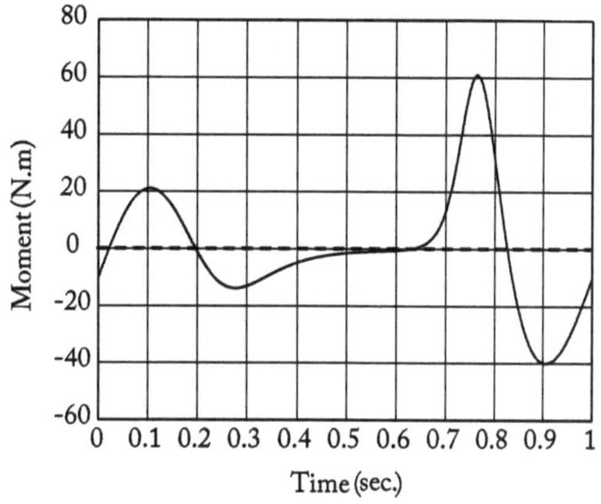

Fig. 5.20 Input torque before (*full line*) and after (*dashed line*) compensation

5.5 Shaking Force and Shaking Moment Balancing of Slider-Crank Mechanisms via Optimal Generation of the Input Crank Rotation

5.5.1 Problem Statement

Let us consider a new technique for the shaking force/shaking moment balancing of the typical axial slider-crank mechanism (Fig. 5.21). We will consider that the mechanism has arbitrary parameters.

The shaking force of a slider-crank mechanism can be computed as the sum of two terms due to the rotational and translational motions:

$$F^{sh}_{(rot)} = (m_2 l_{OS_2} + m_3 r l_{BS_3}/l)(\dot{\varphi}^2 \cos\varphi + \ddot{\varphi}\sin\varphi) \tag{5.60}$$

$$\begin{aligned} F^{sh}_{(tr)} &= -(m_4 + m_3 l_{AS_3}/l)\ddot{x}_B \\ &= (m_4 + m_3 l_{AS_3}/l)[r(\dot{\varphi}^2\cos\varphi + \ddot{\varphi}\sin\varphi) + l(\dot{\psi}^2\cos\psi + \ddot{\psi}\sin\psi)] \end{aligned} \tag{5.61}$$

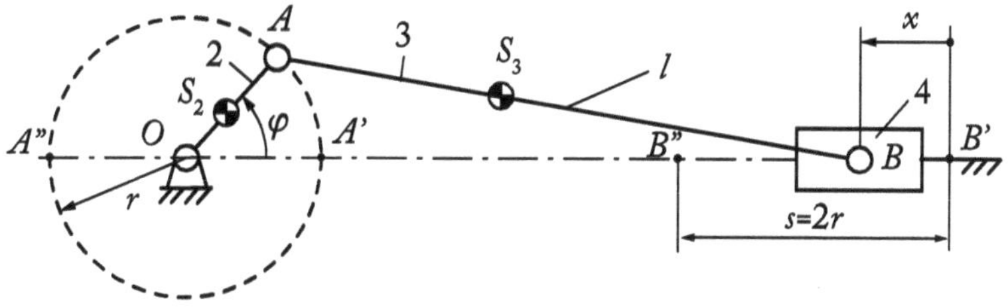

Fig. 5.21 Slider-crank mechanism

where φ is the crank rotation angle; $\psi = \sin^{-1}(-\sin\varphi/\lambda)$ is the connecting rod rotation angle, $\psi = \angle ABO$, $\lambda = l/r$; $\dot{\varphi}$ and $\ddot{\varphi}$ are the angular velocity and acceleration of the crank; $\dot{\psi} = -\dot{\varphi}\cos\varphi/\lambda\cos\psi$ is the angular velocity of the connecting rod; $\ddot{\psi} = (-\dot{\varphi}^2\sin\varphi - \ddot{\varphi}\sin\varphi + \dot{\psi}^2\lambda\sin\psi)/\lambda\cos\psi$ is the angular acceleration of the connecting rod; r is the distance between the centres of the joints O and A; l is the distance between the centres of the joints A and B; l_{OS_2} is the distance of the center of the joint O from the center of masses S_2 of the crank; l_{AS_3} and l_{BS_3} are the distances of the centers of the joints A and B from the center of masses S_3 of the connecting rod; m_2 is the mass of the crank; m_3 is the mass of the connecting rod; m_4 is the mass of the slider.

The shaking force due to the rotational motion can be easily cancelled by adding a counterweight on the input crank. However, the balancing of the shaking force due to the translational rectilinear motion is much more difficult.

From expression (5.61), we can see that the shaking force due to the translational rectilinear motion is minimized, in terms of norm, if the norm $\ddot{x}_B$ of the slider acceleration is minimized during of a full rotation of the crank. This means that if the rotation of the input crank is optimally generated, the shaking force will be minimized. As a result, if the time interval $[t_0, t_f]$ for a full rotation of the crank is fixed, i.e. the time of the stroke is also fixed, the "bang-bang" acceleration law (Fig. 5.22) is the motion profile of the slider displacement that minimizes the value of the maximal acceleration $\ddot{x}_B^{max}$.

Thus, in the present study we would like to reach the following condition

$$F_{(tr)}^{sh} \to \min_{\varphi(t)} \tag{5.62}$$

i.e. to generate such an input rotation angle of the crank which leads to generation of an output acceleration $\ddot{x}_B$ that follows a "bang-bang" motion profile and thus minimizes the norm of $F_{(tr)}^{sh}$.

5.5.2 *Shaking Force and Shaking Moment Minimization*

5.5.2.1 Shaking Force Minimization via Optimal Motion Control of the Input Crank

The problem which has been mentioned above can be formulated as

$$(m_4 + m_3 l_{AS_3}/l)[r(\dot{\varphi}^2\cos\varphi + \ddot{\varphi}\sin\varphi) + l(\dot{\psi}^2\cos\psi + \ddot{\psi}\sin\psi)] \to \min_{\varphi(t)} \tag{5.63}$$

The inverse kinematics of the slider-crank mechanism can be expressed as:

$$\varphi = \cos^{-1}\left[\frac{x_B^2 + r^2 - l^2}{2r\,x_B}\right] \tag{5.64}$$

where $x_B = l + r - x$.

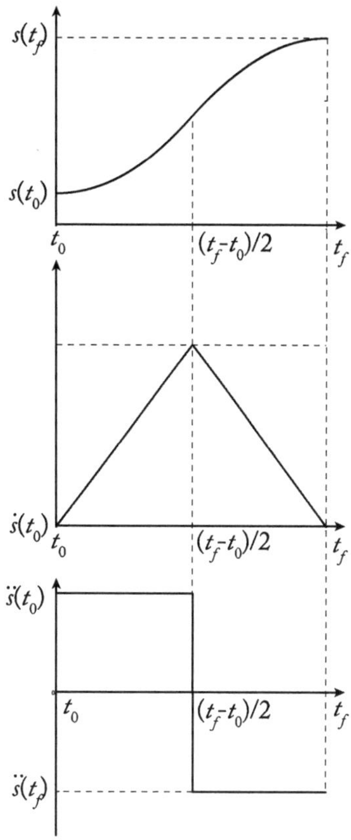

Fig. 5.22 "Bang-bang" acceleration motion profile

Now taking into account that the slider displacement x is generated by a "bang-bang" motion profile, we determine the crank rotation angle. In such a way we ensure that the maximum value of the linear acceleration will be minimal and, as a result, the shaking force will also be small.

5.5.2.2 Shaking Moment Minimization via Optimal Motion Control of the Input Crank

Let us consider the shaking moment balancing of the mechanism. It can be minimized by modification of the axial moment of inertia $I_{S_{1'}}$ of gear 1′ (Fig. 5.23) if it already exists in the initial mechanism or it may be added if such a gear transmission is missing. For this purpose, it is necessary to ensure that $I_{S_{1'}} = I^*_{S_2} i_{2'1'}$, where $I^*_{S_2}$ is the axial moment of inertia of the input crank with added counterweigh and gear 2′, $i_{2'1'}$ is the gear ratio.

Let us consider an illustrative example.

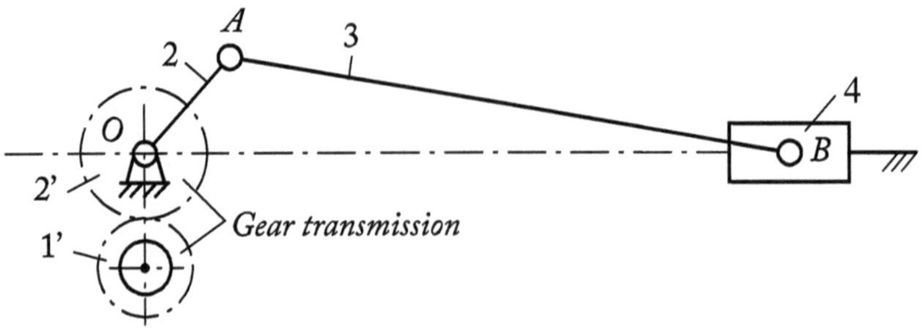

Fig. 5.23 Slider-crank mechanism with gear transmission

Fig. 5.24 Input angle φ (rad)

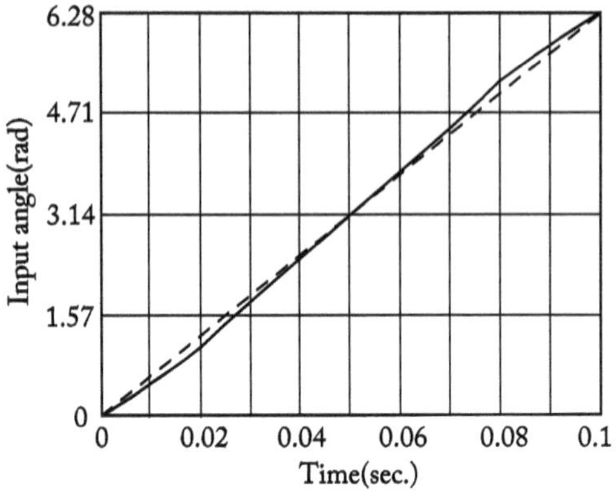

5.5.3 *Illustrative Example*

The parameters of the simulated slider-crank mechanism to be balanced are the following: $r = 0.05$ m, $l = 0.2$ m, $l_{OS_2} = 0.025$ m, $l_{AS_3} = 0.1$ m, $m_2 = 1$ kg, $m_3 = 0.35$ kg, $m_4 = 2$ kg, $I_2 = 0.01$ kg m^2, $I_{S_2} = 0.00137$ kg m^2 and $i_{2'1'} = 1$.

At first, let us determine from expression (5.64) the input angle φ taking into account that $t_f = 0.1$ sec. The variations of the input angle φ and the input angular velocity $\dot{\varphi}$ are shown in Figs. 5.24 and 5.25 by solid lines.

In order to carry out a comparative analysis, a slider-crank mechanism with constant input speed ($\dot{\varphi} = 62.8$ rad/sec) has first been simulated. The variations of the parameters of the slider-crank mechanism with constant input angular velocity are also shown in Figs. 5.24 and 5.25 by dashed lines.

Now we introduce in the software model the input angle calculated from (5.64), i.e. that minimizes the norm of the acceleration $\ddot{x}$. In this case, as was mentioned above, the linear acceleration of the slider should follow a "bang-bang" acceleration motion profile. Indeed, the simulations carried out by using *ADAMS* software showed that the variation of the acceleration of the translational rectilinear motion is a "bang-bang" acceleration motion profile (Fig. 5.26).

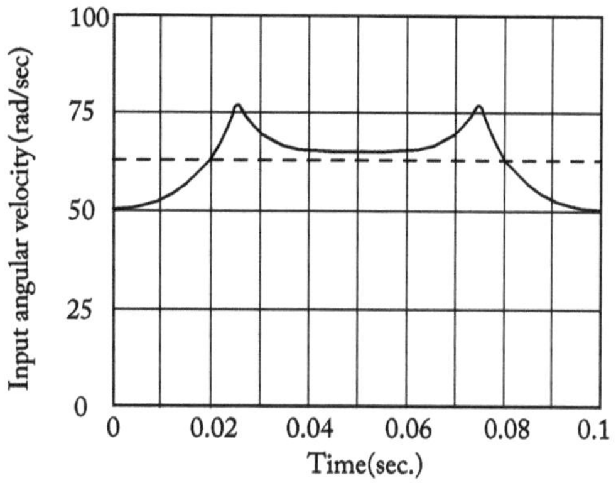

Fig. 5.25 Input angular velocity $\dot{\varphi}$ (rad/sec)

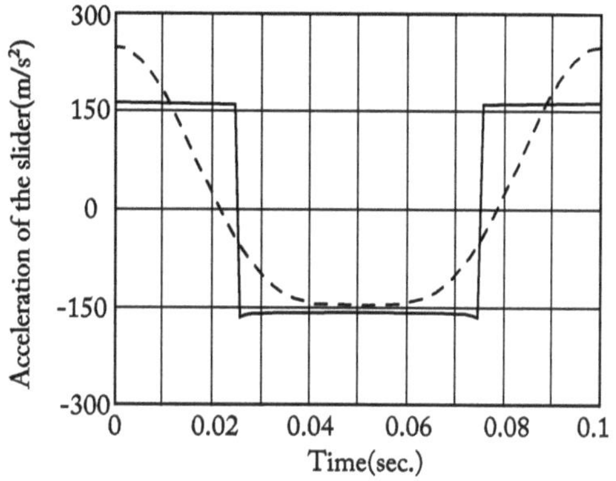

Fig. 5.26 Acceleration of the slider: $\ddot{x}_B$

Thus, the maximal value of the linear acceleration $\ddot{x}_B$ is minimized and, as a result, the shaking force due to the translational motion of the slider is also minimized. For the given example, the minimization of the linear acceleration (shaking force due to the translational motion of the slider) is 36 %.

Thus, only by optimal generation of the input crank rotation 36 % reduction of the translational component of the shaking force has been achieved.

With regard to the rotation component of the shaking force (see expression (5.60)), it can be balanced by a counterweight with static moment:

$$m_{cw} r_{cw} = m_2 l_{OS_2} + m_3\, r\, l_{BS_3}/l \tag{5.65}$$

where m_{cw} is the mass of the counterweigh mounted on the input crank; r_{cw} is the rotation radius of the centre of mass of the counterweight. For the considered mechanism the balancing static moment will be equal to 0.03375 kg m. The variations of the shaking forces are given in Fig. 5.27. The obtained results showed that the reduction of the shaking force due to the rotational and translational motions is 45 %.

The variations of the shaking moment are given in Fig. 5.28.

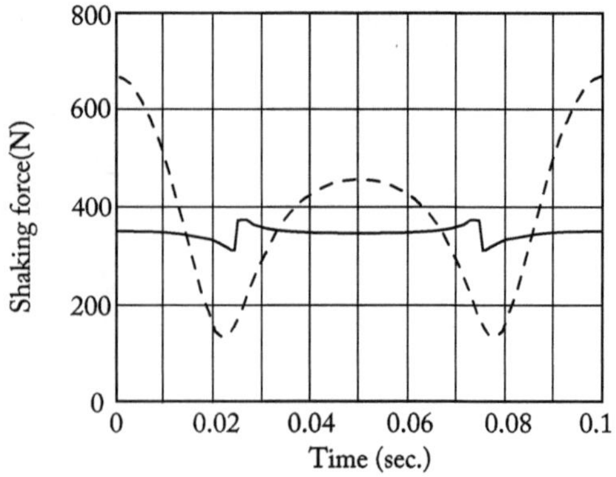

Fig. 5.27 Variations of the shaking force (N)

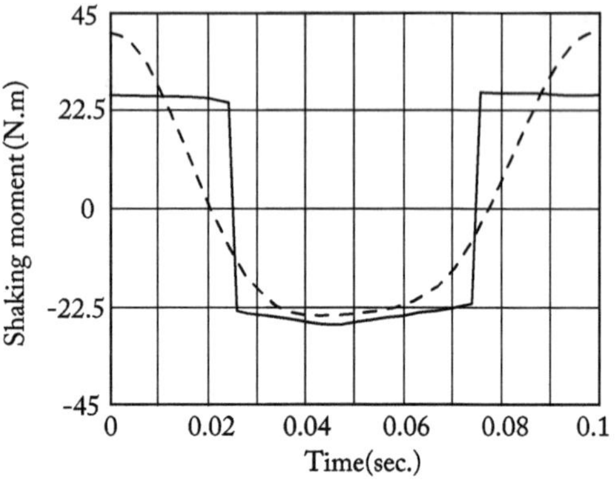

Fig. 5.28 Variations of the shaking moment (N m).

Thus, the proposed balancing method allows the minimisation of the shaking force of the examined slider-crank mechanism up to 45 % and the shaking moment up to 36 %.

Finally, it should be noted that the given "bang-bang" acceleration profile (Fig. 5.22) is based on theoretical considerations. In reality, the actuators are unable to achieve discontinuous efforts. Therefore, this motion profile should be modified by a trapezoidal profile in order to take into account the actuators properties in terms of maximal admissible effort variations. However, our observations showed that the minimizations obtained in the cases of the "bang-bang" and trapezoidal profiles are very close (less than 1 %).

With regard to the motion generation in the slider-crank mechanisms, we would like to note that the controller implementation and control scheme of a slider-crank mechanism executing the input crank rotation by given prescribed variable speed have been minutely examined in (Nagchaudhuri 2002).

Part III
Balancing of Robot Manipulators

Chapter 6
Balancing of Manipulators by Using the Copying Properties of Pantograph Mechanisms

Abstract This Chapter is devoted to the balancing of robot manipulators and hand-operated manipulators by the use of copying properties of pantograph linkages.

The Sect. 6.1 develops a methodology for the balancing of the Delta robot, which is applied as an assisting device for carrying a microscope in surgical operations. It is shown that the use of an auxiliary balancing system consisting of a pantograph linkage can considerably reduce the load on the DELTA robot actuators.

In Sect. 6.2, a new four *dof* parallel manipulator with high-load carrying capacity is presented. This manipulator generates Schoenflies motions (i.e. three translational *dof* and one rotational *dof* around one fixed axis). The particularity of the proposed architecture, which has been patented (Arakelian et al., 2006) is the decoupling of the displacements of the platform in the horizontal plane from its translation along the vertical axis. Such a decoupling is attained thanks to the special design of the robot legs carried out by pantograph linkages. It allows the cancellation of the gravity loads on the actuators which displace the platform in the horizontal plane.

Finally, in Sect. 6.3, the design and balancing of hand-operated manipulators of the pantograph types are proposed. The balancing method is based upon a study concerning the *com* motion. New and more efficient kinematic schemes of hand-operated manipulators are suggested on the basis of this method.

6.1 Design of Balancing Mechanisms for Spatial Parallel Manipulators: Application to the Delta Robot

The Delta robot (Clavel 1990) was developed for high-speed manipulation and it is well known in the electronics, food and pharmaceutical sectors as a reliable system with fast execution of light-duty tasks. However, in recent years, more attention has been paid to the increasing number of possible industrial applications, such as manipulation of medical devices (Fig. 6.1).

In this case, the displacement speed of the platform is not essential because there is no need for productivity acceleration. However, as a result of the increased mass of the platform (about 70 kg), the input torques became important. Thus it became evident that the platform's mass must be balanced. In this case, the traditional approaches with counterweights and springs mounted on the moving links are not applicable.

V. Arakelian, S. Briot, *Balancing of Linkages and Robot Manipulators*,
Mechanisms and Machine Science 27, DOI 10.1007/978-3-319-12490-2_6

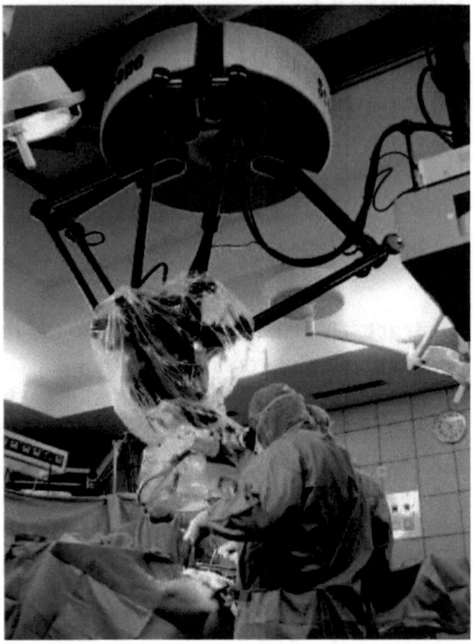

Fig. 6.1 A Delta robot used in the SurgiScope®, a robotized navigation tool-holder designed for neurosurgery and developed by the company ISIS. (ISIS 2014) (courtesy of ISIS)

The Delta robot has a complex structure and after such balancing it becomes either very heavy or a complicated assembly with several complementary articulated dyads. That is why another means for the solution of this problem is proposed. It consists in the addition of a secondary mechanism between the manipulator base and moving platform. This mechanism can create a supplementary vertical force F on the platform to balance the gravitational forces of the robot (Fig. 6.2).

In this context, a new balancing mechanism for the minimization of the input torque of the spatial parallel manipulators with high weight-carrying capacity is developed. This balancing mechanism was presented in (Baradat et al. 2008) and was patented (Baradat et al. 2006).

6.1.1 Description of the Balancing Mechanism

The suggested balancing system includes (Fig. 6.3) a pantograph mechanism mounted on the rotating stand connected with the base.

The input points A and B of the pantograph are located in the horizontal and vertical guides of the rotating stand. So the suggested system has three degrees of freedom: a rotation of the stand about the vertical axis and two translations along the guides. This allows the suggested system to be passive in relation to the Delta robot when the point C is connected with the platform.

Point B is also connected with an actuator which produces a vertical force F_B. This vertical force F_B is used for the balancing of the gravitational forces of the spatial

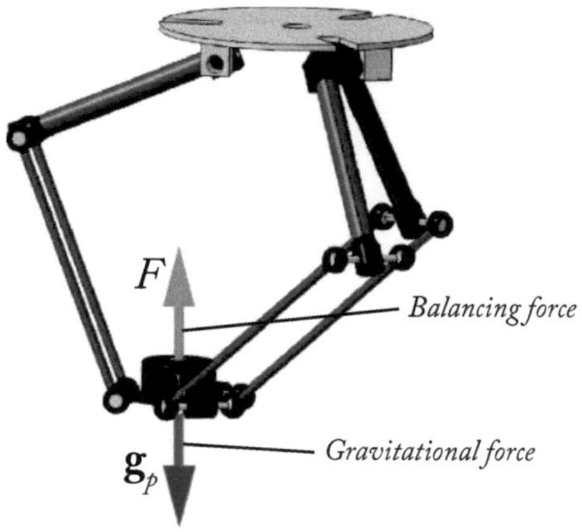

Fig. 6.2 Principle of balancing (without the added balancing mechanism)

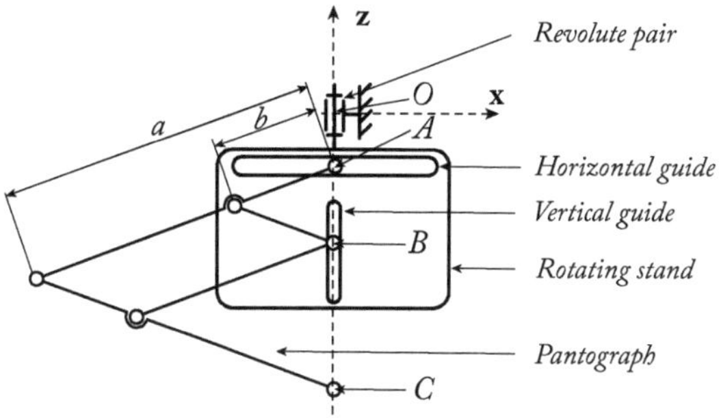

Fig. 6.3 Simplified scheme of the balancing mechanism

parallel robot. It is obvious that the determination of the balancing force $f_B = kf$ takes into account the magnification factor of the pantograph ($k = l_{AC}/l_{AB} = a/b$).

Thus the position of point C is represented by vector $\mathbf{r}_P = [x \;\; y \;\; z]^T$ and the passive motions of the pantograph are represented by $\mathbf{q} = [r \;\; \theta \;\; Z]^T$, where θ is the angle of the rotating stand of the balancing mechanism, r and Z correspond to the displacements in the horizontal and vertical guides of the balancing mechanism.

The kinematic relations between $\mathbf{r}_P$ and $\mathbf{q}$ are the following:

$$\begin{aligned} x &= (1-k)r\cos\theta \\ y &= (1-k)r\sin\theta \\ z &= kZ \end{aligned} \tag{6.1}$$

Differentiating these equations with respect to time, one obtains:

$$\dot{\mathbf{q}} = \mathbf{J}^{-1}\dot{\mathbf{r}}_P \tag{6.2}$$

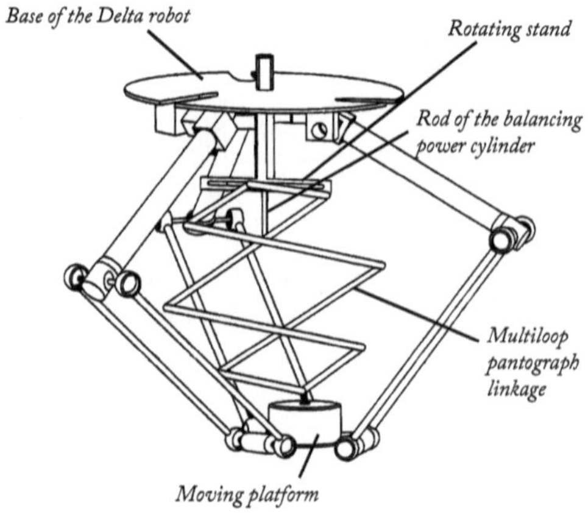

Fig. 6.4 Delta robot with the balancing mechanism

where

$$\begin{aligned}\dot{\mathbf{q}} &= [\dot{r}\ \dot{\theta}\ \dot{Z}]^T \\ \dot{\mathbf{r}}_P &= [\dot{x}\ \dot{y}\ \dot{z}]^T \\ \mathbf{J} &= \begin{bmatrix} (1-k)\cos\theta & (k-1)r\sin\theta & 0 \\ (1-k)\sin\theta & (1-k)r\cos\theta & 0 \\ 0 & 0 & k \end{bmatrix}\end{aligned} \quad (6.3)$$

It is obvious that the added balancing system cannot follow all trajectories of the parallel robot. For example, if the given trajectory of the parallel robot is composed from two mutually perpendicular straight lines, which are intersected at a point of planar coordinates $x = y = 0$, the balancing mechanism cannot execute a continuous motion. In this case, it is necessary to orientate the plane of the pantograph mechanism relative to the **z** axis. Thus, it is obvious that the balancing mechanism must be equipped with a complementary rotating actuator for its orientation in the case of singular trajectories. This complementary actuator may be in operation only for special cases.

It should be noted that in the positions close to the singular trajectories, the secondary mechanism can rotate quickly and it is necessary to carry out an optimal control of the Delta robot with slow-motion.

Figure 6.4 shows the balancing mechanism, which is implemented in the structure of the Delta robot. Multiloop pantograph linkage with several link lengths allows the reduction of the overall size of the balancing mechanism. The size of the pantograph links must be chosen in such a manner that they should not collide with the legs of the Delta robot.

However, it is necessary to note that the suggested balancing mechanism is applicable to many spatial parallel robots from 3 to 6 *dof*.

Figure 6.5 shows an example of such an application for the Gough-Stewart platform with the implemented balancing system.

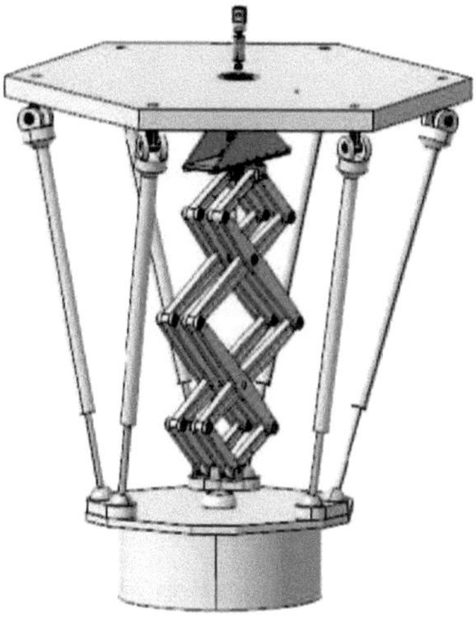

Fig. 6.5 Stewart platform with implemented balancing system

In the proposed design of the Gough-Stewart platform the payload is balanced by the suggested mechanism. So the platform becomes a weightless link and it can be displaced and oriented by low-power linear actuators.

We would like to note some particularity of the balancing of spatial parallel manipulators with $6dof$: if the gravity center of the platform is not situated in the attachment point of the balancing mechanism, a change of orientation of the platform will move its center of mass, which will lead to a complementary moment. However our observations showed that this complementary moment will be incomparable less than initial unbalanced moment.

In the following sub-section, we consider the balancing of the Delta robot by means of the proposed mechanism and we discuss the minimization of the input torques by a constant or a variable force.

6.1.2 Minimization of the Torque by a Constant Force Applied to the Robot Platform

Let us examine two cases: minimization of the torques due to the static loads, i.e. weights of the moving links, and dynamic forces, i.e. inertia forces.

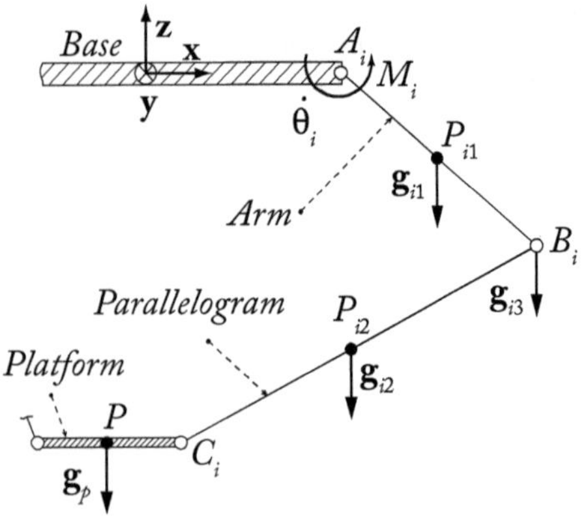

Fig. 6.6 Gravitational forces for the leg i

6.1.2.1 Minimization of the Torques due to Static Loads (Gravitational Forces)

The input torque due to gravity effects of the n-th actuator of the Delta robot can be expressed as:

$$\tau_n^{st} = \sum_{j=1}^{4} \tau_{jn}^{st} \tag{6.4}$$

where τ_{1n}^{st} is the torque due to the gravitational forces of the arms (see Fig. 6.6), τ_{2n}^{st} is the torque due to the parallelograms, τ_{3n}^{st} is the torque due to the joints on points B_i, τ_{4n}^{st} is the torque due to the gravity forces of the platform and medical device. For $n = 1, 2, 3$ and for $j = 1, 2, 3$, τ_n^{st} is equal to:

$$\tau_{jn}^{st}(x, y, z) = \sum_{j=1}^{3} \mathbf{j}_{ijn}^T(x, y, z)\mathbf{g}_{ij} \tag{6.5}$$

and for $n = 1, 2, 3$ and $j = 4$,

$$\tau_{4n}^{st}(x, y, z) = \mathbf{j}_n^T(x, y, z)\mathbf{g}_p \tag{6.6}$$

where, for $i, n = 1, 2, 3$,

- $\mathbf{j}_{i1n}$ is the (3×1) Jacobian matrix relating the velocity of point P_{i1} and the velocity $\dot{\theta}_n$,
- $\mathbf{j}_{i2n}$ is the (3×1) Jacobian matrix relating the velocity of point P_{i2} and the velocity $\dot{\theta}_n$,
- $\mathbf{j}_{i3n}$ is the (3×1) Jacobian matrix relating the velocity of point B_i and the velocity $\dot{\theta}_n$,
- $\mathbf{j}_n$ is the (3×1) Jacobian matrix relating the velocity of point P and the velocity $\dot{\theta}_n$,

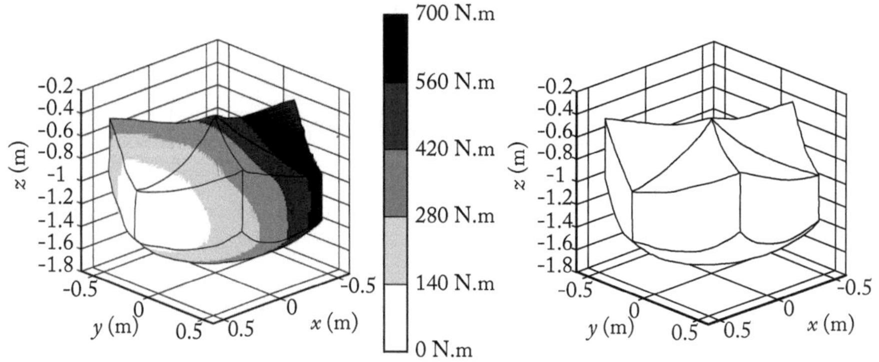

Fig. 6.7 Input torque 1 for unbalanced (*left*) and balanced (*right*) Delta robot (the values of the input torque are differentiated by the contrast intensity)

i.e.

$$\mathbf{j}_{i1n} = \frac{\partial \mathbf{r}_{P_{i1}}}{\partial \theta_n}, \quad \mathbf{j}_{i2n} = \frac{\partial \mathbf{r}_{P_{i2}}}{\partial \theta_n}, \quad \mathbf{j}_{i3n} = \frac{\partial \mathbf{r}_{B_i}}{\partial \theta_n}, \quad \mathbf{j}_n = \frac{\partial \mathbf{r}_P}{\partial \theta_n} \tag{6.7}$$

and $\mathbf{g}_p$ and $\mathbf{g}_{ij}$ are gravity forces whose application points are defined in Fig. 6.6.

Figure 6.7 shows the workspace with the torque of the actuator 1 for each position of the workspace of the Delta robot. It should be noted that, as the Delta robot which we are studying is symmetrical, the values of the input torques for the actuators are also symmetrical but they are situated in different zones (rotations of $\pm 120°$).

Adding the compensation force F on the robot platform exerted by the balancing mechanism, the expression of the robot input torques becomes:

$$\begin{bmatrix} \tau_{1bal}^{st} \\ \tau_{2bal}^{st} \\ \tau_{3bal}^{st} \end{bmatrix} = \begin{bmatrix} \tau_1^{st} \\ \tau_2^{st} \\ \tau_3^{st} \end{bmatrix} + \begin{bmatrix} \mathbf{j}_1^T \\ \mathbf{j}_2^T \\ \mathbf{j}_3^T \end{bmatrix} \begin{bmatrix} 0 \\ 0 \\ F \end{bmatrix} \tag{6.8}$$

where $\tau_{n\,bal}^{st}$ is the optimized torque of actuator n.

The condition for the minimization of the *rms* value of the torques can be expressed as:

$$\sqrt{\frac{\sum_{p=1}^{N} \left(\sum_{i=1}^{3} \left(\tau_i^{st}(x_p, y_p, z_p) + j_{3i} F \right)^2 \right)}{N}} \to \min_F \tag{6.9}$$

where j_{3i} is the third term of the vector $\mathbf{j}_i$, N is the number of calculated positions of the robot, and x_p, y_p, z_p are the coordinates of the p-th calculated position of the robot end-effector.

For the minimization of the *rms* value of the torques, it is necessary to minimize the sum:

$$\Delta = \sum_{p=1}^{N} \left(\sum_{i=1}^{3} \left(\tau_i^{st}(x_p, y_p, z_p) + j_{3i}(x_p, y_p, z_p) F \right)^2 \right) \tag{6.10}$$

For this purpose, we shall achieve the condition $\partial \Delta / \partial F = 0$ from which we determine the force:

$$F = -\frac{\sum_{p=1}^{N} \left(\sum_{i=1}^{3} j_{3i}(x_p, y_p, z_p) \tau_i^{st}(x_p, y_p, z_p) \right)}{\sum_{p=1}^{N} \left(\sum_{i=1}^{3} j_{3i}^2(x_p, y_p, z_p) \right)} \tag{6.11}$$

6.1.2.2 Numerical Example

For the Delta robot of the SurgiScope®, the parameters are the following (see Fig. 6.6):

- $l_{A_i B_i} = 0.75$ m, $l_{B_i C_i} = 0.95$ m, $l_{A_i P_{i1}} = 0.375$ m, $l_{B_i P_{i2}} = 0.475$ m,
- $m_{i1} = 2.3$ kg (mass of the i-th arm with *com* P_{i1}), $m_{i2} = 5.2$ kg (mass of the i-th parallelogram with *com* P_{i2}), $m_{i3} = 3.1$ kg (mass of the joint at point B_i), $m_p = 79$ kg (mass of the platform, joints and medical device, with *com* located at point P).

Thus the value of the added force for the given parameters is $F = 931$ N.

Figure 6.7 shows the variations of the input torques for unbalanced and balanced Delta robot. The obtained results show that the reduction of the *rms* value of the input torque is about 99.5 %. The reduction of the maximum value of the torque is about 92 %.

The purpose of this study is to develop a reliable mechanism for gravitational force balancing of spatial parallel manipulators. Moreover, it is also tempting to consider the minimization of the torques due to the dynamic loads, i.e. inertia forces.

6.1.2.3 Minimization of the Torques due to the Dynamic Loads (Inertia and Gravitational Forces)

The expression of the input torque of the n-th actuator can be obtained using the Lagrange equations with Lagrange multipliers λ:

$$\begin{bmatrix} \tau_1^{dyn} \\ \tau_2^{dyn} \\ \tau_3^{dyn} \end{bmatrix} = \frac{\mathrm{d}}{\mathrm{d}t} \left(\frac{\partial L}{\partial \dot{\mathbf{q}}} \right) - \frac{\partial L}{\partial \mathbf{q}} - \mathbf{B}^T \lambda \tag{6.12}$$

where $L(\mathbf{q}, \dot{\mathbf{q}}, \mathbf{r}_p, \dot{\mathbf{r}}_p)$ is the Lagrangian of the system equal to:

$$L = \frac{1}{2}\left(\left(\sum_{i=1}^{3} m_{i1}\dot{\mathbf{r}}_{P_{i1}}^T\dot{\mathbf{r}}_{P_{i1}} + m_{i2}\dot{\mathbf{r}}_{P_{i2}}^T\dot{\mathbf{r}}_{P_{i2}} + m_{i3}\dot{\mathbf{r}}_{B_i}^T\dot{\mathbf{r}}_{B_i}\right) + m_p\dot{\mathbf{r}}_P^T\dot{\mathbf{r}}_P\right) \tag{6.13}$$

with

$$\dot{\mathbf{r}}_{B_i} = l_{A_iB_i}\dot{\theta}_i\mathbf{Rot}(\gamma_i, \mathbf{z})\begin{bmatrix} -\sin\theta_i \\ 0 \\ \cos\theta_i \end{bmatrix} \tag{6.14}$$

$$\dot{\mathbf{r}}_{P_{i1}} = \frac{l_{A_iP_{i1}}}{l_{A_iB_i}}\dot{\mathbf{r}}_{B_i} \tag{6.15}$$

$$\dot{\mathbf{r}}_{P_{i2}} = \frac{1}{2}\left(\dot{\mathbf{r}}_{B_i} + \dot{\mathbf{r}}_P\right) \tag{6.16}$$

in which

$$\mathbf{Rot}\,(\gamma_i, \mathbf{z}) = \begin{bmatrix} \cos\gamma_i & -\sin\gamma_i & 0 \\ \sin\gamma_i & \cos\gamma_i & 0 \\ 0 & 0 & 1 \end{bmatrix} \tag{6.17}$$

For the Surgiscope®, $\gamma_1 = 0°$, $\gamma_2 = 120°$ and $\gamma_3 = -120°$.

It is noteworthy that in (6.13), the axial moments of inertia of the arm and parallelogram rods are neglected due to their small influence on the input torques of the Surgiscope®.

The Lagrange multipliers can be computed using the following relation:

$$\mathbf{A}^T\lambda = \frac{\mathrm{d}}{\mathrm{d}t}\left(\frac{\partial L}{\partial \dot{\mathbf{r}}_P}\right) - \frac{\partial L}{\partial \mathbf{r}_P} \tag{6.18}$$

In (6.12) and (6.18), the matrices **A** and **B** are matrices that relates the actuator velocities to the end-effector twist, such that:

$$\mathbf{A}\dot{\mathbf{r}}_P + \mathbf{B}\dot{\mathbf{q}} = \mathbf{0} \tag{6.19}$$

With the added force F, and neglecting the inertia of the balancing system, (6.12) can be written as:

$$\begin{bmatrix} \tau_{1bal}^{dyn} \\ \tau_{2bal}^{dyn} \\ \tau_{3bal}^{dyn} \end{bmatrix} = \begin{bmatrix} \tau_1^{dyn} \\ \tau_2^{dyn} \\ \tau_3^{dyn} \end{bmatrix} + \begin{bmatrix} \mathbf{j}_1^T \\ \mathbf{j}_2^T \\ \mathbf{j}_3^T \end{bmatrix}\begin{bmatrix} 0 \\ 0 \\ F \end{bmatrix} \tag{6.20}$$

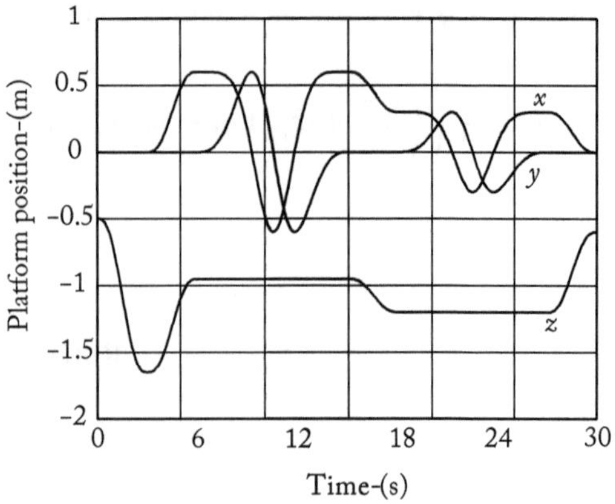

Fig. 6.8 The selected trajectory

Table 6.1 Input torques for unbalanced and balanced robots

Maximal value of the acceleration and velocity	Maximal value of the input torque 1		Balancing force (N)	Gain (%)
	Unbalanced (N.m)	Balanced (N.m)		
$a = 0.1$ m/s^2, $v = 0.26$ m/s	645	57	956	91
$a = 1.05$ m/s^2, $v = 0.79$ m/s	652	110	897	83
$a = 4.13$ m/s^2, $v = 1.57$ m/s	677	356	779	49

We would like to point out that, in the case of the dynamic study, the input torques depend on the velocity and acceleration of the platform displacement and it is impossible to achieve an optimization for the whole workspace of the robot (as it was for the static load minimization). Thus we must define a trajectory on which the input torques will be minimized. The selected trajectory is presented in Fig. 6.8. The kinematic characteristics of the examined motion are given by the maximal values of the acceleration and velocity norms and presented in Table 6.1 (factors 'a' and 'v').

The condition for the minimization of the *rms* value of the torques can be expressed as:

$$\sqrt{\frac{\sum_{p=1}^{N}\left(\sum_{i=1}^{3}\left(\tau_i^{dyn} + j_{3i}F\right)^2\right)}{N}} \to \min_F \tag{6.21}$$

where τ_i^{dyn} is the initial torque of the actuator i and N is the number of calculated positions of the robot.

For the minimization of the *rms* value of the torques, it is necessary to minimize the sum:

$$\Delta = \sum_{p=1}^{N}\left(\sum_{i=1}^{3}\left(\tau_i^{dyn} + j_{3i}(x_p, y_p, z_p)F\right)^2\right) \tag{6.22}$$

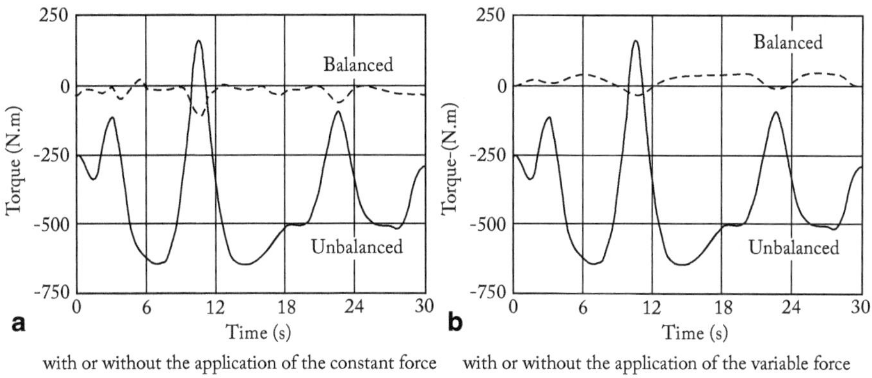

Fig. 6.9 Input torque 1 for unbalanced and balanced Delta robot in a dynamic mode of operation

For this purpose, we shall achieve the condition $\partial\Delta/\partial F = 0$ from which we determine the force:

$$F = -\frac{\sum_{p=1}^{N}\left(\sum_{i=1}^{3} j_{3i}(x_p, y_p, z_p)\tau_i^{dyn}(x_p, y_p, z_p)\right)}{\sum_{p=1}^{N}\left(\sum_{i=1}^{3} j_{3i}^2(x_p, y_p, z_p)\right)} \tag{6.23}$$

For the examined trajectory with $a = 1.05$ m/s^2 and $v = 0.79$ m/s, we determine the external force: $F = 934$ N.

Figure 6.9a shows the variations of the input torque 1 for the unbalanced and balanced Delta robots (for $a = 1.05$ m/s^2 and $v = 0.79$ m/s, when the reduction of the input torques is about 83 %).

6.1.3 Minimization of the Input Torques by a Variable Force Applied to the Platform of the Robot

This section also contains two cases: minimization of the input torques due to the static and dynamic forces.

6.1.3.1 Minimization of the Torques due to the Static Loads (Gravitational Forces)

Trying to minimize the *rms* of (6.8) at any position $\left[x_p, y_p, z_p\right]^T$, the optimization problem becomes:

$$\Delta_p = \sum_{i=1}^{3}\left(\tau_i^{st}(x_p, y_p, z_p) + j_{3i}(x_p, y_p, z_p)F(x_p, y_p, z_p)\right)^2 \rightarrow \min_F \tag{6.24}$$

Thus, $\partial \Delta_p / \partial F = 0$ leads to the optimal value of the force F for the position $[x_p, y_p, z_p]^T$:

$$F(x_p, y_p, z_p) = \frac{\sum_{i=1}^{3} \left(j_{3i}(x_p, y_p, z_p) \tau_i^{st}(x_p, y_p, z_p) \right)}{\sum_{i=1}^{3} j_{3i}^2(x_p, y_p, z_p)} \tag{6.25}$$

It can be demonstrated that for the studied SurgiScope® the difference in torques minimized with a constant or a variable force is very small (about 1 %). For these reasons, no figures are shown as they are similar to those depicted at Fig. 6.7. Thus, for the minimization of the static loads it is better to use constant force. The constant force is easier to create than the variable force.

6.1.3.2 Minimization of the Torques due to the Dynamic Loads (Inertia and Gravitational Forces)

The *rms* of (6.20) can be minimized at any position $[x_p, y_p, z_p]^T$ through the following optimization problem:

$$\Delta_p = \sum_{i=1}^{3} \left(\tau_i^{dyn}(x_p, y_p, z_p) + j_{3i}(x_p, y_p, z_p) F(x_p, y_p, z_p) \right)^2 \to \min_F \tag{6.26}$$

Thus, similarly as in Sect. 6.1.3.1, $\partial \Delta_p / \partial F = 0$ leads to the optimal value of the force F for the position $[x_p, y_p, z_p]^T$:

$$F(x_p, y_p, z_p) = \frac{\sum_{i=1}^{3} \left(j_{3i}(x_p, y_p, z_p) \tau_i^{dyn}(x_p, y_p, z_p) \right)}{\sum_{i=1}^{3} j_{3i}^2(x_p, y_p, z_p)} \tag{6.27}$$

Table 6.2 presents the maximum values of the torques for three examined cases. It should be noted that in this case also, the increase in the velocity and acceleration leads to the reduction of the efficiency of the minimization.

We would like to draw attention to the fact that the added force F is always vertical and cannot compensate for all effects of the inertia forces along the **x** and **y** axes.

Figure 6.9b shows the variations of the input torques for the initial and optimized cases (for $a = 1.05$ m/s^2, $v = 0.79$ m/s, when the reduction of the input torques is 85 %).

Table 6.2 Input torques for unbalanced and balanced robots

Maximal value of the acceleration and velocity	Maximal value of the input torque 1		Gain (%)
	Unbalanced (N.m)	Balanced (N.m)	
$a = 0.1$ m/s^2, $v = 0.26$ m/s	645	40	94
$a = 1.05$ m/s^2, $v = 0.79$ m/s	652	96	85
$a = 4.13$ m/s^2, $v = 1.57$ m/s	677	262	60

It should be noted that these simulations showed that the minimization of the input torques achieved by using a variable force is not very efficient. The difference between the results obtained with constant and variable forces is very small. Taking into account the difficulty involved in the practical realization of the variable force, we can conclude that for the suggested balancing mechanism it is enough to use the constant force.

6.1.4 Prototype and Experimental Validation

6.1.4.1 Prototype

A prototype has been designed and built for validation of the obtained results. It was implemented in the structure of the Delta robot of the SurgiScope® provided by the company ISIS. To design the prototype, the first stage is to find the optimal lengths of the multiloop pantograph linkage taking into account that it should not collide with the legs of the Delta robot. Then the appropriate stiffness characteristics of the multiloop pantograph linkage were found by evolution of the shapes and design concept of links, as well as by successive optimization based on the finite element analysis.

After assembling the prototype, its static balance was verified by placing it vertically and noting that the mechanism is in equilibrium in any of its configurations. In such a manner, the balancing force for developed multiloop pantograph linkage was found through experimentation . The base of the balancing mechanism was then suspended from the fixed structure of the Delta robot and its end was connected to the moving platform (Fig. 6.10). In order to create a balancing force, a counterweight was used. It is obvious that for industrial applications it is better to use pneumatic cylinders or electric actuators with constant moment. However, the validation of the obtained results can also be achieved by a counterweight, which develops the same force as a pneumatic cylinder or electric actuator.

6.1.4.2 Experimental Bench

The experimental bench (Fig. 6.11) is composed of the Delta robot with its control system and a computer to interact with the user and a dSPACE 1103 board. The sampling period is of 1 ms (corresponding sampling frequency $f_e = 1$ kHz).

The Delta robot is composed of three Parvex RX320E DC servo motors with the following main characteristics:

- rated speed: 3000 tr/min and maximum speed: 3900 tr/min,
- rated torque (at slow rotation): 1.08 Nm,
- rated current: 7.8 A and instantaneous maximum current: 20 A,
- 100,000 encoder pulses per revolution (resolution: 0.0036°).

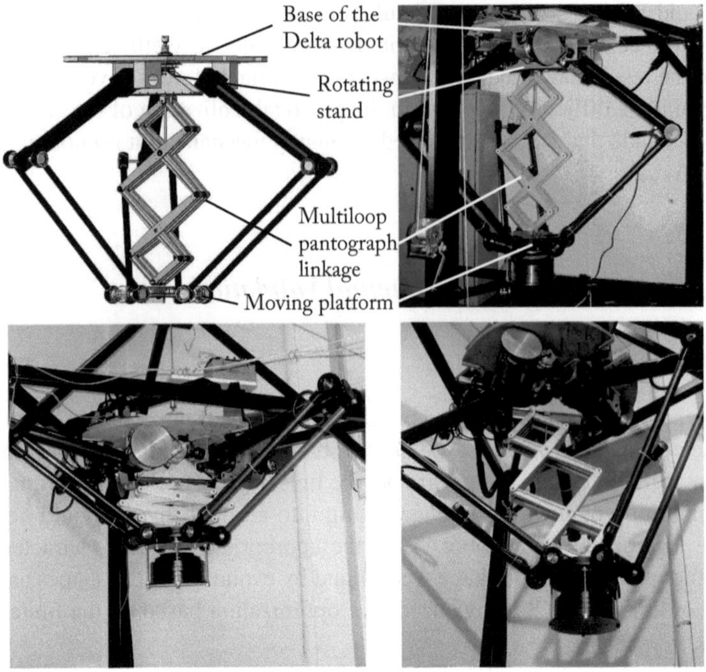

Fig. 6.10 CAD model and prototype of the balancing mechanism implemented in the structure of the Delta robot

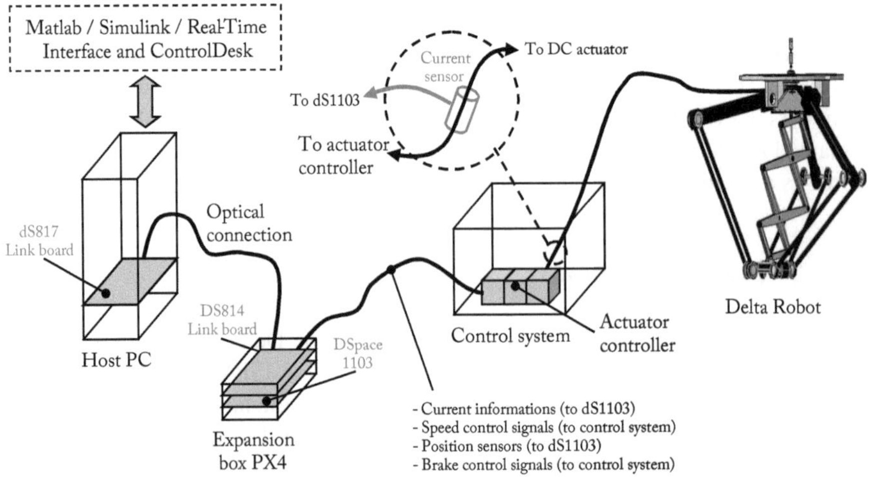

Fig. 6.11 Experimental bench

To show the improvement made in an industrial system by the balancing mechanism, we have preserved the existing industrial control system and used its speed control entries to actuate the motors. The dSPACE 1103 card is the interface between

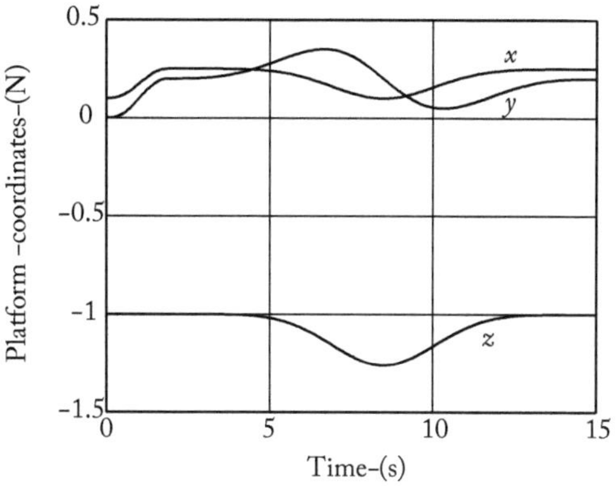

Fig. 6.12 Selected trajectory for experimental validation of torque minimization

the PC and the servosystem. This board allows the control of our three motorized axes and has specific entries to which we have directly connected our three incremental encoders. To control the system the software Matlab/Simulink/RTI/ControlDesk have been used.

The robot is controlled by a classical PID law, already included in the industrial control system. We have observed the torque response of the system to a movement composed of a straight line with respect to **z** axis and a circle in the space (Fig. 6.12).

6.1.4.3 Experimental Results and Comparative Analysis

The input torques measured for each actuator in different cases are denoted by A, B, C and D:

- **Case A**: unloaded robot,
- **Case B**: loaded robot (robot with the load of 690 N),
- **Case C**: load balanced robot (balancing by mechanism with force of 690 N),
- **Case D**: load and mechanical system balanced robot (balancing by mechanism with force of 880 N).

We would like to draw attention to cases C and D. In case C, we have only compensated the load added on the robot platform, to obtain the same result as thus obtained when the robot is unloaded. In case D, we have taken into account the mass of the Delta robot links, which gives the best result.

The measured input torques have been taken for two different speeds of the movement (maximal Cartesian acceleration and maximal Cartesian speed):

- **Experiment 1 (E1)**: 15 % of the maximum capacity of the robot to neglect most of the dynamic effects (we consider it as the static mode of operation),
- **Experiment 2 (E2)**: 100 % of the maximum capacity of the robot to observe the improvement for dynamic mode of operation.

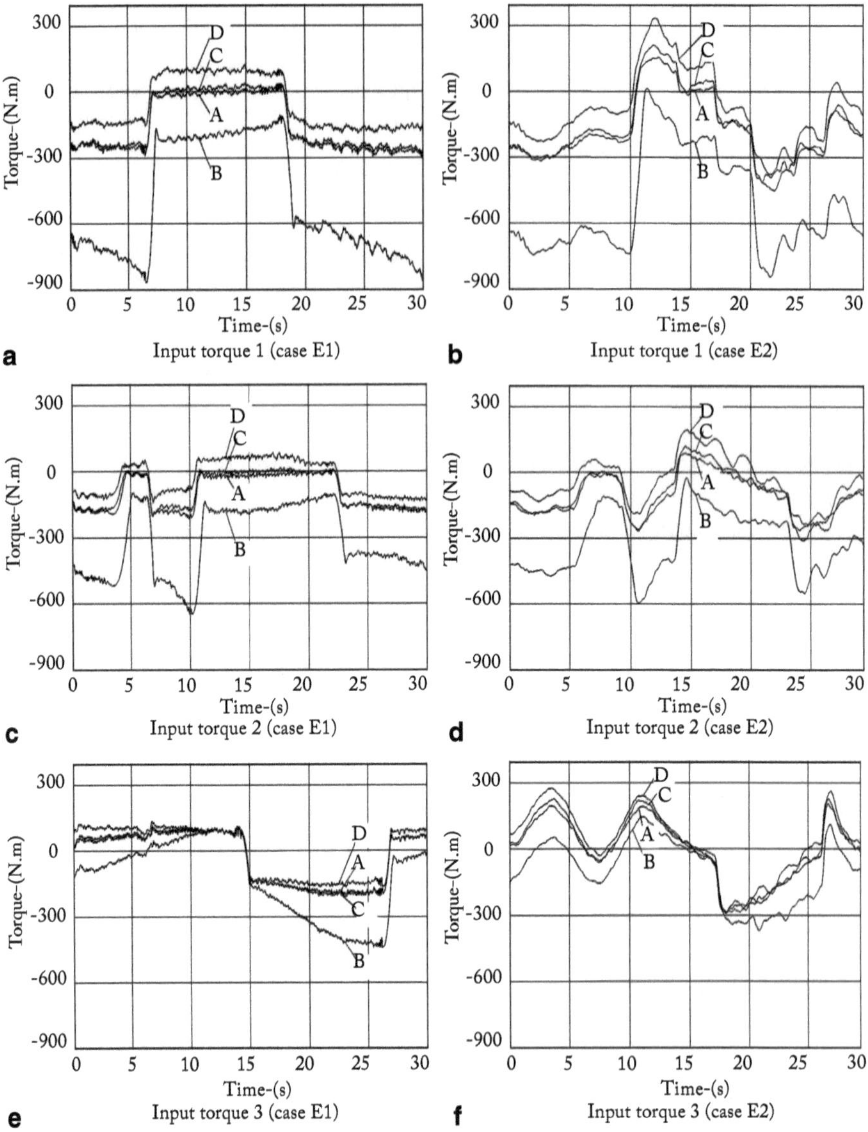

Fig. 6.13 Experimental measures of input torques for three actuators of the Delta robot

The obtained measurements confirm perfectly the theoretical results (Fig. 6.13):

- when balancing is carried out by taking into account only the load on the platform, the results are similar to those obtained for the unloaded robot (cases A and C),
- when balancing is carried out by taking into account the load on the platform and loads of robot links, we obtain the lowest values for the input torques (cases D).

Table 6.3 Input torques (E1: static mode of operation)

						Improvement gain (%)	
		A	B	C	D	C	D
Maximal values of the measured torques (N.m)	Actuator 1	306	882	324	199	63	77
	Actuator 2	217	653	208	145	68	77
	Actuator 3	211	449	221	180	50	59

Table 6.4 Input torques (E2: dynamic mode of operation)

						Improvement gain (%)	
		A	B	C	D	C	D
Maximal values of the measured torques (N.m)	Actuator 1	456	880	502	423	43	52
	Actuator 2	291	608	342	264	44	56
	Actuator 3	313	400	320	328	20	18

Tables 6.3 and 6.4 show the reduction of the input torques for experiments 1 (E1) and 2 (E2). In these tables, the gain in percent represents the reduction of the torques compared to case B.

We can observe that improvement for actuators 1 and 2 in the quasi static movement is 77 % and for actuator 3 is 59 %. For the dynamic mode of operation the improvement for actuator 1 is 52 %, for actuator 2 is 56 % and for actuator 3 is 18 %.

Gains for actuator 3 are quite different from the two others, because this one was less solicited by the given trajectory than the two others, i.e. for the given trajectory the load of the platform on actuator 3 was smaller (see Tables 6.3 and 6.4, case B). So, it is natural that for this actuator we do not observe consequent improvement of its torque by the balancing mechanism.

The experimental validation of the suggested balancing approach showed that satisfactory results are achieved and the developed system is fully operational. The gain is a bit less that predicted, but our mathematical developments were not considering the friction efforts in the actuator gearboxes which cannot be compensated.

Finally, the proposed balancing approach has also interesting properties in terms of increase of the positionning accuracy. It has been shown in (Baradat et al. 2008) that the balancing force allows a reduction of the deformations of the Delta robot under the load of 690 N of more than 90 %.

6.2 Design of Self-Balanced Parallel Manipulators: PAMINSA with 4-*dof*

In this section, a new four *dof* parallel manipulator with high-load carrying capacity is presented.

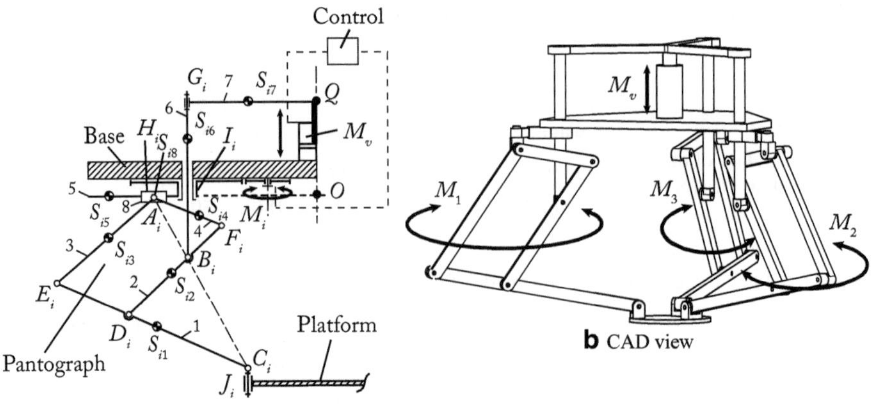

a Kinematic chain of one leg

b CAD view

Fig. 6.14 Schematics of the PAMINSA

6.2.1 *A New Concept for the Design of Partially Decoupled Parallel Manipulators*

6.2.1.1 Description of the Concept

An energetic analysis shows that the gravity work of a body moving in the horizontal plane is equal to zero (the gravitational forces are always perpendicular to the displacements). But the work of the same force moving along the vertical axis is other than zero (the gravitational forces are parallel to the displacements). This phenomenon is used in the design of the hand-operated manipulators (see Sects. 2.3.2 and 6.3), in which the horizontal displacements of the payload are carried out manually and the vertical displacements are actuated. This principle is applied in the design of the new parallel manipulators called PAMINSA (which stands for **PA**rallel **M**anipulator of the **INSA**).

Let us consider the mechanical architecture of the suggested manipulators. The aim of this innovation was to develop a parallel architecture in which the displacements of the platform in the horizontal plane are independent of its vertical motions. For this purpose, the pantograph linkage is used as a leg. The pantograph is a mechanical system with two input points A_i and B_i and one output point C_i (Fig. 6.14a). These input points linearly control the displacement of the output point C_i. Thus, one linear actuator connected with input point B_i can control the vertical displacement of the output point C_i and one other linear actuator with horizontal axis is able to control its horizontal displacements. Please note that these motions are completely decoupled, i.e. they can be carried out independently.

Now let us connect three Scheiner pantograph linkages with the base and the platform as shown in Fig. 6.14b. In the obtained structure, one vertical actuator M_v controls the vertical displacement of point B_i of the pantograph linkages, resulting

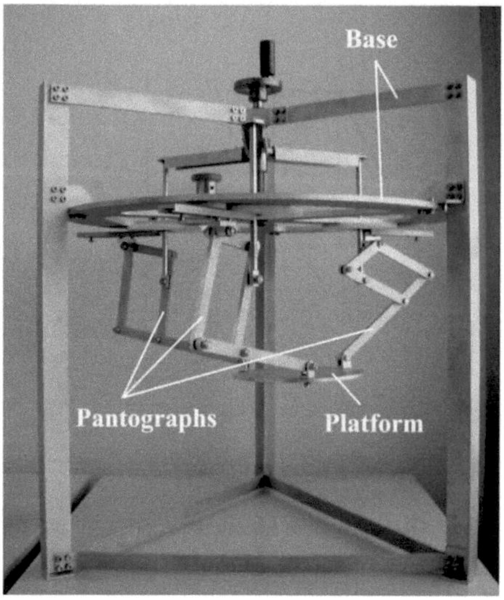

Fig. 6.15 The first model of PAMINSA

in the vertical displacement of joints C_i of the moving platform. The generation of motion in the horizontal plane is achieved by the actuators M_1, M_2 and M_3 moving the input joints A_i.

Figure 6.15 shows the first hand-operated model of PAMINSA. In the suggested architecture, the vertical translation of the platform along **z** axis is decoupled from its displacements in the horizontal plane (translations along **x** and **y** axes and rotation ϕ around **z** axis).

PAMINSA manipulators have numerous advantages, among which it is possible to notice:

1. the decoupling of the control powers into two parts, making it possible to raise an important payload to a fixed altitude with powerful actuators and then to displace it on the horizontal plane with less powerful actuators,
2. a great accuracy in the horizontal positioning because the payload can be locked in the horizontal plane using the mechanical architecture of the manipulator (in other words, if the position of the vertical actuator is fixed, the altitude of the platform cannot change),
3. the cancellation of the loads of gravity on the rotating actuators which displace the platform in the horizontal plane,
4. the simplification of the vertical control which is based on linear input/output relationships.

6.2.1.2 Kinematics and Singularity Analysis

The kinematics of this manipulator is simple because its analysis can be decomposed into two parts. The displacement along the vertical axis is described by the well known

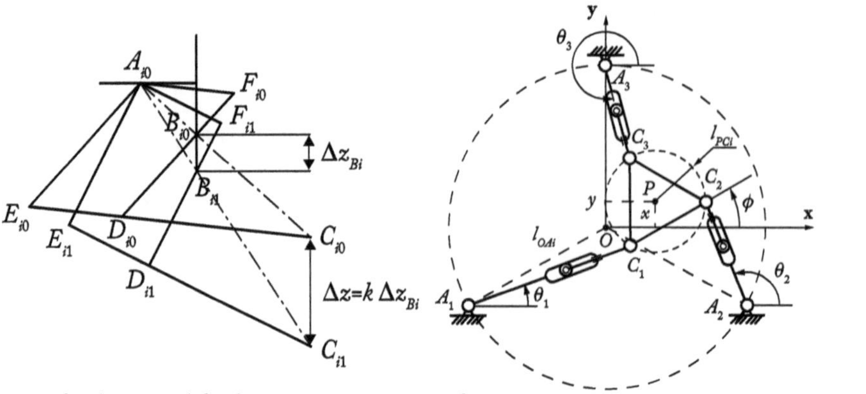

a for the vertical displacements **b** for the displacements in the horizontal plane

Fig. 6.16 Equivalent kinematic models of the PAMINSA

copying properties of the pantograph (Fig. 6.16a):

$$\Delta z = k \Delta z_{B_i} \tag{6.28}$$

where k is the magnification factor of the pantograph, Δz_{B_i} is the displacement of the input point B_i and Δz is the displacement of the output point C_i of the pantograph (Fig. 6.16a). With regard to the displacements in the horizontal plane, the equivalent kinematic model is a 3-$\underline{R}PR$ parallel manipulator (Fig. 6.16b). The position and orientation of the mobile coordinate system of the platform in the fixed coordinate system Oxy can be represented by the vector $\mathbf{x} = [x, y, \phi]^T$, where x and y are the coordinates of the center P of the equilateral triangle $C_1C_2C_3$, ϕ is the angle of rotation around this center. The input parameters are expressed by the vector $\mathbf{q} = [\theta_1, \theta_2, \theta_3]^T$. It must be mentioned that the angle θ_i also corresponds to the angle characterizing the rotation of the pantograph linkage of the leg i with respect to the vertical axis.

The relations between the input $\mathbf{q}$ and output $\mathbf{x}$ parameters can be found from the loop-closure equations (for $i = 1, 2, 3$):

$$\begin{aligned} &\cos\theta_i \left(l_{OA_i} \sin\gamma_i - l_{PC_i} \sin(\phi + \gamma_i) - y \right) \\ &\quad - \sin\theta_i \left(l_{OA_i} \cos\gamma_i - l_{PC_i} \cos(\phi + \gamma_i) - x \right) = 0 \end{aligned} \tag{6.29}$$

where $\gamma_i = \{-5\pi/6, -\pi/6, \pi/2\}$.

The Jacobian matrix $\mathbf{J}$, expressing the relation between the vector of the time derivatives of the input joint coordinates and the time-derivatives of the position of the platform, can be obtained from the differentiation of (6.28) and (6.29) with respect to time (Merlet 2006):

$$\mathbf{J} = \mathbf{A}^{-1}\mathbf{B} \tag{6.30}$$

with

$$\mathbf{A}=\begin{bmatrix}\sin\theta_1 & -\cos\theta_1 & l_{PC_i}\sin(\phi+2\pi/3-\theta_1) & 0\\ \sin\theta_2 & -\cos\theta_2 & l_{PC_i}\sin(\phi-2\pi/3-\theta_2) & 0\\ \sin\theta_3 & -\cos\theta_3 & l_{PC_i}\sin(\phi-\theta_3) & 0\\ 0 & 0 & 0 & 1\end{bmatrix} \tag{6.31}$$

$$\mathbf{B}=\begin{bmatrix}\rho_1 & 0 & 0 & 0\\ 0 & \rho_2 & 0 & 0\\ 0 & 0 & \rho_3 & 0\\ 0 & 0 & 0 & k\end{bmatrix},\quad \rho_i=l_{A_iC_i} \tag{6.32}$$

The actuation singularities of parallel manipulators appears when these matrices becomes rank-deficient (Gosselin and Angeles 1990), i.e. when

$$\det(\mathbf{B})=k\rho_1\rho_2\rho_3=0 \tag{6.33}$$

$$\det(\mathbf{A})=-\frac{\left(l_{PC_i}-l_{OA_i}\cos\phi\right)\left(x^2+y^2-\left(l_{PC_i}^2-l_{OA_i}^2-2l_{PC_i}l_{OA_i}\cos\phi\right)\right)}{\rho_1\rho_2\rho_3}=0 \tag{6.34}$$

These conditions will be reached when:

1. $\rho_1\rho_2\rho_3=0$. These singular positions correspond to manipulator configurations in which the point C_i is aligned with B_i and G_i (Fig. 6.17). In these configurations, infinite number of values for the parameters θ_i correspond to a given position of the platform, i.e. the motion of the input link(s) has no influence on the position of the platform.
2. $l_{PC_i}-l_{OA_i}\cos\phi=0$. This condition is reached if $\phi_{1,2}=\pm\cos^{-1}\left(l_{PC_i}/l_{OA_i}\right)$. Examples of configurations corresponding to this singularity are shown in Fig. 6.18. In these positions, for fixed values of the active-joint variables θ_i, infinitesimal rotations around O^* are possible.
3. $x^2+y^2-\left(l_{PC_i}^2-l_{OA_i}^2-2l_{PC_i}l_{OA_i}\cos\phi\right)=0$. The manipulator configuration corresponding to this singularity is shown in Fig. 6.19. In this configuration, the center P of the platform is on a circle of radius $l_S=\sqrt{l_{PC_i}^2-l_{OA_i}^2-2l_{PC_i}l_{OA_i}\cos\phi}$ and for fixed values of the active-joint variables, the platform may generate infinitesimal rotation around O^*.

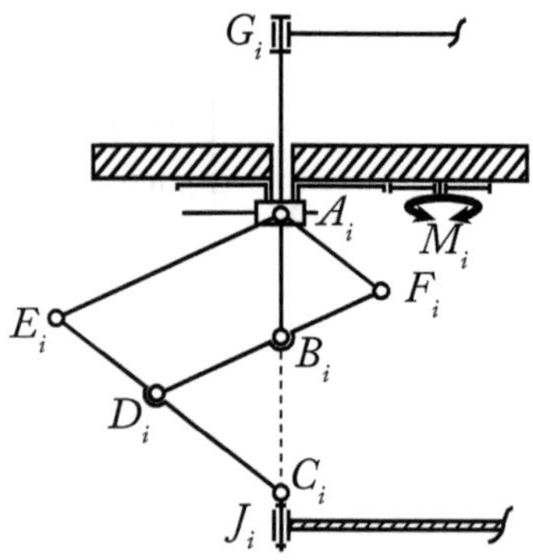

Fig. 6.17 Example of Case 1 singularities

Other types of singularities, which are not described in the approach presented above, are due to the degeneracy of the pantograph linkages. They can be found by analyzing the Jacobian matrix $\mathbf{J}_{leg_i}$, defining the relations between the time derivatives of the coordinates of points C_i (expressed in the plane of the pantograph) and the time derivatives of the angles φ_{1i} and φ_{2i}, where, φ_{1i} is the angle between the directions of the prismatic pair located at H_i and the segment B_iD_i and φ_{2i} is the angle between directions of segments B_iD_i and D_iC_i (Fig. 6.14a):

$$\mathbf{J}_{leg_i} = \begin{bmatrix} -l_{B_iD_i}\sin\varphi_{1i} - l_{D_iC_i}\sin(\varphi_{1i}+\varphi_{2i}) & -l_{D_iC_i}\sin(\varphi_{1i}+\varphi_{2i}) \\ l_{B_iD_i}\cos\varphi_{1i} + l_{D_iC_i}\cos(\varphi_{1i}+\varphi_{2i}) & l_{D_iC_i}\cos(\varphi_{1i}+\varphi_{2i}) \end{bmatrix} \tag{6.35}$$

Thus, the singularity appears when this matrix is rank-deficient, i.e. when

$$\det\left(\mathbf{J}_{leg_i}\right) = l_{B_iD_i} l_{D_iC_i} \sin\varphi_{2i} = 0 \tag{6.36}$$

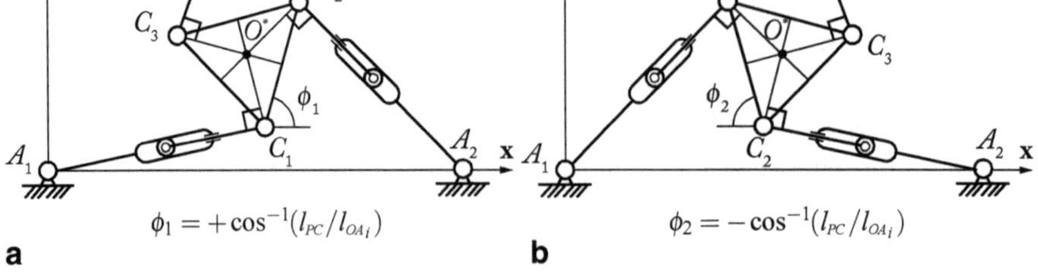

Fig. 6.18 Two singular positions of Case 2 corresponding to

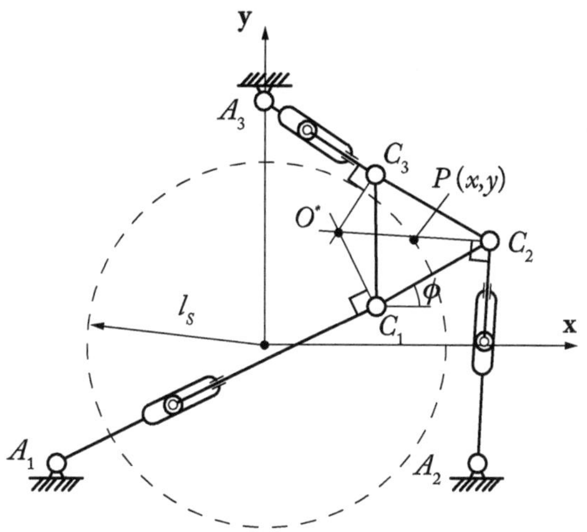

Fig. 6.19 Singularity of Case 3 of the manipulator when the center P of the platform is on a circle of radius $l_S = \sqrt{\left(l_{PC_i}^2 - l_{OA_i}^2 - 2l_{PC_i}l_{OA_i}\cos\phi\right)}$

These conditions will be reached when segment B_iD_i is either fully stretched out or folded back on segment D_iC_i.

It should be noted that there are some particular cases of these singular configurations when finite translations or rotations can be produced (self-motions) (Briot et al. 2008b).

6.2.1.3 The Family of PAMINSA Manipulators

The motion generation of the input point A_i can be carried out by several manners. All architectures shown in Table 6.5 have the same properties as mentioned above, i.e. they have four *dof* and allows the decoupling between the vertical translation and the horizontal displacements. The different schematics for input motion generation can be easily distinguished by their equivalent models for the motions the horizontal plane (the pair M_i' or H_i' corresponds to the displacement of both pair M_i or H_i and pantograph linkage). Changing the equivalent planar model of the robot involves changes in its kinematics relationships, and more particularly, its singularity loci. Especially, by changing the *RPR* equivalent chain by a *PPR* chain, it is possible to create some fully decoupled PAMINSA with three translational *dof* (Fig. 6.20).

We considered above basic structures of PAMINSA with four *dof*. However, on the base of this approach, it is also possible to obtain parallel manipulators from three to six *dof*. Some of them are detailed below.

Table 6.6 shows PAMINSA manipulators from three to six *dof* with a planar projection equivalent to a 3-*RPR* structure. Notation PAMINSA-*iDjL* means that the manipulator has *i dof* and *j* legs. Table 6.6 also represents the output parameters, the actuated joints, as well as the type of connection between the platform and the legs. Such modifications can be easily extended to the other type of kinematic chains represented in Table 6.5.

For each kind of manipulator, the rotations of the legs allow the horizontal displacement of the platform at a given altitude with given inclinations. Note please

Table 6.5 Examples of motion generation of the input point A_j of the pantograph linkage

Kinematic chain	Planar equivalent model		3D representation
	Type	Schematics	
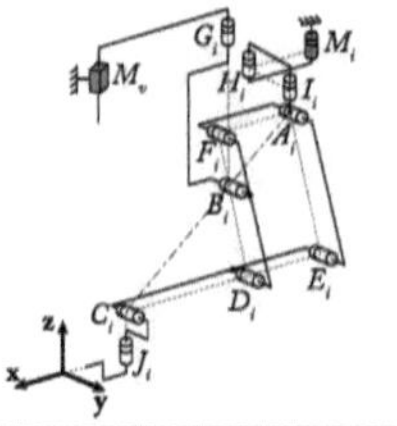	3-$\underline{R}RR$	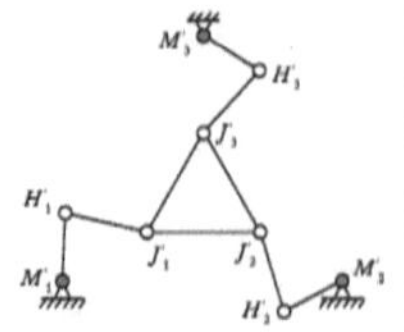	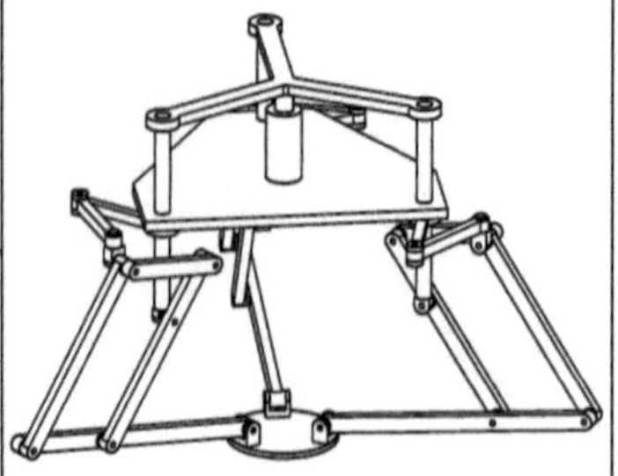
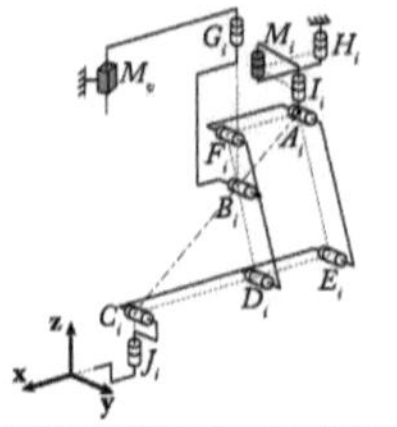	3-$R\underline{R}R$	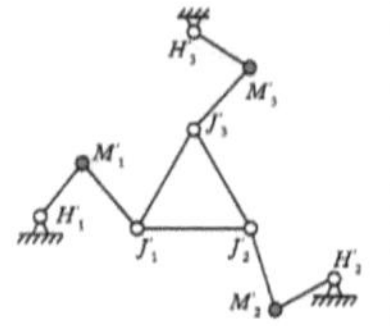	
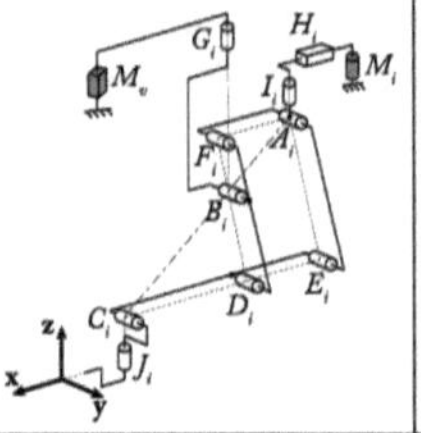	3-$\underline{R}RR$	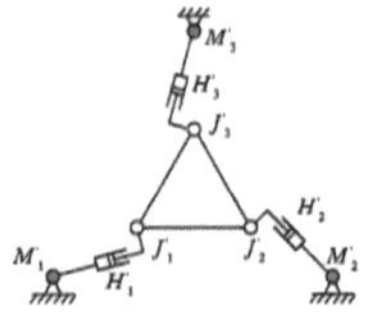	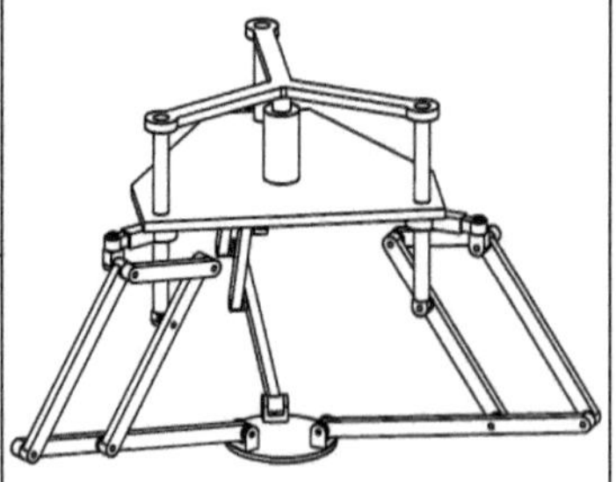
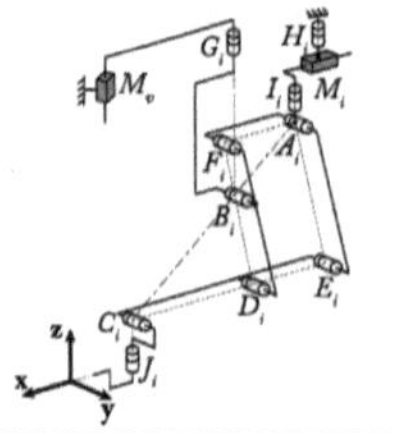	3-$R\underline{R}R$	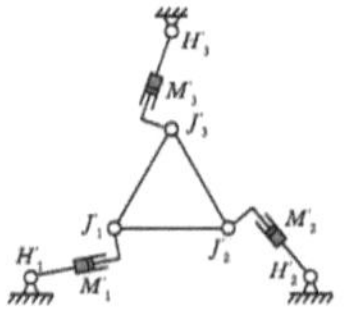	
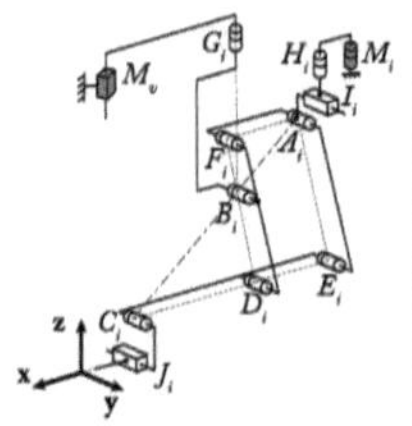	3-$R\underline{R}R$	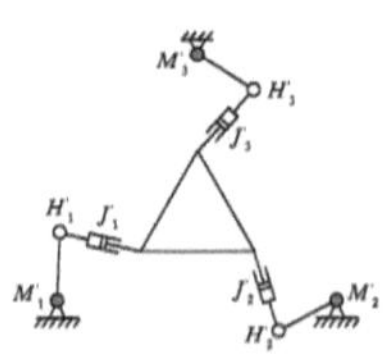	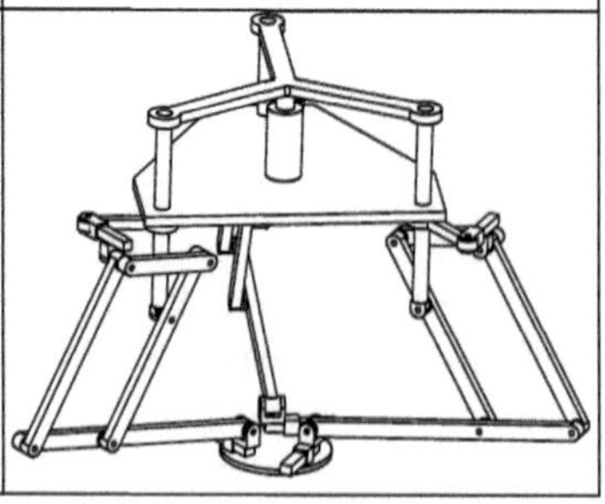

Table 6.5 (Continued)

Kinematic chain	Planar equivalent model		3D representation
	Type	Schematics	
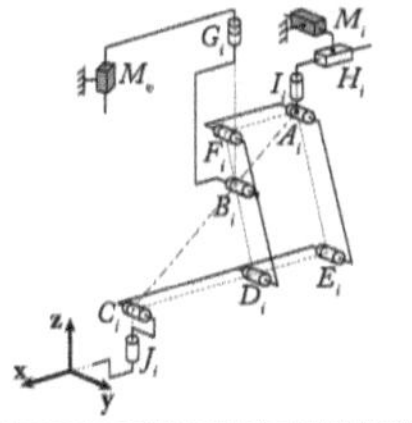	$3\text{-}\underline{R}RR$	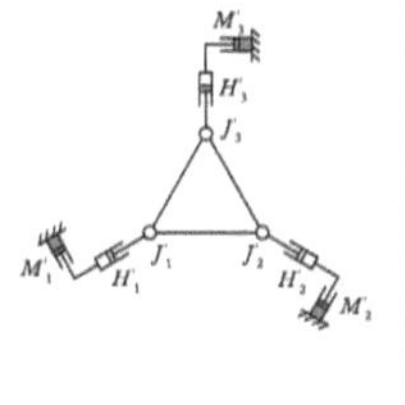	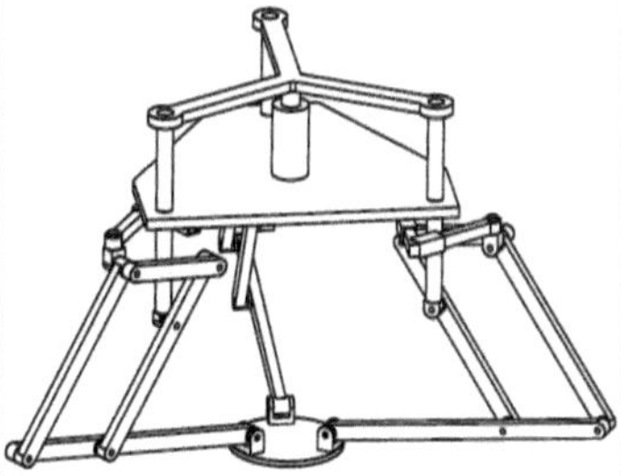
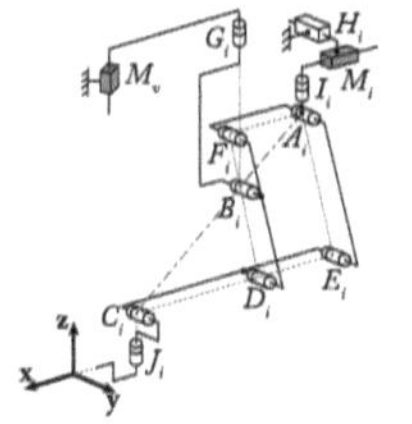	$3\text{-}R\underline{R}R$	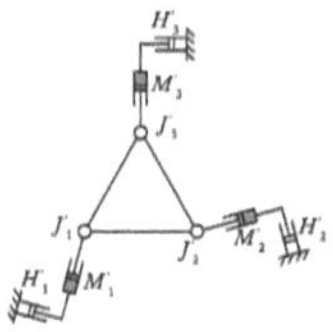	
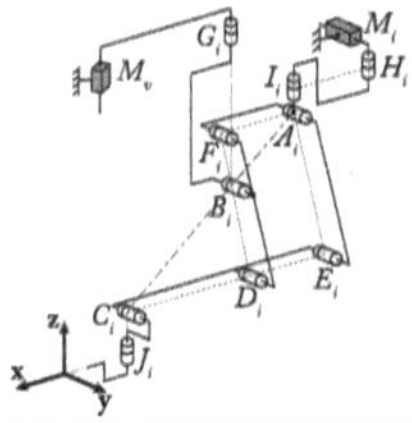	$3\text{-}\underline{R}RR$	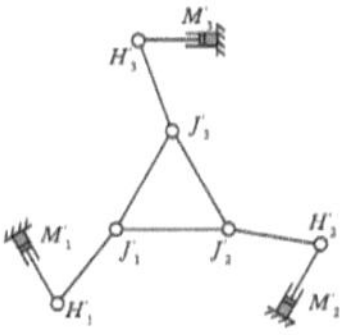	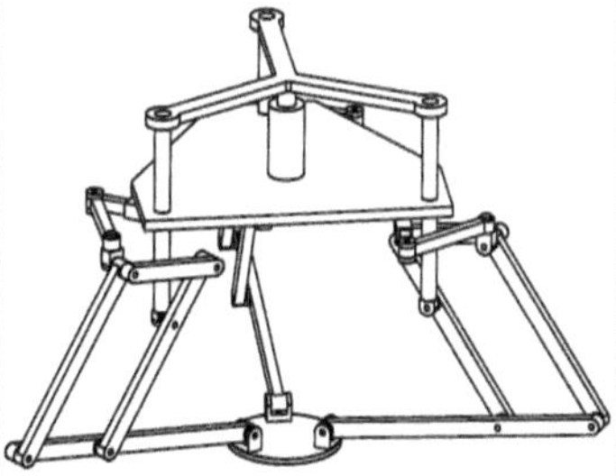
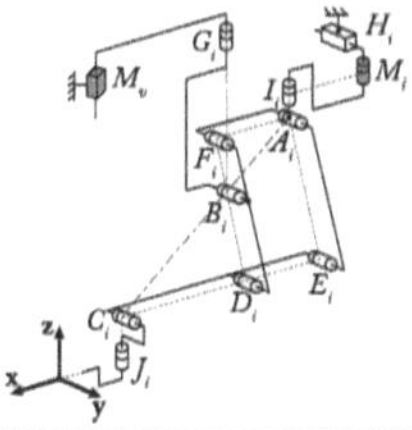	$3\text{-}R\underline{R}R$	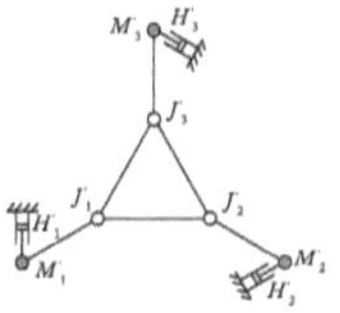	
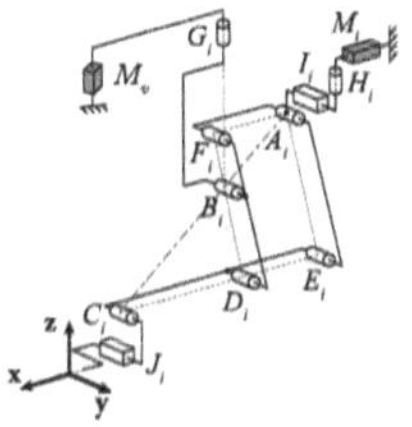	$3\text{-}R\underline{R}R$	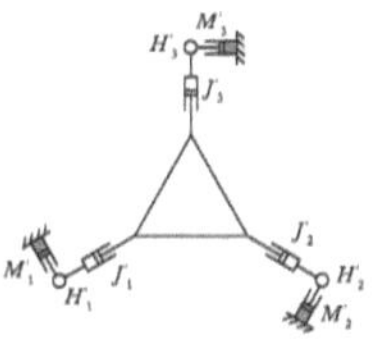	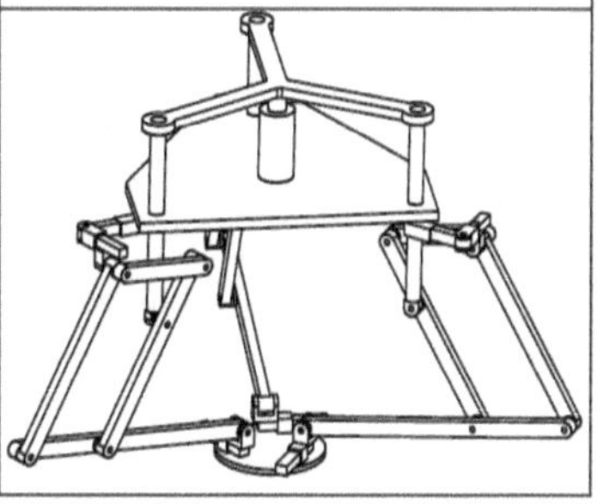

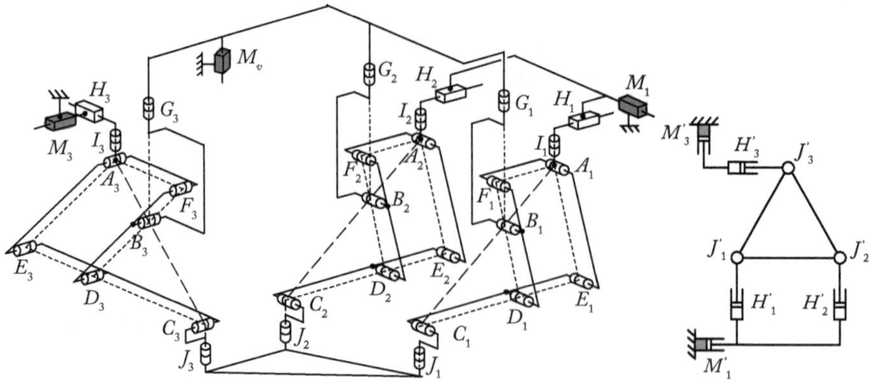

a Kinematic chain **b** Planar equivalent model

Fig. 6.20 Fully decoupled PAMINSA with three translational *dof*

that the inclinations around **x** and **y** axes are obtained by the vertical translations of points B_i.

Each kind of PAMINSA has its advantages and can be used differently. Let us consider the particularity of each architecture. As was mentioned above, the PAMINSA-*4D3L* allows improving the positioning accuracy along the vertical axis because the structure is kinematically locked during the displacement on the horizontal plane. Such a design allows the fixation of an important load in a given altitude, then positioning it on the horizontal plane.

PAMINSA-*4D2L* is able to perform the same task as the PAMINSA-*4D3L* with only two legs. It should be noted that in this case the motorization is quite different.

PAMINSA-*3D3L** can be used in any applications where only three translations along three axes are requested.

PAMINSA-*4D3L** is useful for any task with three translations and one orientation about the **x** or **y** axis.

The PAMINSA-*5D3L* allows the carrying out of all displacements in the horizontal plane with an inclination angle ψ of the platform (Fig. 6.21). The angle of the inclination ψ can be defined as an angle between the normal $\mathbf{n}_p$ to the platform and the normal $\mathbf{n}$ of the plane Oxy. Thus it is possible to move the platform on the horizontal plane with any inclination relative to the horizontal plane. In this case, the inclination is defined by the rotation of the point C_3 about the line C_1C_2.

The PAMINSA-*6D3L* allows for any orientation ϕ of the platform around the **z** axis and the displacements of the platform on the horizontal plane. Two other inclinations of the platform are allowed.

We would like to note that for all versions of presented PAMINSA manipulators, there is a decoupling between the displacements on the horizontal plane and the other displacements.

The kinematic analysis of the whole PAMINSA family is not detailed here but was deeply analyzed in (Briot et al. 2008a, 2009a).

Table 6.6 The family of PAMINSA manipulators from three to six *dof*

<table>
<tr><td colspan="3">PAMINSA-4D2L</td></tr>
<tr><td>dof: 3 Translations and 1 Rotation</td><td>3D view</td><td>Planar equivalent model</td></tr>
<tr><td>Type of connection between the platform and the legs: Universal (Cardan) joints</td><td rowspan="2">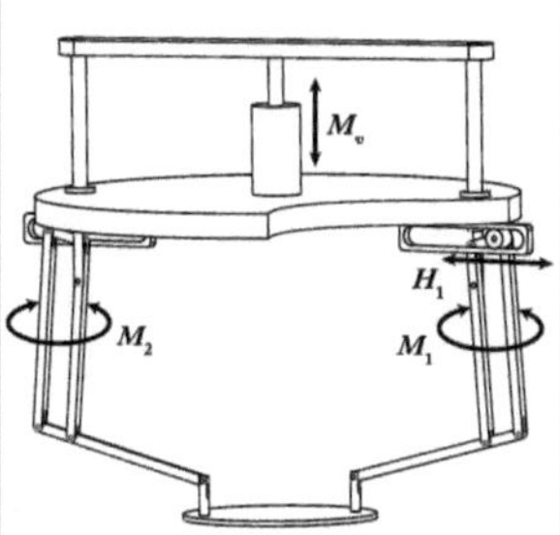
</td><td rowspan="2">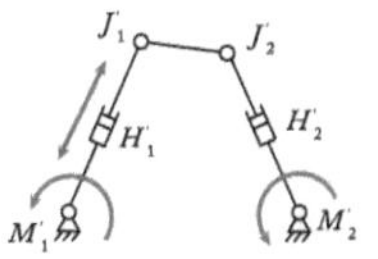
</td></tr>
<tr><td>Actuated joints: M_1, M_2, M_3 and M_v</td></tr>
<tr><td colspan="3">PAMINSA-3D3L[1]</td></tr>
<tr><td>dof: 3 Translations</td><td>3D view</td><td>Planar equivalent model</td></tr>
<tr><td>Type of connection between the platform and the legs: Universal (Cardan) joints</td><td rowspan="2">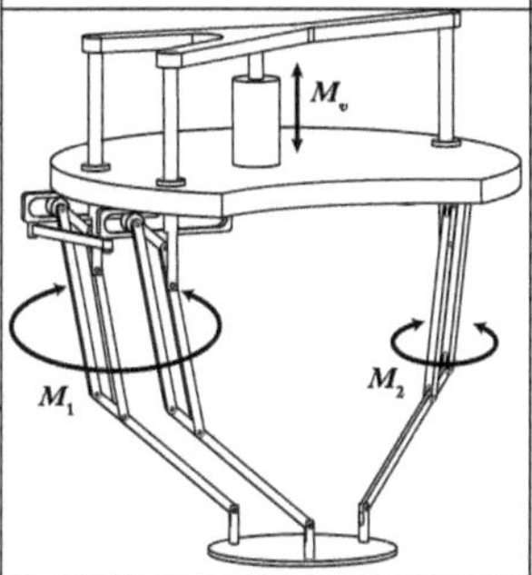
</td><td rowspan="2">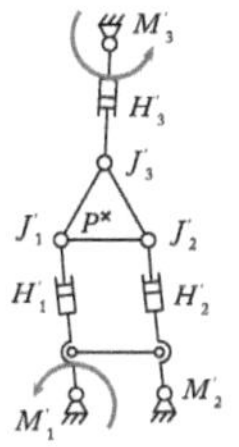
</td></tr>
<tr><td>Actuated joints: M_1, M_2 and M_v</td></tr>
<tr><td colspan="3">PAMINSA-4D3L[1]</td></tr>
<tr><td>dof: 3 Translations and 1 Rotation</td><td>3D view</td><td>Planar equivalent model[2]</td></tr>
<tr><td>Type of connection between the platform and the legs: Spherical pairs</td><td rowspan="2">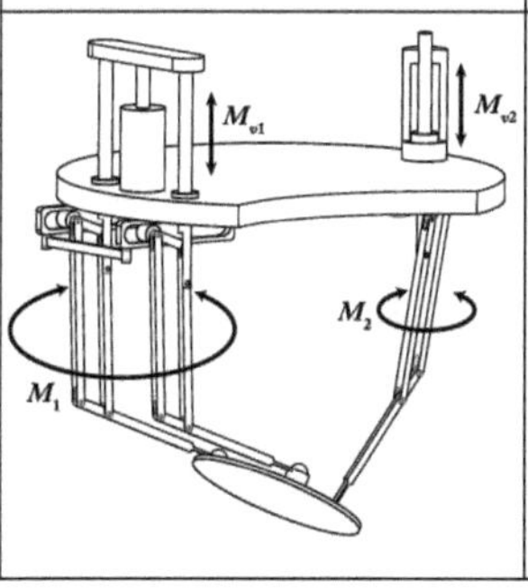
</td><td rowspan="2">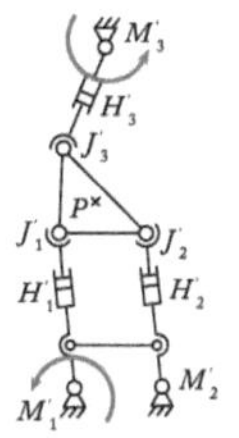
</td></tr>
<tr><td>Actuated joints: M_1, M_2, M_{v1} and M_{v2}</td></tr>
</table>

[1] Two of the three legs of such type of manipulator are actuated with the same motor and stay parallel to each other

[2] The platform shape of the equivalent model for the horizontal displacements changes with the variations of the robot platform orientation

Table 6.6 (Continued)

<table>
<tr><td colspan="3">PAMINSA-4D3L</td></tr>
<tr><td>dof: 3 Translations and 1 Rotation</td><td>3D view</td><td>Planar equivalent model</td></tr>
<tr><td>Type of connection between the platform and the legs: Universal (Cardan) joints</td><td rowspan="2">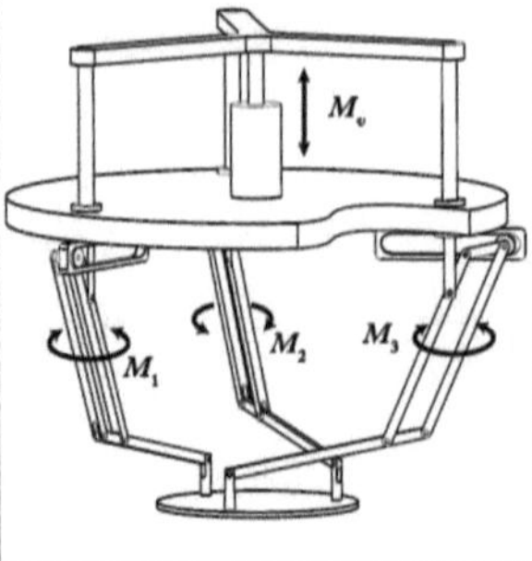
</td><td rowspan="2">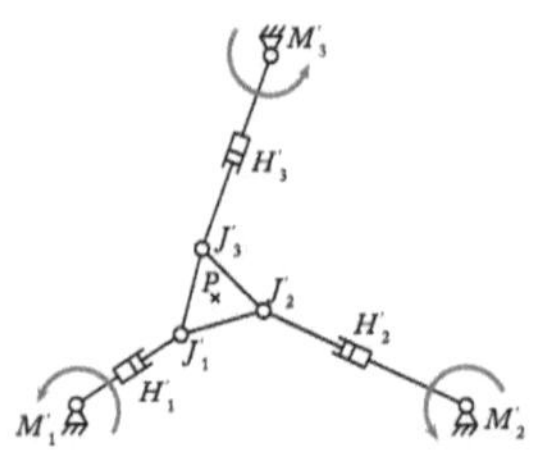
</td></tr>
<tr><td>Actuated joints: M_1, M_2, M_3 and M_v</td></tr>
<tr><td colspan="3">PAMINSA-5D3L</td></tr>
<tr><td>dof: 3 Translations and 2 Rotations</td><td>3D view</td><td>Planar equivalent model[3]</td></tr>
<tr><td>Type of connection between the platform and the legs: Spherical pairs</td><td rowspan="2">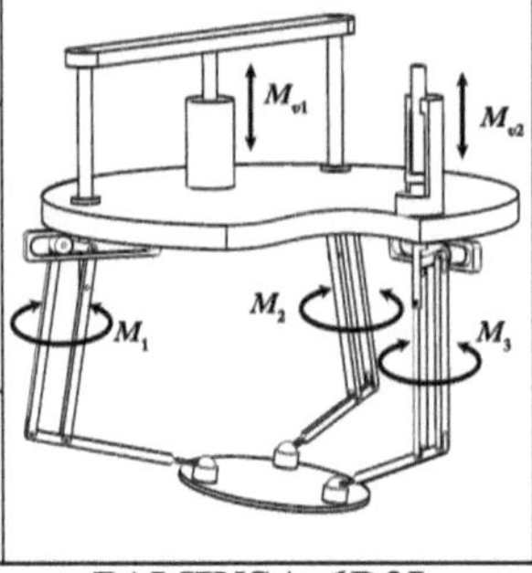
</td><td rowspan="2">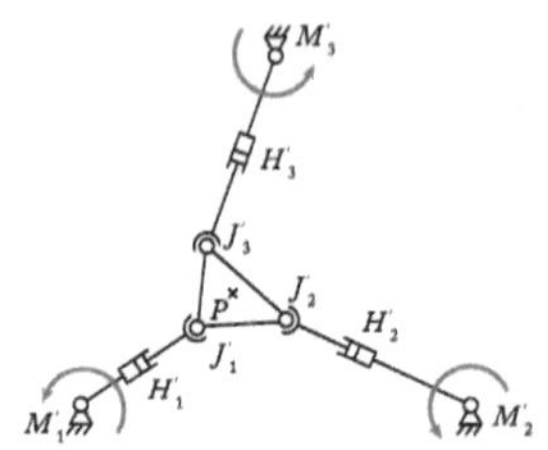
</td></tr>
<tr><td>Actuated joints: M_1, M_2, M_3, M_{v1} and M_{v2}</td></tr>
<tr><td colspan="3">PAMINSA-6D3L</td></tr>
<tr><td>dof: 3 Translations and 3 Rotations</td><td>3D view</td><td>Planar equivalent model[3]</td></tr>
<tr><td>Type of connection between the platform and the legs: Spherical pairs</td><td rowspan="2">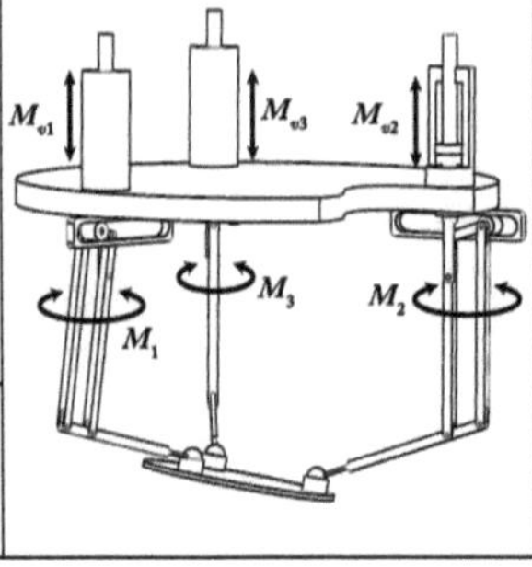
</td><td rowspan="2">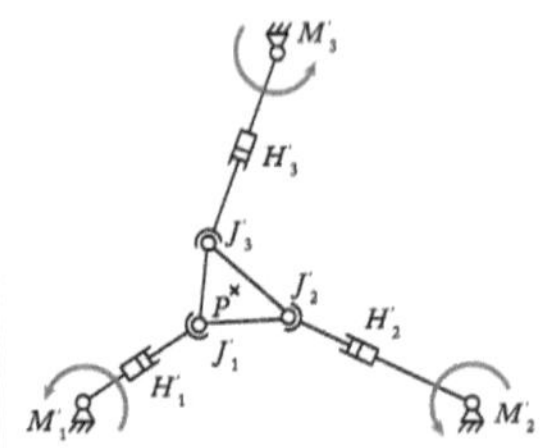
</td></tr>
<tr><td>Actuated joints: M_1, M_2, M_3, M_{v1}, M_{v2} and M_{v3}</td></tr>
</table>

[3] The platform shape of the equivalent model for the horizontal displacements changes with the variations of the robot platform orientation

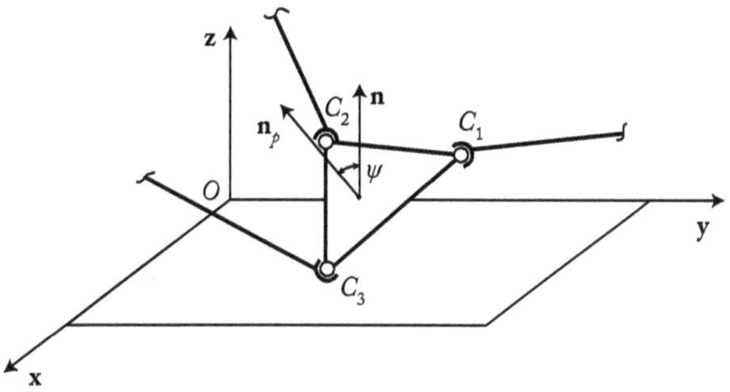

Fig. 6.21 The angle of inclination for the platfor of the PAMINSA-*5D3L*

6.2.2 *Static Analysis of the PAMINSA with 4-dof*

6.2.2.1 Proof of the Concept of Power Decoupling

The static modeling of the basic version PAMINSA-*4D3L* presented in Sects. 6.2.1.1 and 6.2.1.2 can be obtained using the following approach.

The static input torques (or forces) τ_i^{st} applied to the actuator M_i $(i = 1\ldots4)$ due to the force of gravity of links, joints and platform of the studied manipulator can be expressed as:

$$\tau_i^{st} = \tau_{p_i}^{st} + \sum_{j=1}^{3}\sum_{u=1}^{n}\tau_{jui}^{st} \tag{6.37}$$

where τ_{jui}^{st} is the load applied to the actuator i due to the gravity effects on the u-th link or bearings of the j-th leg $(j = 1, 2, 3)$, and $\tau_{p_i}^{st}$ is the load applied to the actuator i due to the gravity effects on the platform.

These loads can be expressed under the form:

$$\tau_{jui}^{st} = \mathbf{j}_{jui}^{T}\begin{bmatrix}0 & 0 & m_{ju}g\end{bmatrix}^{T} \tag{6.38}$$

$$\tau_{p_i}^{st} = m_p g\frac{\partial z}{\partial q_i} \tag{6.39}$$

where m_{ju} is the mass of the u-th link or bearing of the leg j $(j = 1, 2, 3$ and $u = 1\ldots n)$, m_p is the mass of the platform, g is the value of the gravity field, and $\mathbf{j}_{jui}$ is the Jacobian matrix between the *com* S_{ju} of the u-th link or bearing of the leg j and the actuated variables q_i (where $q_i = \theta_i$ for $i = 1, 2, 3$ and $q_4 = z_{B_i}$).

It is easy to see from (6.30) that, for $i = 1, 2, 3$:

$$\tau_{p_i}^{st} = 0 \tag{6.40}$$

i.e. the input torques due to the gravity of the actuators that allows the horizontal displacements of the platform are cancelled because the gravitational forces are always perpendicular to the displacements allowed by these actuators.

It is to be noted that matrices $\mathbf{j}_{jui}$ in (6.38) have the following form:

$$\mathbf{j}_{jui} = \partial \mathbf{r}_{S_{ju}} / \partial q_i \tag{6.41}$$

where, for the joints, S_{ju} are located at the centre of the articulations (Fig. 6.14a), and for the links, considering that they are rectilinear with homogeneous mass distribution,

$$\mathbf{r}_{S_{j7}} = \frac{\mathbf{r}_{G_j}}{2}, \ \mathbf{r}_{S_{j6}} = \frac{\mathbf{r}_{B_j} + \mathbf{r}_{G_j}}{2}, \ \mathbf{r}_{S_{j1}} = \frac{\mathbf{r}_{C_j} + \mathbf{r}_{E_j}}{2} \tag{6.42}$$

$$\mathbf{r}_{S_{j2}} = \frac{\mathbf{r}_{D_j} + \mathbf{r}_{F_j}}{2}, \ \mathbf{r}_{S_{j3}} = \frac{\mathbf{r}_{A_j} + \mathbf{r}_{E_j}}{2}, \ \mathbf{r}_{S_{j4}} = \frac{\mathbf{r}_{A_j} + \mathbf{r}_{F_j}}{2} \tag{6.43}$$

$$\mathbf{r}_{S_{j8}} = \mathbf{r}_{A_j}, \ \mathbf{r}_{S_{j5}} = l_{OS_5} \begin{bmatrix} \cos\theta_j \\ \sin\theta_j \\ 0 \end{bmatrix} \tag{6.44}$$

with

$$\mathbf{r}_{C_j} = \mathbf{r}_P + \begin{bmatrix} l_{C_iP} \cos(\phi + \gamma_i) \\ l_{C_iP} \sin(\phi + \gamma_i) \\ 0 \end{bmatrix}, \ \mathbf{r}_{B_j} = \begin{bmatrix} l_{OA_i} \cos\gamma_i \\ l_{OA_i} \sin\gamma_i \\ z_{C_j}/k \end{bmatrix} \tag{6.45}$$

$$\mathbf{r}_{A_j} = \mathbf{r}_{B_j} + \frac{\mathbf{r}_{C_j} - \mathbf{r}_{B_j}}{(1-k)}, \ \mathbf{r}_{D_j} = \mathbf{r}_{E_j} + \frac{\mathbf{r}_{C_j} - \mathbf{r}_{E_j}}{k} \tag{6.46}$$

$$\mathbf{r}_{F_j} = \mathbf{r}_{A_j} + \frac{\mathbf{r}_{C_j} - \mathbf{r}_{E_j}}{k}, \ \mathbf{r}_{G_j} = \mathbf{r}_{B_j} + \begin{bmatrix} 0 \\ 0 \\ l_{B_iG_i} \end{bmatrix} \tag{6.47}$$

$$\mathbf{r}_{E_j} = \mathbf{r}_{B_j} + \begin{bmatrix} a_{1j} \cos\theta_j \\ a_{1j} \sin\theta_j \\ a_{2j} \end{bmatrix} \tag{6.48}$$

in which

$$\begin{aligned}
a_{1j} &= a_{3j} + a_{4j}a_{2j} \\
a_{2j} &= \left(a_{5j} - a_{6j}\right) / \left(2a_{7j}\right) \\
a_{5j} &= \sqrt{a_{6j} - 4a_{7j}a_{8j}} \\
a_{7j} &= -a_{4j}^2 - 1 \\
a_{6j} &= 2a_{4j}\left(a_{9j} - a_{3j}\right) \\
a_{8j} &= l_{A_iE_i}^2 - a_{9j}^2 + 2a_{3j}a_{9j} - a_{3j}^2 \\
a_{4j} &= z_{C_j} / \left(k\, a_{9j}\right) \\
a_{3j} &= \left(l_{C_iE_i}^2 - l_{A_iE_i}^2 - a_{10j}^2 + a_{9j}^2 - z_{C_j}\right) / \left(2k\, a_{9j}\right) \\
a_{9j} &= -a_{10j} / (k-1) \\
a_{10j} &= \sqrt{\left(x_{C_j} - x_{B_j}\right)^2 + \left(y_{C_j} - y_{B_j}\right)^2}
\end{aligned} \tag{6.49}$$

Thus, developping and simplifying (6.37), it comes that

$$\tau_i^{st} = \left(m_p + \sum_{j=1}^{3} m_{eq1_j}\right) g \frac{\partial z}{\partial q_i} + \sum_{j=1}^{n} m_{eq2_j} g \frac{\partial z_{E_j}}{\partial q_i} \tag{6.50}$$

where

$$m_{eq1_j} = m_{C_j} + \frac{m_{B_j} + m_{D_j} + m_{G_j} + m_{F_j}}{k} + \frac{k\, m_{j3} + 2m_{j1} + m_{j6} + 2m_{j2} + 2m_{j4}}{2k} \tag{6.51}$$

$$m_{eq2_j} = \frac{(k-1)\, m_{D_j} - m_{F_j} + k\, m_{E_j}}{k} + \frac{(k-2)\, m_{j4} - m_{j6} + k\, m_{j3} + k\, m_{j5}}{2k} \tag{6.52}$$

in which m_{B_j} (m_{C_j}, m_{D_j}, m_{E_j}, m_{F_j}, m_{G_j}, resp.) is the mass of the joint located at point B_j (C_j, D_j, E_j, F_j, G_j, resp.).

For the actuators 1–3 that allow the horizontal displacements of the platform, the term $\left(m_p + \sum_{j=1}^{3} m_{eq1_j}\right) g \frac{\partial z}{\partial q_i}$ vanishes, thus their torque is equal to:

$$\tau_i^{st} = \sum_{j=1}^{3} m_{eq2_j} g \frac{\partial z_{E_j}}{\partial q_i} \text{ for } i = 1, 2, 3 \tag{6.53}$$

Finally, it can be demonstrated that the expression (6.53) can be cancelled by the addition of one counterweight by leg, located for example at point F_j with a mass equal to:

$$m_{cw_j} = k\, m_{eq2_j} \tag{6.54}$$

6.2.2.2 Numerical Application

For the numerical application, the following parameters extracted from CAD data of the prototype presented in the following section are used:

- $k = 3$, $l_{OA_i} = l_{OA_i} = 0.35$ m, $l_{C_iP} = 0.1$ m, $g = -9.81$ m/s^2,
- $l_{A_iE_i} = l_{D_iF_i} = k\,l_{B_iF_i} = 0.42$ m, $l_{C_iE_i} = k\,l_{C_iD_i} = k\,l_{A_iF_i} = 0.63$ m, $l_{B_iG_i} =$ 0.442 m, $l_{OS_8} = 0.1$ m,
- $m_{j1} = 1.221$ kg, $m_{j2} = 0.921$ kg, $m_{j3} = 0.672$ kg, $m_{j4} = 0.406$ kg, $m_{j5} =$ 0.403 kg, $m_{j6} = 0.107$ kg, $m_p = 2.301$ kg,
- $m_{B_j} = 0.338$ kg, $m_{C_j} = 0.233$ kg, $m_{D_j} = 0.262$ kg, $m_{E_j} = 0.259$ kg, $m_{F_j} =$ 0.280 kg, $m_{G_j} = 0.214$ kg.

The input torque of actuator 1 before and after mass balancing is plotted in Fig. 6.22 for these parameters and for a platform altitude and orientation equal to $z = -0.6$ m and $\phi = 0°$. It should be noted that, as the studied PAMINSA is symmetrical, the values of the input torques for the actuators 2 and 3 are also symmetrical but they are situated in different zones (rotations of $\pm 120°$).

After complete static balancing, the potential energy of the manipulator is constant for any configuration and zero actuator torques are required. For example, the added counterweights are located at points F_j and their masses are equal to 2.8 kg (to observe the increase in masses after balancing, it should be noted that the mass of each pantograph linkage before balancing was 3.1 kg).

It is obvious that such a balancing is very useful for static mode of operation of the manipulator. However with the increase of the accelerations of moving links, the inertia forces become important and the complete static balancing in dynamic operation can not be optimal. In this context another problem may be formulated: to find such a distribution of movable masses, which allows the minimization of the input torques of the rotating actuators in dynamic mode of operation. This problem was developed in (Briot et al. 2008a, 2009a) but is not detailed here.

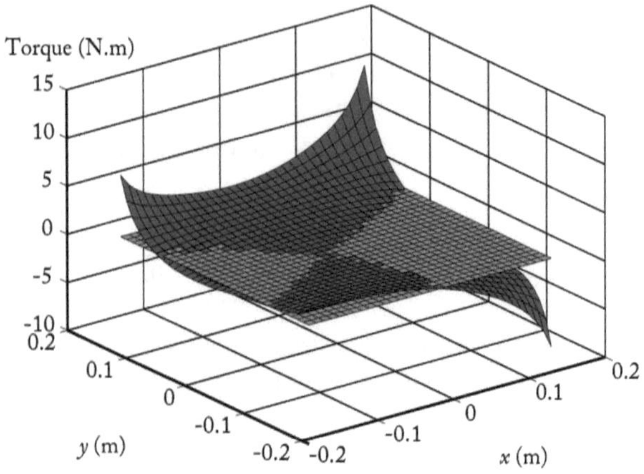

Fig. 6.22 Variations of the actuator torques for $z = -0.6$ m and $\phi = 0°$ before (*dark gray*) and after (*bright gray*) static balancing of legs

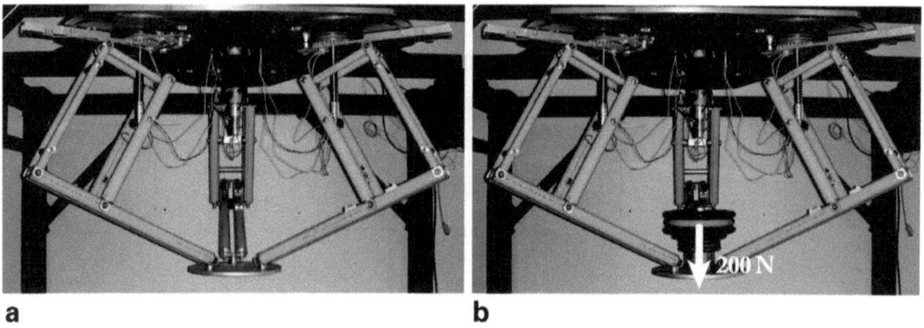

Fig. 6.23 The prototype of PAMINSA developed in the INSA of Rennes. (**a**) without load. (**b**) with a load of 200 N

6.2.3 Prototype and Experimental Validations

6.2.3.1 Description of the Experimental Bench

In order to validate the properties of PAMINSA manipulators, we have developed a prototype of PAMINSA-*4D3L* whose equivalent planar model is a 3-*RPR* manipulator (Fig. 6.23) at the INSA in Rennes. The displacements on the horizontal plane of the developed prototype are obtained by Harmonic Drive motors connected with the legs by means of toothed-belt transmission.

For controlling the linear displacement of the vertical axes of the manipulator's legs, a PARVEX motor system was chosen. The pantograph linkage has been carried out with double rods in order to increase the stiffness of the mechanical system. The pantograph links of the built prototype are hollow tubes with a thickness of 1.5 mm. It is obvious that, for an industrial application, creating a pantograph with single and more rigid links is more attractive.

The experimental bench is detailed in Fig. 6.24. It is composed of the PAMINSA robot with its control system, a computer to interact with the user and a dSPACE 1103 card. To control the system, the softwares Matlab/Simulink/RTI/ControlDesk have been used. The sampling period was 1 ms. For the motions in the horizontal plane, the PAMINSA is composed of three identical Harmonic Drive DC motors (3557012CR) in combination with Fulhaber incremental encoders (HEDS5500) and Fulhaber motion controllers (MCDC2805). A Parvex DC motor, (RX320DR1200) including an encoder and a brake system is used for the translation along the vertical axis, in combination with an e-motion servo-amplifier (SCA-SS-70-10). The dSPACE 1103 carries out the interface between the PC and the servosystems. This card allows the control of the four motorized axes and has specific entries to which we have directly connected the four incremental encoders. We have also connected three LEM (HX10NP) current sensors to observe the torque of motor 1, 2 and 3.

The robot is controlled by the classical PI law directly implemented under Simulink/ControlDesk. In order to make it, each motion controller has been configured to be only used as power amplifiers. Thus, the motor voltage is proportional

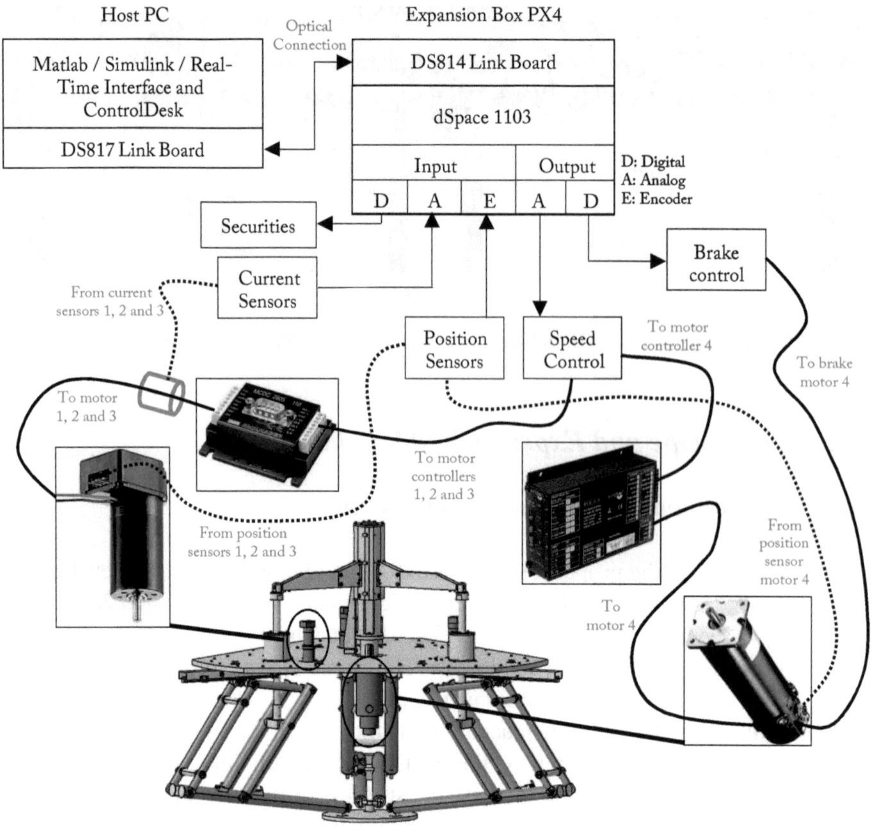

Fig. 6.24 Experimental bench

to the voltage at the analog input of each motion controller. Optics Sensors have been added for the initialization of the robot and for the setting up safety measures.

6.2.3.2 Validation of the Design Concept

In order to validate the numerical simulations with experimental tests, we have measured the input torques/efforts of the actuators with and without a payload of 200 N (Fig. 6.23) for the trajectory given in Fig. 6.25.

The obtained results are presented in Fig. 6.26. The numerical simulations were validated via these experimental tests. The curves with and without payload for the three rotating actuators (Fig. 6.26a–c) are superposed. We can see that they are similar, i.e. the loads on these actuators are cancelled. The small differences might result from friction in the joints, manufacturing errors, elasticity of the links or tracking errors.

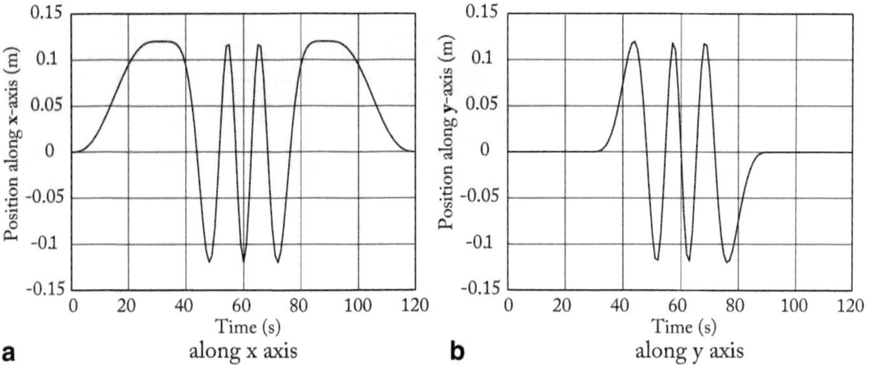

Fig. 6.25 Position of the platform for $z = -0.6$ m and $\phi = 0°$

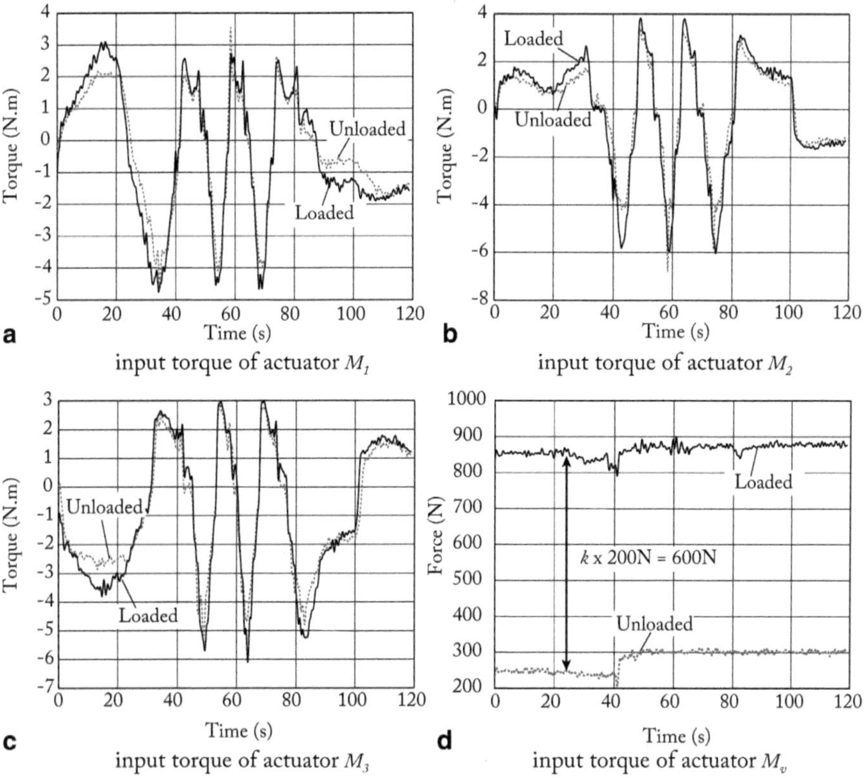

Fig. 6.26 Input torques/effort on the actuators with and without payload of 200 N

Regarding the vertical actuator (Fig. 6.26d), it supports the payload and the increase of the input force is significant.

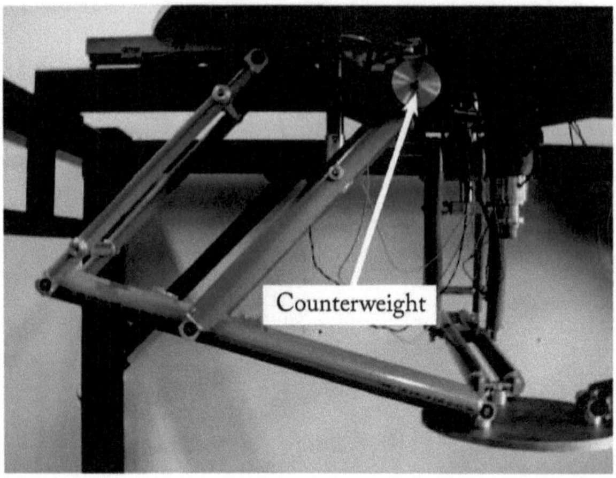

Fig. 6.27 Counterweights mounted into the pantograph linkages

Thus, it can be noted that the obtained measures validate the proposed design approach.

6.2.3.3 Validation of the Leg Static Balancing

The static balancing of the manipulator is experimentally accomplished by adding counterweights of 2.8 kg at the axis F_j of the pantograph linkages (Fig. 6.27).

In order to prove the minimization of input torques before and after balancing, some arbitrary configurations of the manipulator were examined. The tested poses are given in Table 6.7.

For these seven positions of the platform, the maximal absolute values of the input torques of the three rotating actuators before and after complete static balancing are measured (Table 6.8). The reduction of the maximal input torques varies from 74 to 85 %. The input torques are not totally balanced because of friction torques and inaccuracy in the modelling of the robot, but they are drastically decreased.

6.3 Design and Balancing of Hand-operated Manipulators

Hand-operated manipulators are designed such that they can help workers to carry heavy loads. Their principle consist to allows the horizontal displacement of the load without exerting any considerable effort from the operator. A linear actuator allows the carrying of the load and its vertical displacement.

Only few works deal with the problem of the balancing of hand-operated mechanisms. Among them, one can notice (Belyanin 1988b; Popov 1980, 1986) in which, by the application of the principle of virtual displacements, it is demonstrated that for the full balancing of the Scheiner pantograph, it is necessary to apply to the element

Table 6.7 The poses for the experimental validation of the static balancing

Pose	x (m)	y (m)	z (m)	ϕ (deg.)
1	0.124	0.096	−0.6	34.72
2	0.015	0.047	−0.615	−20.23
3	−0.149	0.009	−0.733	4.53
4	0.072	0.129	−0.497	9.23
5	−0.053	0.09	−0.540	33.92
6	−0.134	−0.075	−0.389	−3.5
7	−0.173	−0.042	−0.687	15.64

Table 6.8 The absolute value of the maximal input torques for actuators 1 to 3 before (case 1) and after (case 2) static balancing

Pose	Case 1 (N.m)	Case 2 (N.m)	Reduction (%)
1	1.78	0.46	74
2	1.81	0.26	86
3	1.38	0.34	76
4	3.31	0.47	86
5	3.23	0.59	82
6	1.93	0.35	82
7	2.4	0.55	77

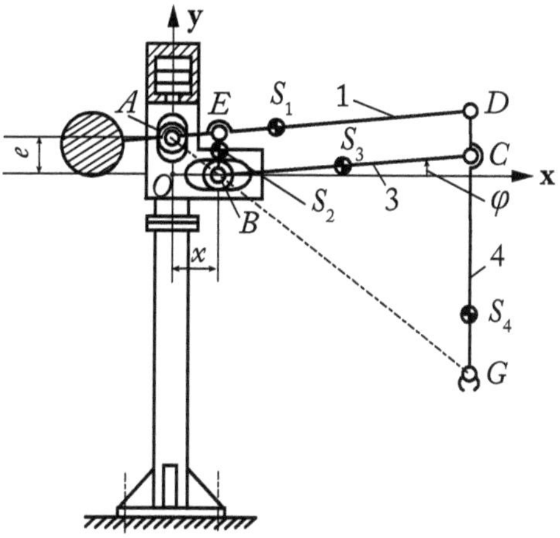

Fig. 6.28 Hand-operated manipulator using a pantograph mechanism

AD (Fig. 6.28) a torque $M_{eq} = M \sin\varphi$, where M is a constant torque depending of the pantograph links mass and dimension.

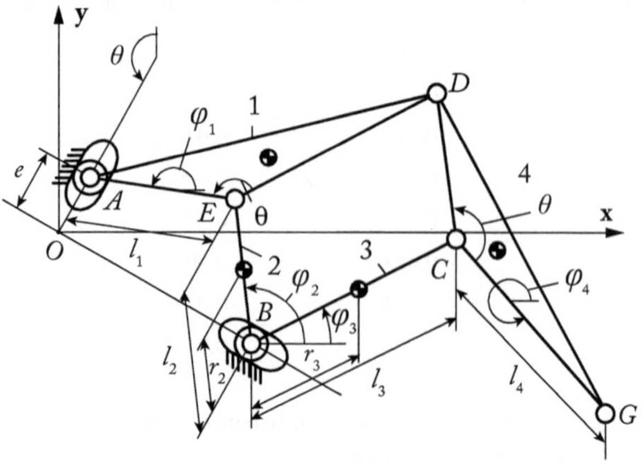

Fig. 6.29 A general pantograph mechanism

In this section, a method based on the study of the *com* displacement for the balancing of hand-operated manipulators designed via the use of pantographs is proposed.

6.3.1 Methodology

The motion of the global *com* of the moving links of the pantograph mechanism from Fig. 6.29 along **y** axis can be written as:

$$y_S = \frac{m_{1A}y_A + m_{1E}y_E + (m_{1D} + m_{4D})\,y_D + m_2y_2 + m_3y_3 + m_{4C}y_C + m_{4G}y_G}{\sum_{i=1}^{4} m_i} \tag{6.55}$$

where

$$\begin{aligned}
y_2 &= l_1 \sin\varphi_1 + (l_2 - r_2)\sin\varphi_2 + e\cos\theta \\
y_3 &= l_1 \sin\varphi_1 + l_2\sin\varphi_2 - r_3\sin\varphi_3 + e\cos\theta \\
y_A &= e\cos\theta \\
y_C &= l_1 \sin\varphi_1 + l_2\sin\varphi_2 - l_3\sin\varphi_3 + e\cos\theta \\
y_D &= l_1 \sin\varphi_1 - l_3\sin\varphi_3 + e\cos\theta \\
y_E &= l_1 \sin\varphi_1 + e\cos\theta \\
y_G &= l_1 \sin\varphi_1 + l_2\sin\varphi_2 - l_3\sin\varphi_3 - l_4\sin\varphi_4 + e\cos\theta
\end{aligned} \tag{6.56}$$

Thus, (6.55) can be rewritten under the form:

$$y_S = \frac{\sum_{i=1}^{4} \alpha_i \sin\varphi_i + \sum_{i=1}^{4} m_i e\cos\theta}{\sum_{i=1}^{4} m_i} \tag{6.57}$$

where

$$\begin{aligned}\alpha_1 &= (m_{1E} + m_{1D} + m_2 + m_3 + m_4)\, l_1 \\ \alpha_2 &= m_2 (l_2 - r_2) + (m_3 + m_{4C} + m_{4G})\, l_2 \\ \alpha_3 &= -m_3 r_3 - (m_4 + m_{1D})\, l_3 \\ \alpha_4 &= -m_G l_4\end{aligned} \tag{6.58}$$

In order to the operator be able to move the load horizontally, acting against friction forces only, it is necessary that the total work of the gravity forces applied on the moving links vanish when the manipulator is moving horizontally. Thus, the *com* of the pantograph mechanism must only move horizontally (or must be fixed), i.e. $\alpha_1 = \alpha_2 = \alpha_3 = \alpha_4 = 0$.

The Scheiner pantograph is generally used for hand-operated manipulator when $\theta = \pi$. For this mechanism, (6.57) becomes:

$$y_S = \frac{\alpha \sin\varphi + \left(\sum_{i=1}^{4} m_i - \beta\right) e \cos\theta}{\sum_{i=1}^{4} m_i} \tag{6.59}$$

where

$$\alpha = m_1 r_1 + m_{2E} l_1 + m_3 r_3 + m_{4D}\,(l_1 + l_3) \tag{6.60}$$

and

$$\beta = m_2 (l_2 - r_2) / l_2 + m_4 r_4 / l_2 + m_3 \tag{6.61}$$

In such a case, in order to balance the manipulator, it is necessary that $\alpha = 0$, which can be obtained through the addition of one counterweight.

Another request for hand-operated manipulators is to make the load carried by the actuator invariant. This load must be independent of the position of the payload within the workspace. In this case, the condition $\left(\sum_{i=1}^{4} m_i - \beta\right) = 0$ must be achieved.

When these two conditions are obtained, the *com* of the manipulator coincide with point B (Fig. 6.28). When the payload moves vertically, the actuator input effort is proportional to the payload. The coefficient of proportionality is equal to the magnification factor of the pantograph.

6.3.2 Application

Table 6.9 describes the kinematic architectures of the most currently used hand-operated manipulators. It also gives the formulas for calculating their balancing. The analysis of the actuation forces for the different balancing methods shows that balancing is optimal when the *com* coincides with the centre of the joint A, which

Table 6.9 Kinematic schemes of the most known hand-operated manipulators and formulas for their balancing

Kinematic scheme	Notations	Formulas for balancing
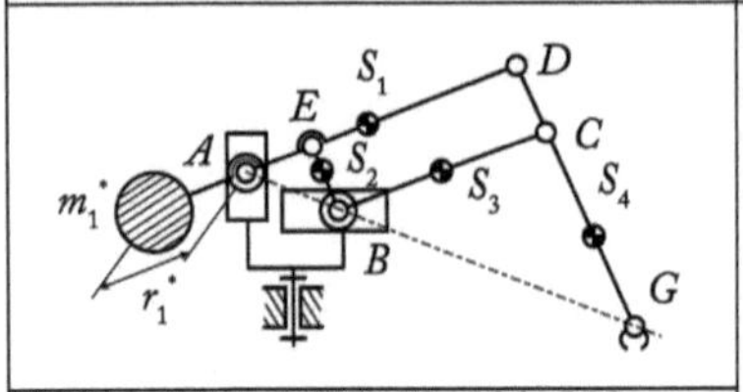 KHC-160M1; MBR-63; 120; 250	$l_{BC}=l_{ED}=a$; $l_{BE}=l_{DC}=b_1$; $l_{AE}=a_1$; $l_{CG}=b$; $l_{AS_1}=r_1$; $l_{ES_2}=r_2$; $l_{BS_3}=r_3$; $l_{DS_4}=r_4$	$m_1^*=m_2(1-r_2/b)$ $+m_4(1-r_4/b_1)$ $r_1^*=1/m_1^*(m_2a_1+$ $m_3r_3+m_4(a+a_1)$ $-m_2r_2a_1/b_1$ $-m_4r_4a_1b_1)$
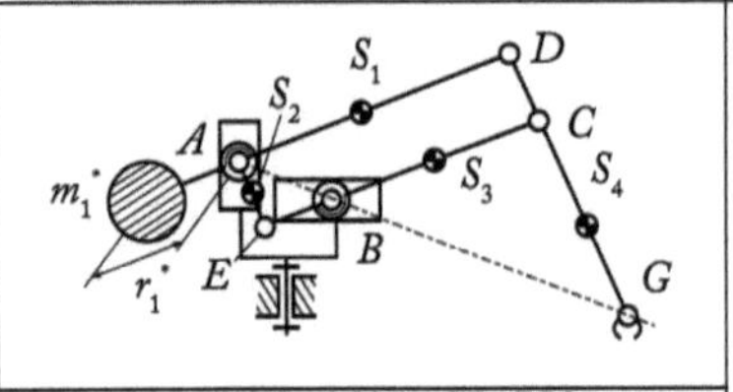 HBL-75A; HBL-150A; SL-750	$l_{AE}=l_{DC}=b_1$; $l_{EB}=a_1$; $l_{BC}=a$; $l_{CG}=b$; $l_{AD}=a+a_1$; $l_{AS_1}=r_2$; $l_{BS_3}=r_3$; $l_{DS_4}=r_4$	$m_1^*=m_2(1-r_2/b_1)$ $+m_4(1-r_4/b_1)$ $r_1^*=1/m_1^*(m_3r_3+$ $m_4(a+a_1)-m_2r_2a_1/b_1$ $-m_4r_4a_1/b_1$
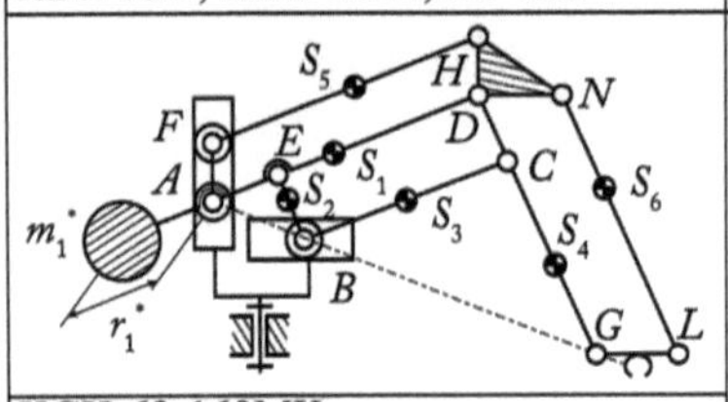 KCH-63-160MK	$l_{BC}=l_{ED}=a$; $l_{BE}=l_{DC}=b_1$; $l_{AE}=a_1$; $l_{CG}=b$; $l_{FH}=a+a_1$; $l_{NL}=b+b_1$ $l_{AS_1}=r_1$; $l_{ES_2}=r_2$; $l_{BS_3}=r_3$; $l_{DS_4}=r_4$; $l_{FS_5}=r_5$; $l_{NS_6}=r_6$	$m_1^*=m_2(1-r_2/b)$ $+m_4(1-r_4/b_1)$ $+m_6(1-r_6/b_1)$ $r_1^*=1/m_1^*(m_2a_1+$ $m_3r_3+m_4(a+a_1)$ $-(m_2r_2+m_4r_4)a_1/b_1$ $m_5r_5-m_6r_6a_1b_1)$

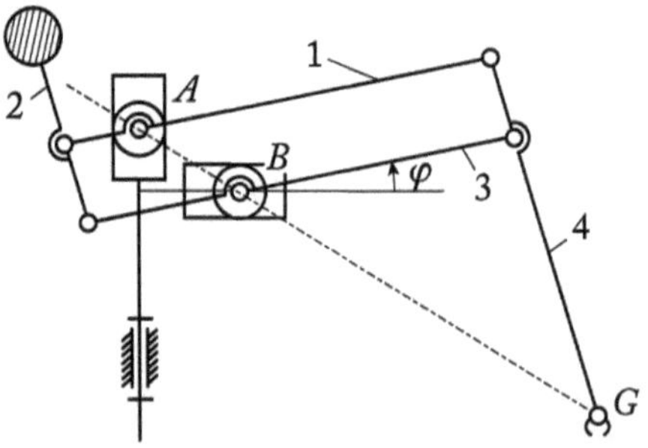

Fig. 6.30 Kinematic scheme of a pantograph mechanism ensuring an optimal balancing of the hand-operated manipulator via the use of only one counterweight

is linked with the linear actuator. In such case, the reaction wrench in the joint B is minimal which leads to the reduction of the effort applied by the operator. However, the design of a mechanism with two counterweights is not interesting. This the reason why it is more convenient to use another pantograph mechanism such as the one presented in Fig. 6.30. Such an architecture allows the coincidence of the global *com* with point A, with the use of only one counterweight.

Figure 6.31 shows a diagram of the hand-operated manipulator without motor system designed to assist a machine-tool. It must ensure the production of a batch

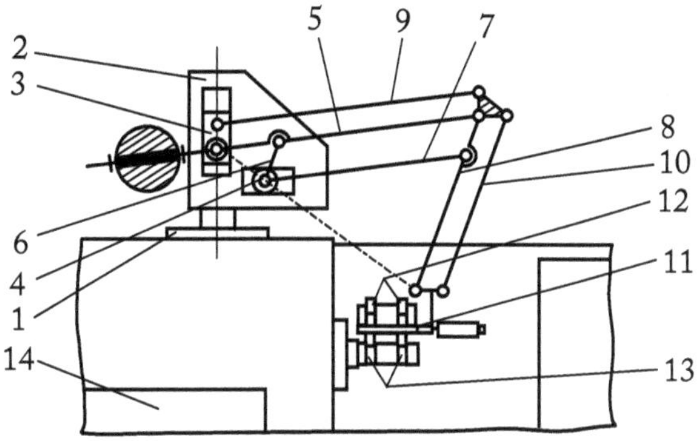

Fig. 6.31 Kinematic scheme of a pantograph mechanism ensuring an optimal balancing of the hand-operated manipulator via the use of only one counterweight

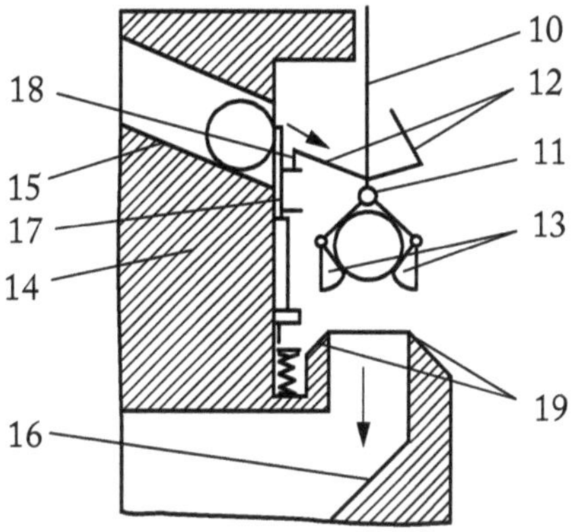

Fig. 6.32 Device for the loading and unloading of the manipulator

of workpieces. After completing the machining of the machine-tool, the operator makes the clamping device down (the gripper of the preformed workpiece 12 with which it is loaded). Using the gripper 13, the operator retrieves the machined part of the mandrel. During this operation, and as a result of the joint action of the weight of the preformed workpiece and of the machined workpiece, the arm imbalance and falls to the position set by the stop (mounted in the groove of the vertical movement and not shown in Fig. 6.31). Here, the clamping axis of the preformed workpiece gets in coincidence with the axis of the mandrel. In this position, the operator mounts the piece outlined in the mandrel, after which, the balancing of the manipulator is restored. It manually brings the actuator 14 to the device for the loading and unloading of the manipulator (Fig. 6.32). To approach, the actuator of the device 14 must rotate the device 14 from 90° and lower it. With the lowering of the gripping, the element 18 moves the interceptor 17. By an inclined plane, the preformed workpiece is brought to the gripper 12. As a result of the joint action of the weight of the preformed workpiece and the machined workpiece, the actuator drops out of balance. The gripper itself goes down to touch the jaws, thus abandoning the workpiece bit higher above the tank 16 for receiving the parts and thus restoring the balancing of the manipulator.

The operator then manually goes back the gripper and features it on the mandrel. After completing the machining process, the cycle is repeated.

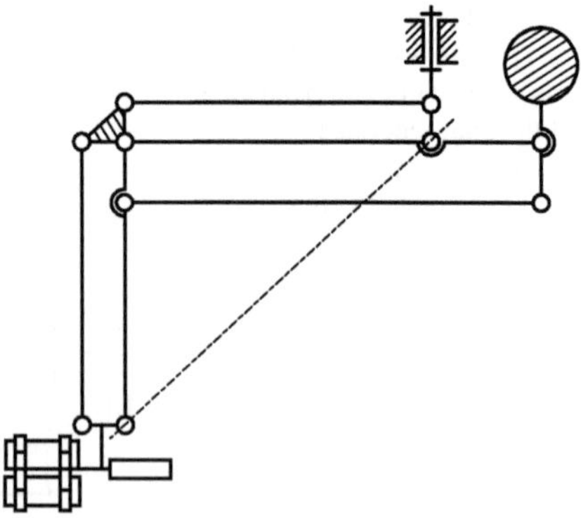

Fig. 6.33 Hand-operated manipulator with actuation system applying only torques

The exclusion of the manipulator motor system is made possible through this balancing method by which the parameters of the actuator of the manipulator are determined from the conditions for balancing the weight of the elements of the actuator with the weight of the manipulated device:

$$m_5r_5 + m_6r_5 + m_8l_9 + m_7r_7 + m_9r_9 + ml_9 - m_6r_6\frac{l_9 - l_7}{l_6} \\ -m_8r_8\frac{l_9 - l_7}{l_6} - m_{10}r_{10}\frac{l_9 - l_7}{l_6} - ml_9\frac{l_9 - l_7}{l_6} = 0 \tag{6.62}$$

$$\frac{m_6r_6 + m_8r_8 + m_{10}r_{10} + ml_8}{l_6} = \sum_{i=5}^{13} m_i + m - m_7 \tag{6.63}$$

where $m = 0.5(m_p + m_{pr})$ is the mass of the manipulated device which is equal to the half of the sum of the masses from the machined workpiece (m_p) and the preformed workpiece (m_{pr}); m_i, l_i and r_i are the mass, the length and the distance from the *com* of the i-th element to the "left" joint ($i = 5, 7, 9$) or to the "upper" joint ($i = 6, 8, 10$) (Fig. 6.31).

These relations describe the conditions for the self balancing of the system "arm-manipulated device" that, if carried out, make it possible to manually displace the manipulated device in the horizontal place as well as in the vertical plane.

Thus, by overcoming a small non-compensated force which is equal to 5–8 % of the weight of the manipulated device, the operator will have the possibility to load and unload the machine-tool.

A practical application shows that moving the hand-operated manipulator is more easy when the actuator can apply torques instead of forces. In such a case, the friction effects are the lowest and, consequently, the manual effector exerted by the operator is lower. A kinematic scheme of such a manipulator is shown in Fig. 6.33.

Chapter 7
Shaking Force and Shaking Moment Balancing of Robot Manipulators

Abstract This Chapter is devoted to the shaking force and shaking moment balancing of robot manipulators.

The Sect. 7.1 discusses the development of reactionless 3-*RRR* planar parallel manipulators, which apply no reaction forces or moments to the mounting base during motion. Design equations and techniques are proposed which allow for the dynamic substitution of the mass of the moving platform of a parallel manipulator by three concentrated masses. The dynamic model of the moving platform consequently represents a weightless link with three concentrated masses. This allows for the transformation of the problem of the design of a reactionless manipulator into a problem of balancing pivoted legs carrying concentrated masses. The total angular momentum of the manipulator is reduced to zero using two approaches: (i) on the basis of counter-rotations and (ii) using an inertia flywheel rotating with a prescribed angular velocity.

Section 7.2 deals with the complete shaking force and shaking moment balancing of planar parallel manipulators with prismatic pairs. The cancellation of the dynamic loads transmitted to the ground is a challenge for these types of manipulators. It is obvious that the classical methods based on the optimal redistribution of movable masses and additional counter-rotations can be used to cancel shaking force and shaking moment. However, such a balancing of parallel manipulators with prismatic pairs is only attained via a considerably complicated design. This study shows that it is possible to balance planar parallel mechanisms using Scott-Russell mechanisms. Such an approach enables a division of the number of counter-rotations by two.

Then we consider in Sect. 7.3 a simple and effective balancing method, which allows the considerable reduction of the shaking force of non-redundant manipulators without adding counterweights. It is based on the optimal control of the acceleration of the total *com* of moving links. The full shaking force and shaking moment balancing of robots using an optimal motion control is also used in last Section (Sect. 7.4) of this Chapter. This strategy takes advantage of the particularities of position/orientation decoupled robots for which the motion of the wrist is decoupled from the motion of the position finder device. Such an approach is illustrated via two applicative examples: the SCARA serial robot and the position/orientation decoupled PAMINSA parallel manipulator.

V. Arakelian, S. Briot, *Balancing of Linkages and Robot Manipulators*,
Mechanisms and Machine Science 27, DOI 10.1007/978-3-319-12490-2_7

7.1 Complete Shaking Force and Shaking Moment Balancing of 3-*dof* 3-*RRR* Parallel Manipulators

7.1.1 3-dof 3-RRR Planar Parallel Manipulator and Dynamic Model with Concentrated Masses

The moving platform of a planar 3-*dof* 3-*RRR* parallel manipulator is connected to its legs by three revolute joints located at points P_i ($i = 1, 2, 3$—Fig. 7.1). Each leg comprises two links connected by revolute joints located at A_i and they are mounted on the frame by revolute joints located at O_i. The input parameters of such a manipulator are defined by the joint angles θ_i of each leg and the output parameters by the pose of the moving platform, i.e. its orientation ϕ and position of one point of the moving platform, by example, the *com* of the moving platform with coordinates $[x, y]^T$.

Note that all axes of revolute joints are parallel, i.e. this is a mechanism in which all points of the links describe paths located in parallel planes.

The conditions for dynamic substitution of the mass of the platform (Fig. 7.2) by three concentrated masses situated on the axis of joints P_i are the followings (Seyferth 1974; Wu and Gosselin 2007):

$$\sum_{i=1}^{3} m_i = m_p, \ \sum_{i=1}^{3} m_i x_i = \sum_{i=1}^{3} m_i y_i = 0, \ \sum_{i=1}^{3} m_i \left(x_i^2 + y_i^2\right) = I_p \tag{7.1}$$

where m_i are point masses located on the joint axis, x_i and y_i are coordinates of the point masses with respect to the platform frame, m_p is the mass of the moving platform and I_p is the axial moment of inertia of the moving platform with respect to the *com*.

Thus, if we have a platform with such a redistribution of masses, when the conditions (7.1) are fulfilled, the mass of the platform can be dynamically substituted by

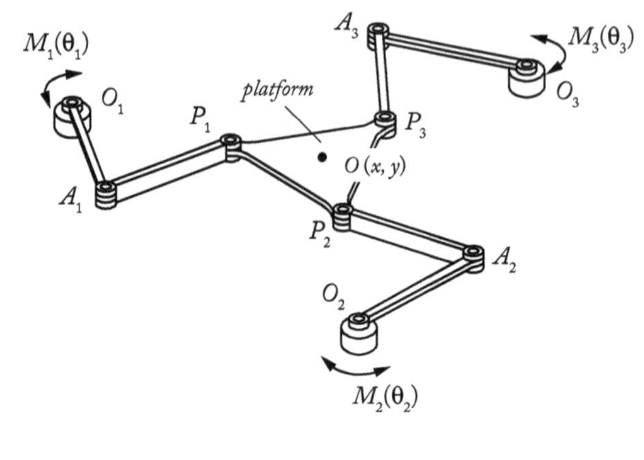

Fig. 7.1 Planar 3-*dof* 3-*RRR* parallel manipulator

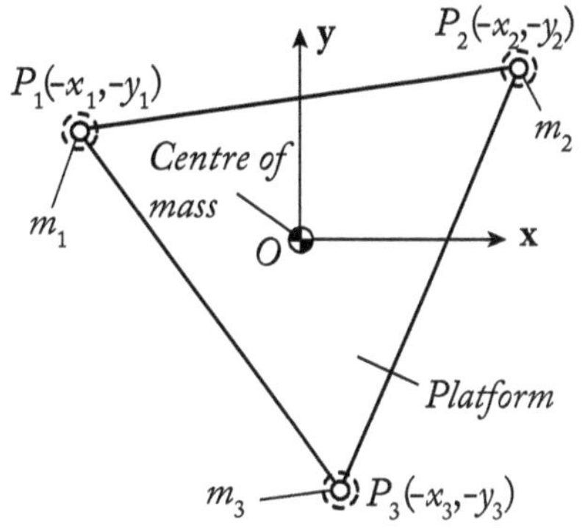

Fig. 7.2 Moving platform of the parallel manipulator and the point masses m_i $(i = 1, 2, 3)$

three concentrated masses, i.e. the platform can be considered as a weightless link with three point masses attached.

A design example of such a platform is now considered. Figure 7.3 shows a platform of a parallel manipulator, which represents a cylinder of radius R. The axial moment of inertia of this platform with respect to the centre of the mass is equal to $I_p = m_p R^2/2$. If it is desired to substitute dynamically the mass of the platform by three point masses disposed on the vertices of an isosceles triangle $P_1P_2P_3$, it is necessary to situate the revolute joints P_i at distances $r = R/\sqrt{2}$ from the centre O. In this case the axial moment of inertia of the three point masses and the moving platform are identical. This is an example of conceivable shape, but it is obviously possible to find several examples which allow the dynamic substitution of the mass of the platform by three concentrated masses.

Thus, we can replace the effect of mass and inertia of the moving platform by three point masses, which are at the end of each leg. This model allows the transformation of the manipulator balancing problem into one of balancing the legs. The latter is much simpler than the former.

7.1.2 Balancing of Legs

7.1.2.1 Balancing by Counter-rotations

The suggested balancing method is based on balancing of the inertia forces by means of counterweights mounted on the links and balancing of the total angular momentum by means of counterweights with planetary gear trains to generate the counter-rotations.

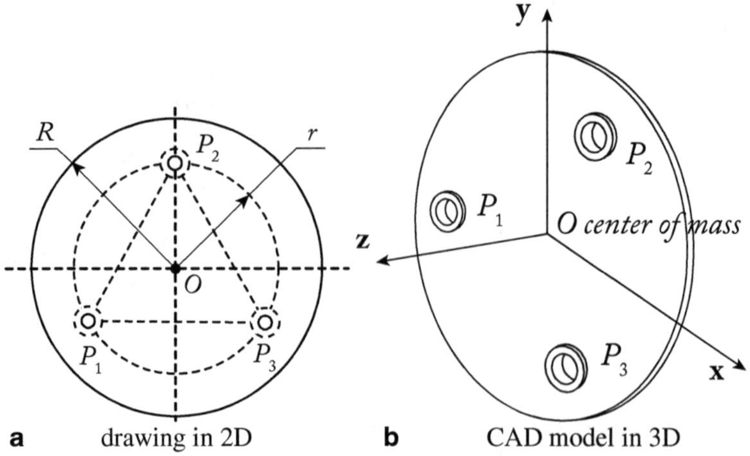

Fig. 7.3 A special shape of the moving platform

Shaking Force Balancing

In order to achieve the dynamic balancing of the manipulator, we first have to ensure that it is force-balanced, i.e. statically balanced. As mentioned above, the mass of the moving platform is substituted by three equivalent point masses located at the legs, i.e. each leg of the manipulator can be balanced independently.

The location of the *com* S_i of each leg relative to its base joint located at O_i (Fig. 7.4) can be found by the expressions:

$$\begin{bmatrix} x_{S_i} \\ y_{S_i} \end{bmatrix} = \frac{1}{m_{2i} + m_{3i} + m_{P_i}} \begin{bmatrix} m_{2i}x_{2i} + m_{3i}x_{3i} + m_i x_{P_i} \\ m_{2i}y_{2i} + m_{3i}y_{3i} + m_i y_{P_i} \end{bmatrix} \tag{7.2}$$

where

$$\begin{aligned} x_{2i} &= l_{O_iS_{2i}} \cos\theta_i \\ y_{2i} &= l_{O_iS_{2i}} \sin\theta_i \\ x_{3i} &= l_{O_iA_i} \cos\theta_i + l_{A_iS_{3i}} \cos\theta_{i+3} \\ y_{2i} &= l_{O_iA_i} \sin\theta_i + l_{A_iS_{3i}} \sin\theta_{i+3} \\ x_{P_i} &= l_{O_iA_i} \cos\theta_i + l_{A_iP_i} \cos\theta_{i+3} \\ y_{P_i} &= l_{O_iA_i} \sin\theta_i + l_{A_iP_i} \sin\theta_{i+3} \end{aligned} \tag{7.3}$$

with m_{2i} and m_{3i} the masses of links $2i$ and $3i$.

It is clear that the motion of the *com* of each leg is generated by the change in the two angles θ_i and θ_{i+3}. Thus, for the position of the *com* to remain constant and located at the point O_i, it is sufficient that

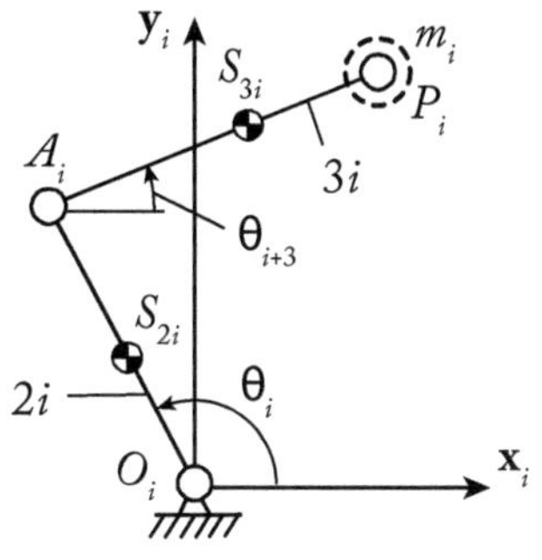

Fig. 7.4 Modeling of leg i of the planar parallel manipulator ($i = 1, 2, 3$)

$$m_{3i}\overrightarrow{A_iS_{3i}} + m_i\overrightarrow{A_iP_i} = \mathbf{0} \tag{7.4}$$
$$m_{2i}\overrightarrow{O_iS_{2i}} + (m_i + m_{3i})\overrightarrow{O_iA_i} = \mathbf{0}$$

The conditions (7.4) can be satisfied by adding two counterweights mounted on links $2i$ and $3i$, such that vectors

- $\overrightarrow{A_iS_{3i}}$ and $\overrightarrow{A_iP_i}$, and
- $\overrightarrow{O_iS_{2i}}$ and $\overrightarrow{O_iA_i}$

have opposite directions.

After such a redistribution of masses, all moving masses of the manipulator can be replaced by three fixed masses $m_{O_i} = m_{2i} + m_{3i} + m_{P_i}$ located at the point O_i and the *com* of the manipulator is located at the centre of these three fixed masses. Thus, the *com* of the manipulator remains motionless for any motion of links and hence, the manipulator transmits no inertia loads to its base.

Shaking Force and Shaking Moment Balancing

Now that the inertia force balancing is achieved, we have to consider the cancellation of the shaking moment. As in the first case, we consider the balancing of the manipulator legs. There are several approaches for complete shaking moment balancing of articulated dyads with two revolute joints (Arakelian and Smith 1999; Berkof 1973; Esat and Bahai 1999; Gao 1991; Ye and Smith 1991). The balancing method applied in this case is based on the shaking moment balancing by means of counterweights with planetary gear trains carrying out the counter-rotations (Arakelian and Smith 1999). The dynamic balancing scheme of each leg is designed in the following manner. The gear 3_{GRi} (Fig. 7.5) is mounted on the rotation axis of input link $2i$ and is linked kinematically with $3i$ through belt transmission $6i$. It meshes also with gear $4i$ mounted on the base. The gear 2_{GRi} is mounted on input link $2i$ and meshes with gear $5i$ mounted on the rotation axis passing through B_i. It should be noted that the joint between the gear 3_{GRi} and the ground is different from the joint located at O_i and it will be designated as the joint located at $O_{3_{GRi}}$.

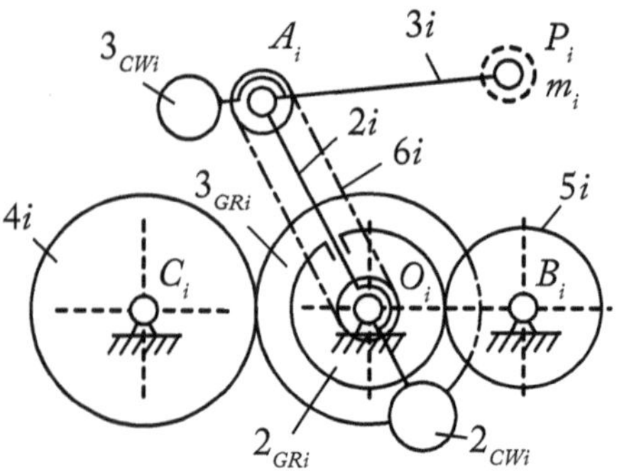

Fig. 7.5 Dynamic balancing scheme of leg i $(i = 1, 2, 3)$

Thus the shaking moment may be balanced by the moment of inertia of gears $4i$ and $5i$ taking into account that the angular velocities of links are the following: $\dot{\theta}_{4i} = -\dot{\theta}_{i+3}$ and $\dot{\theta}_{5i} = -\dot{\theta}_i$.

After shaking force balancing, the shaking moment applied on the base is constant relative to any point, i.e. for a given position of the manipulator it has the same value for any point of the base and can be expressed as:

$$M^{sh} = \sum_{i=1}^{3} M_i^{sh} = \sum_{i=1}^{3} \frac{\mathrm{d}H_{O_i}}{\mathrm{d}t} \tag{7.5}$$

where H_{O_i} is the angular momentum of the moving links of each leg with respect to point O_i.

In order to have a shaking moment equal to zero for all trajectories, the sum of the angular momentum of the legs must be constant over time.

The angular momentum for each leg with added planetary gear trains can be written as:

$$\begin{aligned} H_{O_i} &= m_{2i}\left(x_{2i}\dot{y}_{2i} - y_{2i}\dot{x}_{2i}\right) + \left(I_{S_{2i}} + I_{2GRi}\right)\dot{\theta}_i + I_{S_{5i}}\dot{\theta}_{5i} \\ &\quad + m_{3i}\left(x_{3i}\dot{y}_{3i} - y_{3i}\dot{x}_{3i}\right) + \left(I_{S_{3i}} + I_{3GRi}\right)\dot{\theta}_{i+3} + I_{S_{4i}}\dot{\theta}_{4i} + m_i\left(x_{P_i}\dot{y}_{P_i} - y_{P_i}\dot{x}_{P_i}\right) \end{aligned} \tag{7.6}$$

where $I_{S_{2i}}$ and $I_{S_{3i}}$ are the moments of inertia of links $2i$ and $3i$ about the *com* of the links (axial moment of inertia), and $I_{S_{4i}}$ and $I_{S_{5i}}$ are the axial moments of inertia of the added gears.

We substitute (7.3) and their derivatives into (7.6) and, taking into account condition (7.4), we obtain the following expression of the angular momentum for each leg:

$$\begin{aligned} H_{O_i} &= \left(I_{S_{3i}} + I_{3GRi} + m_{3i}l_{A_iS_{3i}}^2 + m_i l_{A_iP_i}^2\right)\dot{\theta}_{i+3} \\ &\quad + I_{S_{4i}}\dot{\theta}_{4i} + \left(I_{S_{2i}} + I_{2GRi} + m_{2i}l_{O_iS_{2i}}^2 + (m_i + m_{3i})\, l_{O_iA_i}^2\right)\dot{\theta}_i + I_{S_{5i}}\dot{\theta}_{5i} \end{aligned} \tag{7.7}$$

from which we obtain the conditions of shaking moment balancing:

$$I_{S_{4i}} = I_{S_{3i}} + I_{3GRi} + m_{3i} l^2_{A_i S_{3i}} + m_i l^2_{A_i P_i}$$
$$I_{S_{5i}} = I_{S_{2i}} + I_{2GRi} + m_{2i} l^2_{O_i S_{2i}} + (m_i + m_{3i}) l^2_{O_i A_i} \tag{7.8}$$

Hence, any 3-*dof* 3-*RRR* parallel manipulator satisfying (7.1), (7.4) and (7.7) will be dynamically balanced, i.e. reactionless.

The disadvantage of the suggested balancing scheme is the need for the connection of gears to the oscillating links. The oscillations of the links of the manipulator will create noise unless expensive anti-backlash gears are used.

7.1.2.2 Numerical Example and Simulation Results

We shall now examine the ground bearing forces and the ground bearing moments of a 3-*dof* 3-*RRR* parallel manipulator which is fully force and moment balanced. The geometry and mass distribution parameters of the links are listed in Table 7.1 (parameters x_{P_i} and y_{P_i} show the selected configuration of the manipulator).

The platform of the examined parallel manipulator represents a cylinder with radius $R = 0.082$ m and $m_p = 3$ kg. As a result, $I_p = 0.01$ kg m^2 (so that $m_i = 1$ kg and $r = R/\sqrt{2} = 0.058$ m). The actuator motions are given by the expressions (Ma and Angeles 1989): $\theta_i = a_i\pi + b_i\,(2\pi t/T - \sin(2\pi t/T))$, where $a_1 = 1/3$, $a_2 = 4/3$, $a_3 = 10/3$, $b_1 = 1/6$, $b_2 = -1/6$, $b_3 = 1/12$ and $T = 0.3$ s. The angles θ_i of the input links are measured with respect to the global **x** axis. The proposed functions give zero velocity and acceleration at the start and end of the motion.

Figure 7.6 shows the resultant bearing forces of the balanced planar 3-*dof* 3-*RRR* manipulator along the **x** and **y** axes.

In Fig. 7.7 are presented the variations of the moment of the ground bearing forces and the reactions of the input torques.

Table 7.1 Parameters of the balanced manipulator (SI units)

Parameter	Leg 1	Leg 2	Leg 3	Parameter	Leg 1	Leg 2	Leg 3
x_{O_i}	0	0.46	0.22	$l_{O_i S_{2i}}$	0.09	0.09	0.09
y_{O_i}	0	0	0.4	$l_{A_i S_{3i}}$	0.07	0.07	0.07
$l_{O_i A_i}$	0.18	0.18	0.18	m_{2i}	7.2	7.2	7.2
$l_{A_i P_i}$	0.18	0.18	0.18	m_{3i}	2.6	2.6	2.6
x_{P_i} at $t = 0$	0.18	0.28	0.22	$I_{S_{2i}}$	0.02	0.02	0.02
y_{P_i} at $t = 0$	0	0	0.087	$I_{S_{3i}}$	0.017	0.017	0.017
m_i	1	1	1	$I_{S_{4i}}$	0.068	0.068	0.068
r_i	0.058	0.058	0.058	$I_{S_{5i}}$	0.2	0.2	0.2

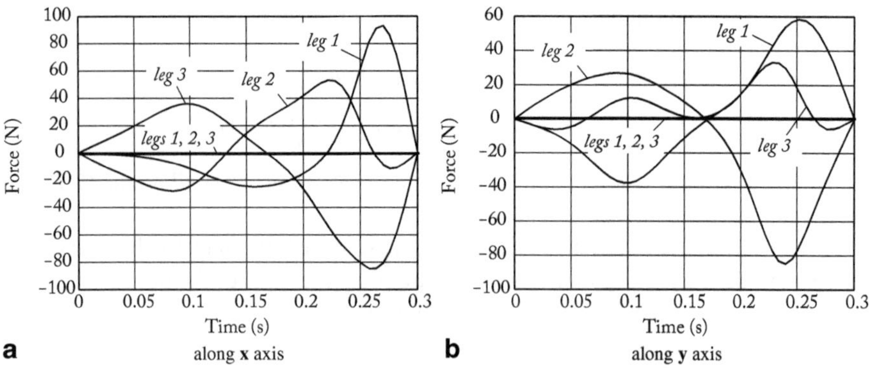

Fig. 7.6 Variations of the ground bearing forces of the balanced manipulator

Fig. 7.7 Variations of the moment of the ground bearing forces and the reactions of the input torques of the balanced manipulator

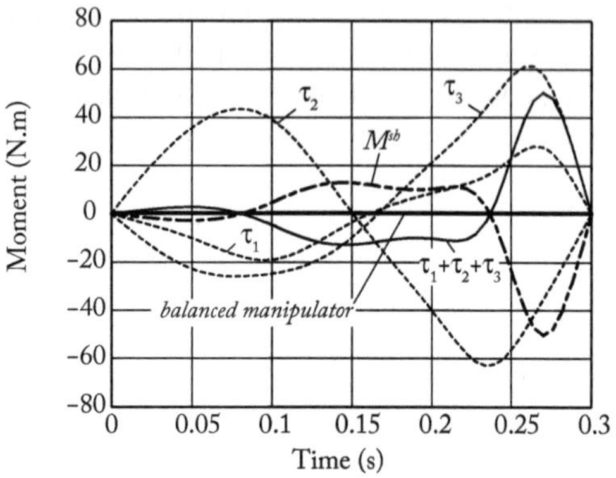

7.1.3 *Balancing of the 3-RRR Robot by Using an Inertia Flywheel*

7.1.3.1 Balancing by Inertia Flywheel

In this section we consider the shaking moment cancellation of the fully-force balanced 3-*dof* 3-*RRR* parallel manipulator by an inertia flywheel with prescribed rotation. It is evident that this solution is constructively more efficient.

Figure 7.8 shows the fully force-balanced 3-*dof* 3-*RRR* parallel manipulator and balancing inertia flywheel, which is mounted on the base of the manipulator. The conditions for balancing the shaking moment of the manipulator are determined from the following consideration.

Please note that, when an inertia flywheel is added, it is not necessary that the platform has an axial inertia moment which can be dynamically substituted by three concentrated masses, i.e. $\sum_{i=1}^{3} m_i \left(x_i^2 + y_i^2\right) \neq I_p$.

Thus, the angular momentum for the fully force-balanced manipulator can be deduced from (7.7) by cancelling the axial moment of inertia of the counter-rotations

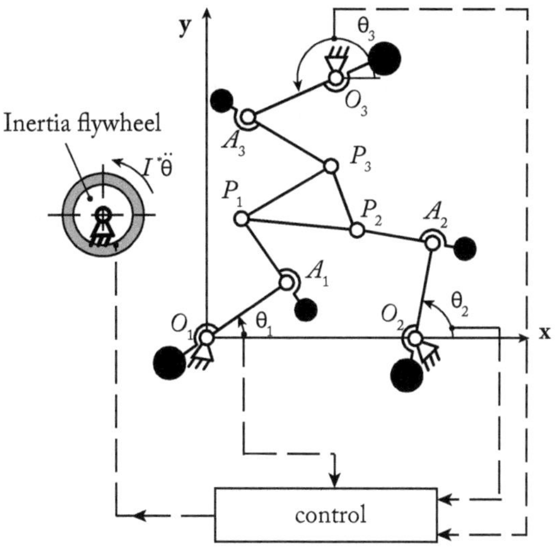

Fig. 7.8 Shaking moment balancing of fully force-balanced 3-*dof* 3-*RRR* parallel manipulator by an inertia flywheel

and of the gears:

$$H = \sum_{i=1}^{3} H_{O_i} + \Delta I_p \dot{\phi} = \sum_{i=1}^{3} \left(I_{S_{3i}} + m_{3i} l_{A_i S_{3i}}^2 + m_i l_{A_i P_i}^2 \right) \dot{\theta}_{i+3} \quad (7.9)$$

$$+ \sum_{i=1}^{3} \left(I_{S_{2i}} + m_{2i} l_{O_i S_{2i}}^2 + (m_i + m_{3i}) l_{O_i A_i}^2 \right) \dot{\theta}_i + \Delta I_p \dot{\phi}$$

where $\Delta I_p = I_p - \sum_{i=1}^{3} m_i \left(x_i^2 + y_i^2 \right)$.

Hence, the shaking moment is the following:

$$M^{sh} = \frac{d}{dt} \left(\sum_{i=1}^{3} H_{O_i} + \Delta I_p \dot{\phi} \right) = \sum_{i=1}^{3} \left(K_i \ddot{\theta}_i + K_{i+3} \ddot{\theta}_{i+3} \right) + \Delta I_p \ddot{\phi} \quad (7.10)$$

where $K_i = I_{S_{2i}} + m_{2i} l_{O_i S_{2i}}^2 + (m_i + m_{3i}) l_{O_i A_i}^2$ and $K_{i+3} = I_{S_{3i}} + m_{3i} l_{A_i S_{3i}}^2 + m_i l_{A_i P_i}^2$.

To balance the shaking moment, an inertia flywheel with axial inertia moment I^* can be used. The angular acceleration of this inertia flywheel driven by a complementary actuator 4 is the following:

$$\ddot{\theta} = M^{sh} / I^* = \frac{\sum_{i=1}^{3} \left(K_i \ddot{\theta}_i + K_{i+3} \ddot{\theta}_{i+3} \right) + \Delta I_p \ddot{\phi}}{I^*} \quad (7.11)$$

It should be noted that the axial inertia moment of the flywheel must be selected in such a manner that its rotation with prescribed acceleration will be feasible. Therefore, the reaction of the balancing inertia flywheel on the frame cancels the shaking moment due to the parallel manipulator. In other words, the actuator, which rotates the balancing inertia flywheel with a prescribed angular acceleration $\ddot{\theta}$ has a reaction on the frame which is similar but opposite to the shaking moment of the parallel manipulator. Thus, full shaking moment is cancelled.

The angular velocity $\dot{\theta}(t)$ and angular displacements $\theta(t)$ can be determined by simple integration of the obtained values of $\ddot{\theta}(t)$.

7.1.3.2 Numerical Example and Simulation Results

Let us consider a numerical example for computer simulation. As a model we could use the previous example with the link parameters given in Table 7.1. However, for the best illustration of the suggested balancing approach, we change the value of the axial moment of inertia of the platform: $I_p = 0.015\ \mathrm{kg\,m^2}$, i.e. the mass of the platform cannot be dynamically substituted by three concentrated masses. It should be noted that in this case we do not need this condition. Thus, by substituting statically the mass of the platform by three concentrated masses, the shaking force balancing is carried out in the same way as in the previous case. Then the angular accelerations of the movable links are determined taking into account that the drivers are given by the expressions (Ma and Angeles 1989): $\theta_i = a_i\pi + b_i\,(2\pi t/T - \sin(2\pi t/T))$, where $a_1 = 1/3$, $a_2 = 4/3$, $a_3 = 10/3$, $b_1 = 1/6$, $b_2 = -1/6$, $b_3 = 1/12$ and $T = 0.3$ sec.

Now, by determining the shaking moment from (7.10) and taking as the axial moment inertia of the flywheel $I^* = 0.01\ \mathrm{kg\,m^2}$, we determine the angular acceleration of the balancing flywheel, which gives complete shaking moment balancing of the manipulator. Figure 7.9 shows the obtained law of rotation of the flywheel, which produces complete shaking moment balancing.

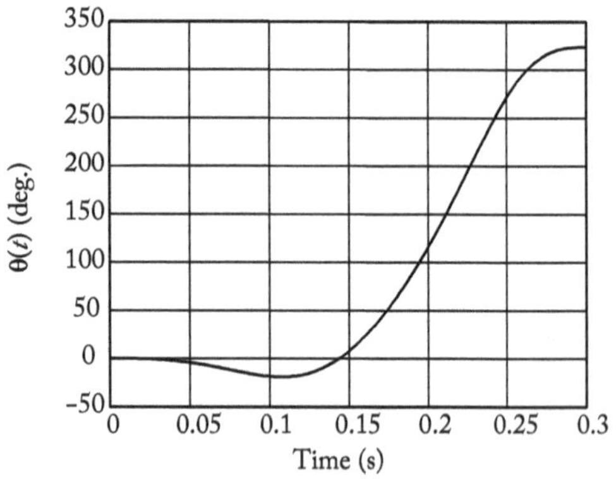

Fig. 7.9 Law of rotation of the balancing flywheel

7.2 Complete Shaking Force and Shaking Moment Balancing of Planar Parallel Manipulators with Prismatic Pairs

7.2.1 Complete Shaking Moment and Shaking Force Balancing by Adding an Idler Loop Between the Base and the Platform

In this sub-section, the complete shaking force and shaking moment balancing of planar manipulators by adding an idler loop is discussed. The added balancing loop is mounted between the base and the platform of the mechanism. We illustrate the suggested balancing technique on a 3-*RPR* mechanism (Fig. 7.10). Please note that we do not mention the type of actuation of the mechanism as it has no influence on the balancing.

Firstly, let us analyze the cancellation of the dynamic reactions of the 3-*RPR* planar parallel mechanism (Fig. 7.10a). Such a mechanism has 3 *dof* (two translations in the (Oxy) plane and one rotation of the moving platform around an axis perpendicular to (Oxy)) and is composed of three identical legs, each being composed of a revolute joint attached to the base at point A_i (in the following of Sect. 7.2, $i = 1 \cdots 3$), one moving prismatic guide, located at point B_i, and another revolute joint attached to the platform at C_i. The base and platform triangles, denoted $A_1A_2A_3$ and $C_1C_2C_3$, are equilateral. On this manipulator, typically, the actuated joints are the first revolute joint at A_i or the linear guide at B_i.

Considering that the **x** axis is directed along the line A_1A_2, the **y** axis being perpendicular to the **x** axis and the origin of the base frame located at point O, the centre of the circumcircle of triangle $A_1A_2A_3$, one can define the coordinate x, y

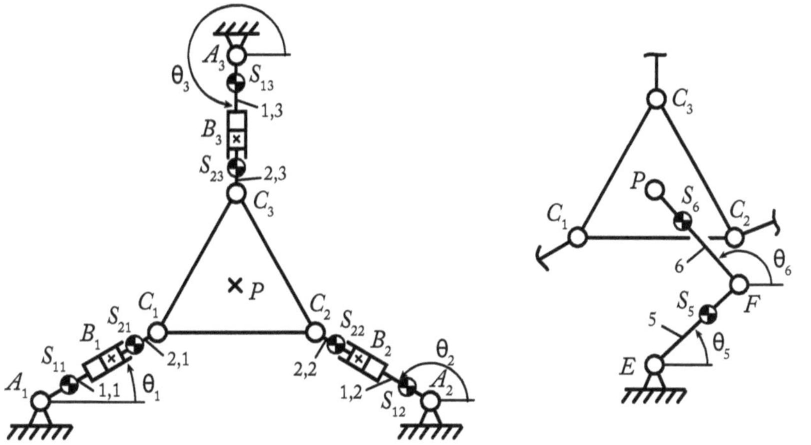

a Schematics of the 3-*RPR* mechamism **b** Schematics of the added idler loop (*RRR* chain)

Fig. 7.10 Schematic of the 3-*RPR* robot under study

and ϕ of the platform, as being respectively the coordinates of point P along the **x** and **y** axes and the angle between the line C_1C_2 and A_1A_2.

Let us denote as S_{ij} the *com* of link ij ($j = 1, 2$), which has a mass m_j and an axial moment of inertia I_{S_j}. The *com* of the platform is located at point P. The mass of the platform is m_p and its axial moment of inertia I_p.

In order to cancel the shaking forces and shaking moment of the manipulator, an idler loop is added between the base and the platform (Fig. 7.10b). The *com* of elements 5 and 6 of the idler loop are located at S_5 and S_6, respectively. Their masses are denoted as m_5 and m_6 and their axial moments of inertia as I_{S_5} and I_{S_6}, respectively. The positions of the *coms* are defined such that $\overrightarrow{A_iS_{1i}} = r_1 l_{B_iC_i}\mathbf{u}_i$, $\overrightarrow{C_iS_{2i}} = (r_2 1) l_{B_iC_i}\mathbf{u}_i$, $\overrightarrow{ES_5} = r_5\overrightarrow{EF}$, and $\overrightarrow{FS_{6i}} = r_6\overrightarrow{FP}$, r_1, r_2, r_5, and r_6 being dimensionless coefficients, and $\mathbf{u}_i$ a unit vector directed along $\overrightarrow{B_iC_i}$.

The expression of the shaking force $\mathbf{f}^{sh}$ transmitted by the robot to the ground is:

$$\mathbf{f}^{sh} = \left(\sum_{i=1}^{3}\sum_{j=1}^{2} m_j \ddot{\mathbf{r}}_{S_{ij}}\right) + m_p\ddot{\mathbf{r}}_P + m_5\ddot{\mathbf{r}}_{S_5} + m_6\ddot{\mathbf{r}}_{S_6} \tag{7.12}$$

where $\ddot{\mathbf{r}}_{S_{ij}}$, $\ddot{\mathbf{r}}_P$, $\ddot{\mathbf{r}}_{S_5}$ and $\ddot{\mathbf{r}}_{S_6}$ are the accelerations of the *coms* S_{ij}, of P, S_5 and S_6, respectively.

Developing (7.12), it can be demonstrated that the shaking force $\mathbf{f}^{sh}$ can be expressed as:

$$\begin{aligned}\mathbf{f}^{sh} = &(m_1 r_1 - m_2(1 - r_2))\sum_{i=1}^{3}\mathbf{a}_i + \left(3m_2 + m_p + m_6 r_6\right)\mathbf{a}_6 \\ &+ \left(3m_2 + m_p + m_5 r_5 + m_6\right)\ddot{\mathbf{r}}_F\end{aligned} \tag{7.13}$$

with

$$\mathbf{a}_i = l_{B_iC_i}\left(\ddot{\theta}_i\begin{bmatrix}-\sin\theta_i \\ \cos\theta_i\end{bmatrix} - \dot{\theta}_i^2\begin{bmatrix}\cos\theta_i \\ \sin\theta_i\end{bmatrix}\right) \tag{7.14}$$

$$\mathbf{a}_6 = l_{FP}\left(\ddot{\theta}_6\begin{bmatrix}-\sin\theta_6 \\ \cos\theta_6\end{bmatrix} - \dot{\theta}_6^2\begin{bmatrix}\cos\theta_6 \\ \sin\theta_6\end{bmatrix}\right) \tag{7.15}$$

At this step, only five counterweights are needed in the cancellation of the shaking force, but it could be demonstrated after more derivations that three others are necessary for the cancellation of the shaking moment. Therefore, we propose directly adding three supplementary counterweights (Fig. 7.11). The positions of the eight added counterweights are $\overrightarrow{A_iM_{1i}} = r_{cw_1} l_{B_iC_i}\mathbf{u}_i$, $\overrightarrow{B_iM_{2i}} = (r_{cw_2} - 1) l_{B_iC_i}\mathbf{u}_i$, $\overrightarrow{EM_5} = r_{cw_5}\overrightarrow{EF}$, and $\overrightarrow{FM_6} = r_{cw_6}\overrightarrow{FP}$, r_{cw_1}, r_{cw_2}, r_{cw_5}, and r_{cw_6} being dimensionless

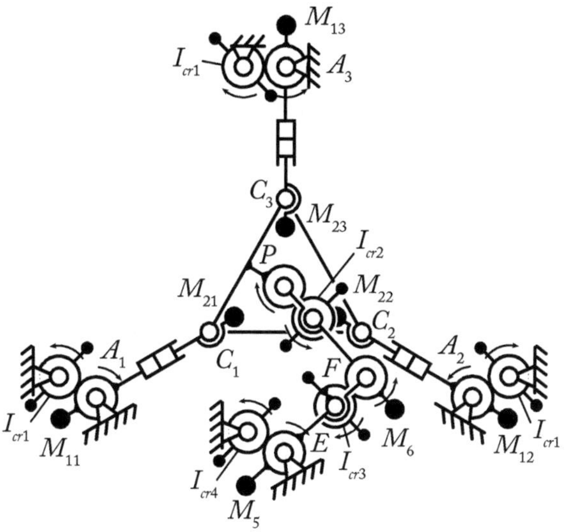

Fig. 7.11 Schematics of the 3-RPR mechanism with the added RRR chain used for the cancellation of the shaking force and shaking moment

coefficients. Their masses are respectively denoted m_{cw_1}, m_{cw_2}, m_{cw_5} and m_{cw_6}. With the addition of the counterweights, the shaking force becomes:

$$\mathbf{f}^{sh^*} = \mathbf{f}^{sh} + \left(m_{cw_1} r_{cw_1} - m_{cw_2}\left(1 - r_{cw_2}\right)\right) \sum_{i=1}^{3} \mathbf{a}_i + m_{cw_6} r_{cw_6} \mathbf{a}_6 + \left(m_{cw_5} r_{cw_5} + m_{cw_6}\right) \ddot{\mathbf{r}}_F \tag{7.16}$$

Thus, the shaking force is cancelled if:

$$\begin{aligned} m_{cw_1} &= -\frac{m_1 r_1}{r_{cw_1}} \\ m_{cw_2} &= -\frac{m_2(1 - r_2)}{1 - r_{cw_2}} \\ m_{cw_6} &= -\frac{3\left(m_2 + m_{cw_2}\right) + m_p + m_6 r_6}{r_{cw_6}} \\ m_{cw_5} &= -\frac{3\left(m_2 + m_{cw_2}\right) + m_p + m_5 r_5 + m_6 + m_{cw_6}}{r_{cw_5}} \end{aligned} \tag{7.17}$$

The expression of the shaking moment M_O^{sh} of the modified structure (expressed at point O) can be written as:

$$M_O^{sh} = \frac{d}{dt} H_O \tag{7.18}$$

where H_O is the angular momentum of the leg (expressed at point O). Thus, in order to cancel the shaking moment, the angular momentum is held constant over time.

The expression of the angular momentum H_O is equal to:

$$
\begin{aligned}
H_O =& \sum_{i=1}^{3}\sum_{j=1}^{2}\left(m_j\left(x_{S_{ij}}\dot{y}_{S_{ij}} - y_{S_{ij}}\dot{x}_{S_{ij}}\right) + m_{cw_j}\left(x_{M_{ij}}\dot{y}_{M_{ij}} - y_{M_{ij}}\dot{x}_{M_{ij}}\right) + I_{S_j}\dot{\theta}_i\right) \\
&+ I_p\dot{\phi} + \sum_{j=5}^{6}\left(m_j\left(x_{S_j}\dot{y}_{S_j} - y_{S_j}\dot{x}_{S_j}\right) + m_{cw_j}\left(x_{M_j}\dot{y}_{M_j} - y_{M_j}\dot{x}_{M_j}\right) + I_{S_j}\dot{\theta}_j\right)
\end{aligned} \tag{7.19}
$$

where x_Q, y_Q, $\dot{x}_Q$ and $\dot{y}_Q$ are the position and velocities of any point Q along **x** and **y** axes, respectively (Q being either point S_{ij}, M_{ij}, $(j = 1, 2)$, S_j or M_j $(j = 5, 6)$).

Developing and introducing (7.17) into (7.19) yields

$$
\begin{aligned}
H_O =& \sum_{i=1}^{3}\left(I_{S_1} + I_{S_2} + \left(m_1 r_1^2 + m_{cw_1} r_{cw_1}^2 + m_2\left(1 - r_2\right)^2\right.\right. \\
&\left.\left. + m_{cw_2}\left(1 - r_{cw_2}\right)^2\right) l_{B_iC_i}^2\right)\dot{\theta}_i \\
&+ \left(I_{S_6} + \left(m_6 r_6^2 + m_{cw_6} r_{cw_6}^2 + m_p + 3\left(m_2 + m_{cw_2}\right)\right) l_{FP}^2\right)\dot{\theta}_6 \\
&+ \left(I_{S_5} + \left(m_5 r_5^2 + m_{cw_5} r_{cw_5}^2 + m_6 + m_{cw_6} + m_p + 3\left(m_2 + m_{cw_2}\right)\right) l_{EF}^2\right)\dot{\theta}_5 \\
&+ \left(I_p + 3\left(m_2 + m_{cw_2}\right) l_{C_iP}^2\right)\dot{\phi}
\end{aligned} \tag{7.20}
$$

After such modifications of the *RRR* chain, the angular momentum of the legs of the mechanism and of the *RRR* chain can be balanced using six counter-rotations (Fig. 7.11), which have an axial moment of inertia equal to:

$$
\begin{aligned}
I_{cr_1} &= I_{S_1} + I_{S_2} + \left(m_1 r_1^2 + m_{cw_1} r_{cw_1}^2 + m_2\left(1 - r_2\right)^2 + m_{cw_2}\left(1 - r_{cw_2}\right)^2\right) l_{B_iC_i}^2 \\
I_{cr_2} &= I_p + 3\left(m2 + m_{cw_2}\right) l_{C_iP}^2 \\
I_{cr_3} &= I_{S_6} + \left(m_6 r_6^2 + m_{cw_6} r_{cw_6}^2 + m_p + 3\left(m_2 + m_{cw_2}\right)\right) l_{FP}^2 + 2I_{cr_2} \\
I_{cr_4} &= I_{S_5} + \left(m_5 r_5^2 + m_{cw_5} r_{cw_5}^2 + m_6 + m_{cw_6} + m_p + 3\left(m_2 + m_{cw_2}\right)\right) l_{EF}^2 + 2I_{cr_3}
\end{aligned} \tag{7.21}
$$

Numerical Application

Let us illustrate the suggested balancing approach using numerical simulations carried out with ADAMS software. For this purpose, non balanced and balanced 3-*RPR* parallel manipulators will be compared.

The chosen trajectory for simulations is a straight line of the controlled point of the platform, achieved in $t_f = 0.25$ s, between $P_0 = [x_0\ \ y_0]^T = [0.05\ m\ \ 0\ m]^T$ and $P_f = \left[x_f\ \ y_f\right]^T = [0.2\ m\ \ 0\ m]^T$ with a rotation of the platform from $\phi_0 = 0°$. to $\phi_f = 30°$. For the displacement of the mechanism, fifth order polynomial laws are used and therefore the trajectory is defined by the following expressions:

$$\begin{cases} x(t) = x_0 + s(t)(x_f - x_0) \\ y(t) = 0 \\ \phi(t) = \phi_0 + s(t)(\phi_f - \phi_0) \end{cases} \tag{7.22}$$

with

$$s(t) = \frac{10}{t_f^3}t^3 - \frac{15}{t_f^4}t^4 + \frac{6}{t_f^5}t^5 \tag{7.23}$$

The parameters used for the simulations are the followings:

- $l_{OE} = 0$ m, $l_{OA_i} = 0.35$ m, $l_{B_iC_i} = 0.05$ m, $l_{C_iP} = 0.1$ m, $l_{EF} = 0.15$ m and $l_{FP} = 0.1581$ m,
- $r_1 = 2, r_2 = r_5 = r_6 = 0.5$,
- $m_1 = 0.75$ kg, $m_2 = 0.37$ kg, $m_5 = 0.42$ kg, $m_6 = 0.47$ kg; $m_p = 1$ kg,
- $I_{S_1} = 0.00344$ kg m^2, $I_{S_2} = 0.00025$ kg m^2, $I_{S_5} = 0.00122$ kg m^2, $I_{S_6} = 0.00146$ kg m^2, $I_p = 0.00436$ kg m^2.

For such parameters and such a trajectory, the shaking force and shaking moment are computed using ADAMS software and are presented in Fig. 7.12 (solid line). Then, we add the counterweights and the idler loop *EFP* to the mechanism. The position coefficients of the counterweights are all equal to $r_{cw_j} = -0.5$ $(j = 1, 2, 5, 6)$. Therefore, the added masses are equal to $m_{cw_1} = 0.75$ kg, $m_{cw_2} = 0.37$ kg, $m_{cw_5} = 6.92$ kg, $m_{cw_6} = 21.66$ kg. The new values of the shaking force and moment are presented in Fig. 7.12 (dashed line). It is possible to see that with the added counterweights the shaking efforts are cancelled, while the maximal value of shaking moment is increased by a factor 17. Finally, we add the counter-rotations. Their values are equal to $I_{cr_1} = 0.01917$ kg m^2, $I_{cr_2} = 0.02665$ kg m^2, $I_{cr_3} = 0.18169$ kg m^2, $I_{cr_4} = 0.72781$ kg m^2. With such counter-rotations, the shaking moment is balanced (in gray line in Fig. 7.12c).

7.2.2 *Complete Shaking Force and Shaking Moment Balancing Using Scott-Russell Mechanism*

In this sub-section another approach for complete shaking force and shaking moment balancing is developed, which consists of adding Scott-Russell mechanisms to the initial architecture of a manipulator. This approach enables a reduction in the number of counter-rotations.

7.2.2.1 Properties of the Scott-Russell Mechanism

Let us observe a simple slider-crank mechanism (Fig. 7.13). The *com* of link i ($i = 1, 2, 3$) is denoted as S_i. Link i has a mass m_i and an axial moment of inertia I_{S_i}. The positions of the centres of mass are $\overrightarrow{AS_1} = r1\overrightarrow{AB}$, $\overrightarrow{BS_2} = r_2\overrightarrow{BC}$, $\overrightarrow{CS_3} = l_3 r_3 \mathbf{x}$, r_1, r_2 and r_3 being dimensionless coefficients, and l_3 being a constant length.

It is known that the compete shaking force and shaking moment balancing of a general slider-crank mechanism can be obtained by adding two counterweights mounted on the links and two pairs of counter-rotations. However, it is possible to balance this mechanism without counter-rotation if it has specific geometrical parameters, as in Scott-Russell mechanisms ($a = 0$ m, $l_{AB} = l_{BC}$ – Fig. 7.13).

Let us consider the balancing of this mechanism. The expression of the shaking force $\mathbf{f}^{sh}$ of a slider-crank mechanism can be written as:

$$\mathbf{f}^{sh} = \sum_{i=1}^{3} m_i \ddot{\mathbf{r}}_{S_i} = (m_1 r_1 + m_2)\,\ddot{\mathbf{r}}_B + (m_2 r_2 + m_3)\,\mathbf{a} \tag{7.24}$$

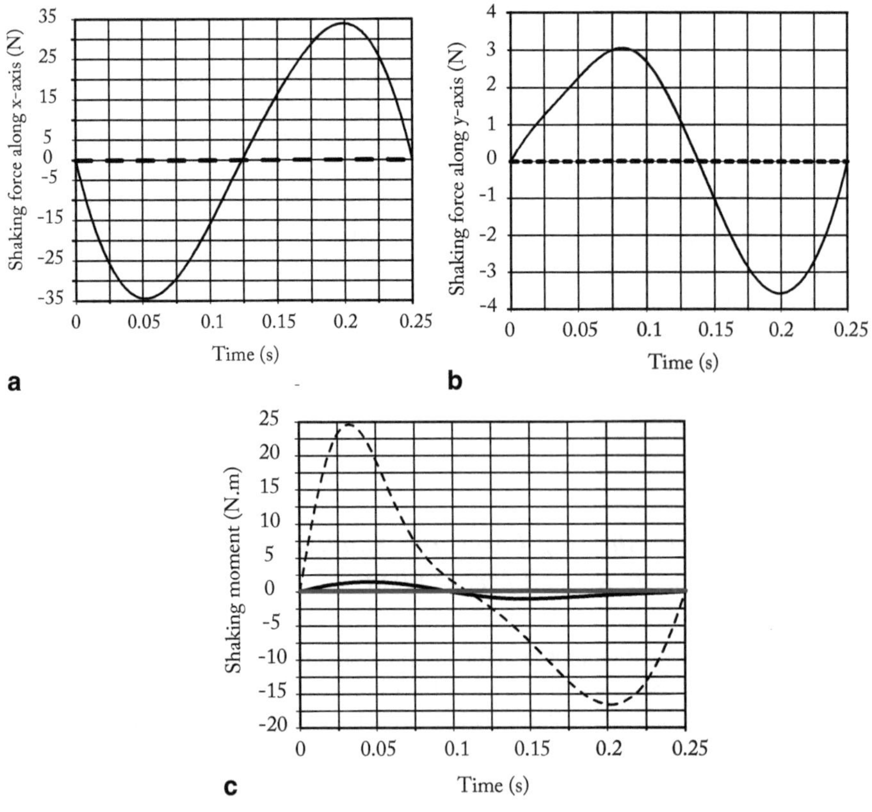

Fig. 7.12 Shaking force and shaking moment before (*solid line*) and after (*dashed line*) the addition of the counterweights, and after the addition of the counter-rotations (*gray line*)

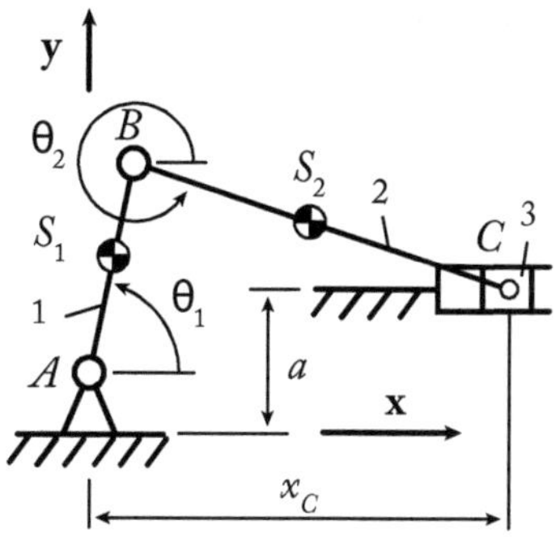

Fig. 7.13 A general slider-crank mechanism

with

$$\mathbf{a} = l_{BC}\left(\ddot{\theta}_2\begin{bmatrix}-\sin\theta_2\\ \cos\theta_2\end{bmatrix} - \dot{\theta}_2^2\begin{bmatrix}\cos\theta_2\\ \sin\theta_2\end{bmatrix}\right) \tag{7.25}$$

$\ddot{\mathbf{r}}_B$ being the acceleration of point B.

The constant terms of (7.24) can be cancelled by the addition of two counterweights located at M_j, $(j = 1, 2)$ (Fig. 7.13), whose masses are denoted as m_{cw_j}. Their positions are equal to: $\overrightarrow{AM_1} = r_{cw_1}\overrightarrow{AB}$, $\overrightarrow{BM_2} = r_{cw_1}\overrightarrow{BC}$, r_{cw_1} and r_{cw_2} being dimensionless coefficients. With the addition of the counterweights, the shaking force becomes:

$$\mathbf{f}^{sh^*} = \mathbf{f}^{sh} + \left(m_{cw_1}r_{cw_1} + m_{cw_2}\right)\ddot{\mathbf{r}}_B + m_{cw_2}r_{cw_2}\mathbf{a} \tag{7.26}$$

Thus, the shaking force is balanced if:

$$m_{cw_2} = -\frac{m_2r_2 + m_3}{r_{cw_2}} \text{ and } m_{cw_1} = -\frac{m_1r_1 + m_2 + m_{cw_2} + m_3}{r_{cw_1}} \tag{7.27}$$

The expression of the angular momentum H_A (expressed at point A) is:

$$H_A = \sum_{j=1}^{3}\left(m_j\left(x_{S_j}\dot{y}_{S_j} - y_{S_j}\dot{x}_{S_j}\right)\right) + \sum_{j=1}^{2}\left(m_{cw_j}\left(x_{M_j}\dot{y}_{M_j} - y_{M_j}\dot{x}_{M_j}\right) + I_{S_j}\dot{\theta}_j\right) \tag{7.28}$$

where x_Q, y_Q, $\dot{x}_Q$ and $\dot{y}_Q$ are the position and velocities of any point Q along **x** and **y** axes, respectively (Q being either point S_j or M_j, $(j = 1, 2, 3)$).

Developing and introducing (7.27) into (7.28),

$$\begin{aligned} H_A = &\left(I_{S_1} + \left(m_1r_1^2 + m_{cw_1}r_{cw_1}^2 + m_2 + m_{cw_2} + m_3\right)l_{AB}^2\right)\dot{\theta}_1 \\ &+ \left(I_{S_2} + \left(m_2r_2^2 + m_{cw_1}r_{cw_1}^2 + m_3\right)l_{BC}^2\right)\dot{\theta}_2 \end{aligned} \tag{7.29}$$

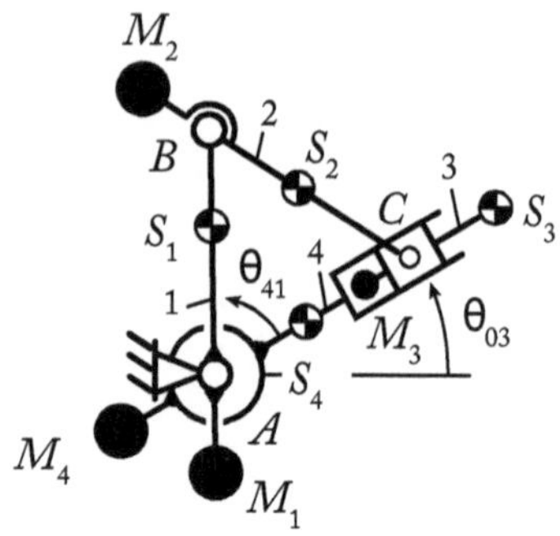

Fig. 7.14 A manipulator leg with added Scott-Russell mechanism

with

$$\dot{\theta}_2 = -\frac{\dot{y}_B\,(x_C - x_B) + (a - y_B)\,(\dot{x}_C - \dot{x}_B)}{l_{BC}^2} \tag{7.30}$$

where x_B, y_B, x_C are the coordinates along **x** and **y** axes of points B and C, respectively, and $\dot{x}_B$, $\dot{y}_B$, $\dot{x}_C$ their velocities.

In order to cancel the shaking moment M_A^{sh}, the angular momentum has to be constant or null. Developing (7.29), M_A^{sh} can be cancelled if $a = 0$ and $l_{AB} = l_{BC}$ (in such a case, $\dot{\theta}_1 = -\dot{\theta}_2$) and if:

$$\begin{aligned} &I_{S_1} + \left(m_1 r_1^2 + m_{cw_1} r_{cw_1}^2 + m_2 + m_{cw_2} + m_3\right) l_{AB}^2 \\ &- I_{S_2} - \left(m_2 r_2^2 + m_{cw_1} r_{cw_1}^2 + m_3\right) l_{BC}^2 = 0 \end{aligned} \tag{7.31}$$

7.2.2.2 Balancing of a Manipulator's Leg Using a Scott-Russell Mechanism

Now let us consider a manipulator's leg with an added Scott-Russell mechanism (Fig. 7.14). Let us denote as S_4 the centre of mass of link 4, which has a mass m_4 and an axial moment of inertia I_{S_4}. The position of S_4 is such that: $\overrightarrow{AS_4} = l_3 r_4 \mathbf{u}$, r_4 being a dimensionless coefficient and **u** a unit vector along $\overrightarrow{CS_3}$.

Now the shaking force becomes:

$$\begin{aligned} \mathbf{f}^{sh} = &\left(m_1 r_1 + m_{cw_1} r_{cw_1} + m_2 + m_{cw_2} + m_3\right) \ddot{\mathbf{r}}_B + (m_3 r_3 + m_4 r_4)\,\mathbf{a}_1 \\ &+ \left(m_2 r_2 + m_{cw_2} r_{cw_2} + m_3\right) \mathbf{a}_2 \end{aligned} \tag{7.32}$$

with

$$\mathbf{a}_1 = l_3 \left(\ddot{\theta}_{03} \begin{bmatrix} -\sin\theta_{03} \\ \cos\theta_{03} \end{bmatrix} - \dot{\theta}_{03}^2 \begin{bmatrix} \cos\theta_{03} \\ \sin\theta_{03} \end{bmatrix} \right) \tag{7.33}$$

$$\mathbf{a}_2 = l_{AB} \left(\left(\ddot{\theta}_{03} - \ddot{\theta}_{41}\right) \begin{bmatrix} -\sin\left(\theta_{03} - \theta_{41}\right) \\ \cos\left(\theta_{03} - \theta_{41}\right) \end{bmatrix} - \left(\dot{\theta}_{03} - \dot{\theta}_{41}\right)^2 \begin{bmatrix} \cos\left(\theta_{03} - \theta_{41}\right) \\ \sin\left(\theta_{03} - \theta_{41}\right) \end{bmatrix} \right)$$

At this step, only one supplementary counterweight is necessary for the cancellation of the shaking force, but it could be demonstrated after more derivations that another is necessary for the cancellation of the shaking moment. Therefore, we propose adding this additional counterweight directly. The two counterweights are located at points M_3 and M_4, defined such that: $\overrightarrow{CM_3} = r_{cw_3} l_3 \mathbf{u}$, $\overrightarrow{AM_4} = r_{cw_4} l_3 \mathbf{u}$, r_{cw_3} and r_{cw_4} being dimensionless coefficients. Their masses are respectively denoted m_{cw_3} and m_{cw_4}. With the addition of the counterweights, the shaking force becomes:

$$\mathbf{f}^{sh^*} = \mathbf{f}^{sh} + m_{cw_3}\ddot{\mathbf{r}}_B + m_{cw_3}\mathbf{a}_1 + \left(m_{cw_3} r_{cw_3} + m_{cw_4} r_{cw_4}\right)\mathbf{a}_2 \tag{7.34}$$

Thus, the shaking force is cancelled if:

$$\begin{aligned} m_{cw_4} &= -\frac{m_4 r_4}{r_{cw_4}} \\ m_{cw_3} &= -\frac{m_3 r_3}{r_{cw_3}} \\ m_{cw_2} &= -\frac{m_2 r_2 + m_3 + m_{cw_3}}{r_{cw_2}} \\ m_{cw_1} &= -\frac{m_1 r_1 + m_2 + m_{cw_2} + m_3 + m_{cw_3}}{r_{cw_1}} \end{aligned} \tag{7.35}$$

Developing and simplifying, the expression of the angular momentum is equal to:

$$H_A = I_{eq_1}\dot{\theta}_{03} + I_{eq_2}\dot{\theta}_{41} \tag{7.36}$$

with

$$\begin{aligned} I_{eq_1} =& \sum_{i=1}^{4} I_{S_i} + \left(m_1 r_1^2 + m_{cw_1} r_{cw_1}^2 + m_2 (1 - r_2)^2 + m_{cw_2}\left(1 - r_{cw_2}\right)^2\right) l_{AB}^2 \\ &+ \left(m_3 r_3^2 + m_{cw_3} r_{cw_3}^2 + m_4 r_4^2 + m_{cw_4} r_{cw_4}^2\right) l_3^2 \end{aligned} \tag{7.37}$$

$$\begin{aligned} I_{eq_2} =& I_{S_1} + \left(m_1 r_1^2 + m_{cw_1} r_{cw_1}^2 + m_2 + m_{cw_2} + m_3\right) l_{AB}^2 \\ &- I_{S_2} - \left(m_2 r_2^2 + m_{cw_2} r_{cw_2}^2 + m_3\right) l_{BC}^2 \end{aligned} \tag{7.38}$$

From (7.31), $I_{eq_2} = 0$. Therefore, the shaking moment of the leg can be cancelled using a simple counter-rotation I_{cr} with an axial moment of inertia equal to I_{eq_1}.

7.2.2.3 Shaking Moment and Shaking Force Balancing of the 3-*RPR* Manipulator

Now, let us apply such an approach to the 3-*RPR* mechanism. First of all, let us substitute the platform mass by three points masses located at C_1, C_2 and C_3, with

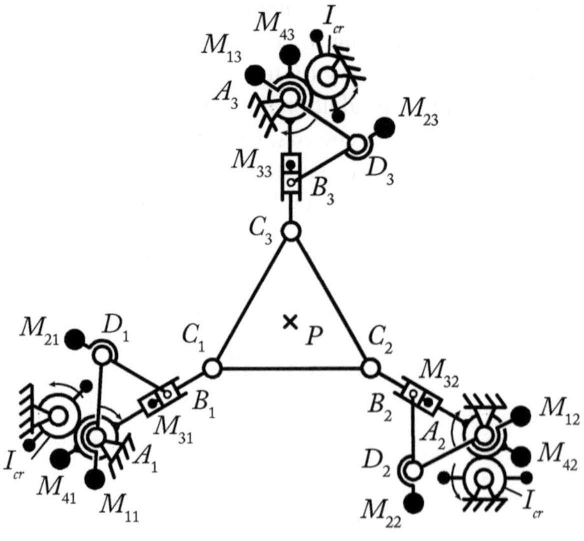

Fig. 7.15 Schematics of a shaking force and shaking moment balanced 3-RPR mechanism

the values of mass equal to m_{p_1}, m_{p_2} and m_{p_3} respectively (Arakelian and Smith 1999; Baradat et al. 2008; Seyferth 1974). Such a condition can be obtained if:

$$m_{p_i} = m_p/3 \text{ and } I_p = 3m_{p_i} l_{C_i P}^2 \tag{7.39}$$

Such a decomposition of the platform enables us to consider the shaking force and shaking moment balancing of each leg of the mechanism. Then, modifying each leg in order to obtain a mechanism similar to a Scott-Russel linkage, the shaking force and shaking moment are cancelled if:

$$\begin{aligned}
m_{cw_4} &= -\frac{m_4 r_4}{r_{cw_4}} \\
m_{cw_3} &= -\frac{m_3 r_3 + m_{p_i}}{r_{cw_3}} \\
m_{cw_2} &= -\frac{m_2 r_2 + m_3 + m_{cw_3} + m_{p_i}}{r_{cw_2}} \\
m_{cw_1} &= -\frac{m_1 r_1 + m_2 + m_{cw_2} + m_3 + m_{cw_3} + m_{p_i}}{r_{cw_1}} \\
0 &= I_{S_1} + \left(m_1 r_1^2 + m_{cw_1} r_{cw_1}^2 + m_2 + m_{cw_2} + m_3\right) l_{A_i B_i}^2 \\
&\quad - I_{S_2} - \left(m_2 r_2^2 + m_{cw_2} r_{cw_2}^2 + m_3\right) l_{B_i C_i}^2, \text{ and} \\
I_{cr} &= \sum_{i=1}^{4} I_{S_i} + \left(m_1 r_1^2 + m_{cw_1} r_{cw_1}^2 + m_2 (1 - r_2)^2 + m_{cw_2} \left(1 - r_{cw_2}\right)^2\right) l_{A_i B_i}^2 \\
&\quad + \left(m_3 r_3^2 + m_{cw_3} r_{cw_3}^2 + m_4 r_4^2 + m_{cw_4} r_{cw_4}^2 + m_{p_i}\right) l_3^2
\end{aligned} \tag{7.40}$$

taking into account that I_{cr} is the axial moment of inertia of the counter-rotations (Fig. 7.15).

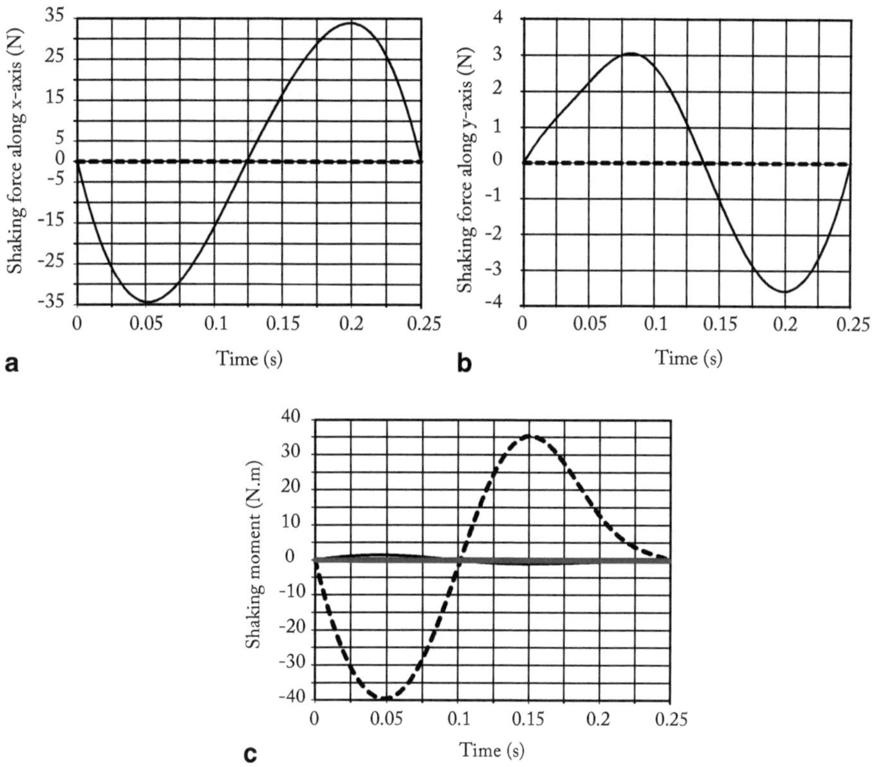

Fig. 7.16 Shaking force and shaking moment before (*solid line*) and after (*dashed line*) the addition of the counterweights, and after the addition of the counter-rotations (*gray line*)

Thus, with this approach it is possible to create a fully-balanced shaking force and shaking moment 3-*RPR* mechanism with only three counter-rotations (Fig. 7.15), i.e., this method enables a reduction in the number of counter-rotations by a factor of two.

Numerical Application

The parameters used for the simulations are the followings:

- $l_{OA_i} = 0.35$ m, $l_{A_iB_i} = l_{B_iC_i} = 0.25$ m, $l_{C_iP} = 0.1$ m and $l_3 = 0.025$ m,
- $r_1 = r_2 = 0.5$, $r_3 = 0$ and $r_4 = 4$,
- $m_1 = 1.09$ kg, $m_2 = 1.1$ kg, $m_3 = 0.37$ kg, $m_4 = 0.75$ kg; $m_p = 1$ kg,
- $I_{S_1} = 0.00738$ kg m^2, $I_{S_2} = 0.58389$ kg m^2, $I_3 = 0.00344$ kg m^2, $I_{S_4} = 0.00025$ kg m^2, and $I_p = 0.01$ kg m^2.

For these new parameters and for the trajectory used in Sect. 7.2.1, taking into account that the position coefficients of the counterweights are equal to $r_{cw_j} = -0.5$ ($j = 1, 3, 4$), $r_{cw_2} = -1$, the new values of the counterweights are: $m_{cw_1} = 3.17$ kg,

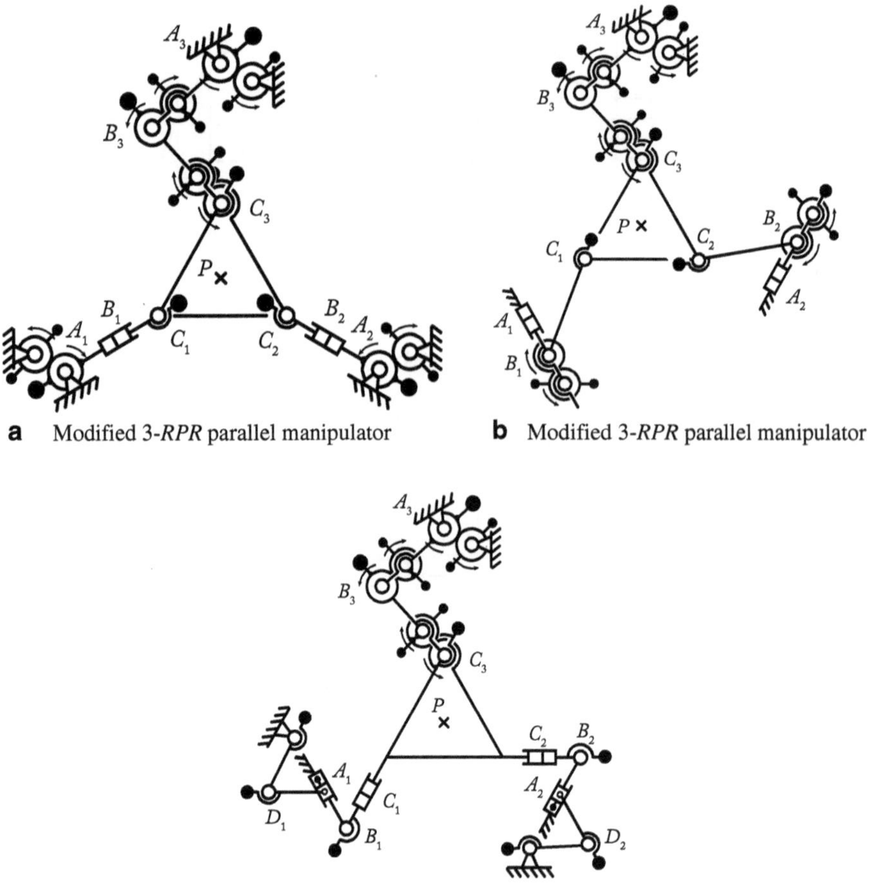

a Modified 3-*RPR* parallel manipulator **b** Modified 3-*RPR* parallel manipulator

c Modified 3-*RPR* parallel manipulator

Fig. 7.17 Complete shaking force and shaking moment balancing of planar manipulators with prismatic pairs via structural modification of one leg

$m_{cw_2} = 11.71$ kg, $m_{cw_3} = 0.33$ kg, $m_{cw_4} = 0.75$ kg. The shaking force and shaking moment are then computed (dashed line in Fig. 7.16). It is possible to see that, with the counterweights, the shaking efforts are cancelled, while the maximal value of the shaking moment is increased by a factor 28. Finally, we add the counter-rotations. Their values are equal to $I_{cr} = 1.56907$ kg m^2. With such counter-rotations, the shaking moment is balanced (gray line in Fig. 7.16c).

Finally, it should be noted that the combination of the proposed two techniques of balancing enables the creation of fully balanced parallel manipulators with modified legs. As examples, different structures of balanced manipulators are presented in Fig. 7.17 (3-*RPR*, 3-*PRR* and 3-*PRP*) in which one leg with a prismatic pair is replaced by a leg with only revolute joints. Such a modification allows the displacement of the *com* of the manipulator to C_3 and then to balance the manipulator via the modified leg $C_3B_3A_3$.

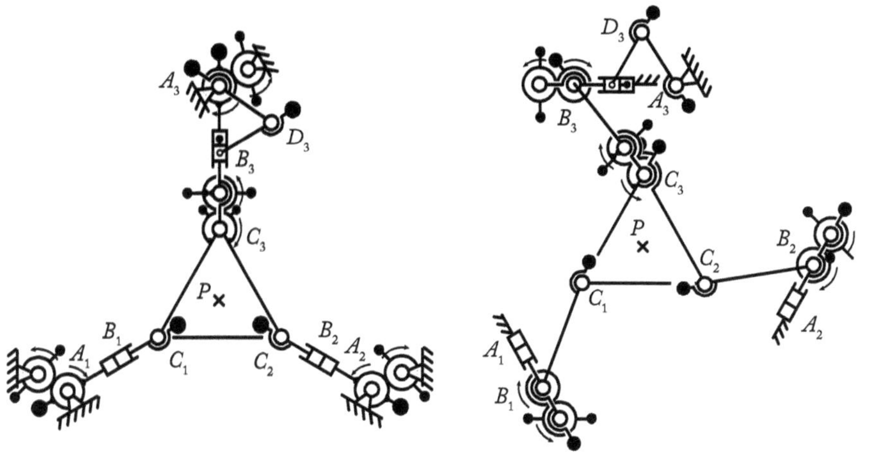

a Balacing of 3-*RPR* parallel manipulator **b** Balacing of 3-*RPR* parallel manipulator

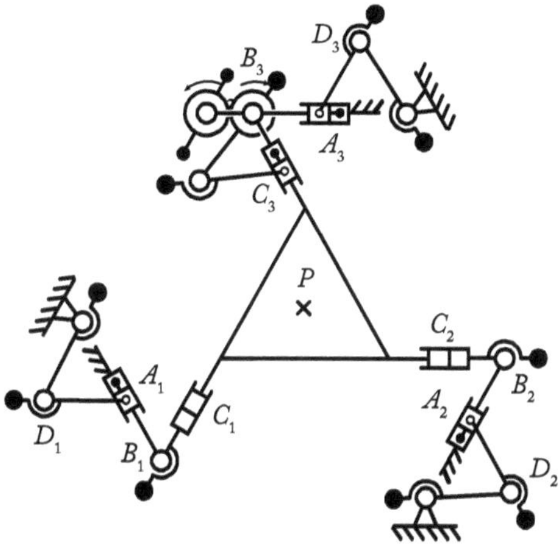

c Balacing of 3-*RPR* parallel manipulator

Fig. 7.18 Complete shaking force and shaking moment balancing of planar manipulators with prismatic with reduced number of Scott-Russell mechanisms

In the same way, it is possible to balance a parallel manipulator with prismatic pairs by adding fewer Scott-Russell mechanisms. The balancing schemes for several parallel manipulators are presented in Fig. 7.18.

Thus, it has been presented new balancing schemes for the shaking force and shaking moment of planar parallel manipulators whose legs are made of prismatic pairs.

Usually, the balancing of parallel manipulators with prismatic pairs is only attained via a considerably complicated design. This section showed that it is possible to balance planar parallel mechanisms using Scott-Russell mechanisms. Such an approach enables a division of the number of counter-rotations by two. Numerical simulations carried out using ADAMS software validated the obtained results and illustrated that the suggested balancing enables the creation of a parallel manipulator transmitting no inertia load to its base.

7.3 Shaking Force Minimization of High-speed Robots via Centre of Mass Acceleration Control

7.3.1 Minimization of the Shaking Forces via an Optimal Motion Planning of the Total Mass Centre of Moving Links

7.3.1.1 Definition of the optimal trajectory

The shaking forces $\mathbf{f}^{sh}$ of a manipulator can be written in the form:

$$\mathbf{f}^{sh} = \sum m_i \ddot{\mathbf{r}}_S \tag{7.41}$$

where $\sum m_i$ is the total mass of the moving links of the manipulator and $\ddot{\mathbf{r}}_S$ is the acceleration of the *com* S of the robot. The classical balancing approach consists in adding counterweights in order to keep the total *com* of moving links stationary. In this case, $\ddot{r}_S = 0$ for any configuration of the mechanical system. But, as a consequence, the total mass of the manipulator is considerably increased. Thus, in order to avoid this drawback, in the present Section, a new approach is proposed, which consists of the optimal control of the total *com* of the robot. Such an optimal motion planning allows the reduction of the total mass centre acceleration and, consequently, the reduction of the shaking force.

Classically, manipulator displacements are defined considering either joint coordinates $\mathbf{q}$ or Cartesian variables $\mathbf{x}$. Knowing the initial and final manipulator configurations at time t_0 and t_f, denoted as $\mathbf{q}_0 = \mathbf{q}(t_0)$ and $\mathbf{q}_f = \mathbf{q}(t_f)$, or $\mathbf{x}_0 = \mathbf{x}(t_0)$ and $\mathbf{x}_f = \mathbf{x}(t_f)$, in the case of the control of the Cartesian variables, the classical displacement law may be written in the form:

$$\mathbf{q}(t) = s_q(t)(\mathbf{q}_f - \mathbf{q}_0) + \mathbf{q}_0 \text{ or } \mathbf{x}(t) = s_x(t)(\mathbf{x}_f - \mathbf{x}_0) + \mathbf{x}_0 \tag{7.42}$$

where $s_q(t)$ and $s_x(t)$ may be polynomial (of order 3, 5 and higher), sinusoidal, bang-bang, etc. motion profiles (Khalil and Dombre 2002).

From expression (7.41), it is possible to see that the shaking force, in terms of norm, is minimized if the norm $\|\ddot{r}_S\|$ of the *com* acceleration is minimized along the trajectory. This means that if the displacement $\mathbf{r}_S$ of the manipulator *com* is optimally controlled, the shaking force will be minimized. As a result, the first problem is to define the optimal trajectory for the displacement $\mathbf{r}_S$ of the manipulator *com*.

For this purpose, let us consider the displacement $\mathbf{r}_S$ of a point S in the Cartesian space. First, in order to minimize the *com* acceleration, the length of the path followed by S should be minimized, i.e. point S should move along a straight line passing through its initial and final positions, denoted as $\mathbf{r}_{S_0}$ and $\mathbf{r}_{S_f}$, respectively.

Then, the motion profile used on this path should be optimized. It is assumed that, at any moment during the displacement, the norm of the maximal admissible acceleration the point S can reach is constant and denoted as $\ddot{r}_S^{max}$. Taking this maximal value for the acceleration into consideration, it is known that the motion profile that minimize the time interval $\{t_0, t_f\}$ for going from position $\mathbf{r}_{S_0}$ to position $\mathbf{r}_{S_f}$ is the bang-bang acceleration profile (Khalil and Dombre 2002), given by (Fig. 7.19a),

$$\begin{cases} \mathbf{r}_S(t) = s(t)(\mathbf{r}_{S_f} - \mathbf{r}_{S_0}) + \mathbf{r}_{S_0} \\ \dot{\mathbf{r}}_S(t) = \dot{s}(t)(\mathbf{r}_{S_f} - \mathbf{r}_{S_0}) \\ \ddot{\mathbf{r}}_S(t) = \ddot{s}(t)(\mathbf{r}_{S_f} - \mathbf{r}_{S_0}) \end{cases} \tag{7.43}$$

where

$$\ddot{s}(t) = \frac{1}{\|\mathbf{r}_{S_f} - \mathbf{r}_{S_0}\|} \begin{cases} \ddot{r}_S^{max} & \text{for } t \leq (t_f - t_0)/2 \\ -\ddot{r}_S^{max} & \text{for } t \geq (t_f - t_0)/2 \end{cases} \tag{7.44}$$

Consequently, if the time interval $\{t_0, t_f\}$ for the displacement between positions $\mathbf{r}_{S_0}$ and $\mathbf{r}_{S_f}$ is fixed, the bang-bang acceleration profile is the trajectory that minimizes the value of the maximal acceleration $\ddot{r}_S^{max}$. Thus, in order to minimize $\|\ddot{r}_S\|$ for a displacement during the fixed time interval $\{t_0, t_f\}$, the bang-bang acceleration profile must be applied on the displacement $\mathbf{r}_S$ on the manipulator total *com*.

7.3.1.2 Observations About the Modification of the Optimal Trajectory for Taking into Account the Actuators Properties

It should be mentioned that the given bang-bang acceleration profile (Fig. 7.19a) is based on theoretical considerations. In reality, the actuators are unable to achieve discontinuous efforts. Therefore, this motion profile should be modified by a trapezoidal acceleration profile (Fig. 7.19b) in order to take into account the actuators properties in terms of maximal admissible effort variations.

For a given time interval $\{t_0, t_f\}$, the trapezoidal profile, as we define it, is characterized by two parameters: t_1 and t_2 (Fig. 7.19b). In order to find the optimal values for t_1 and t_2, the following problem should be considered:

$$\ddot{r}_S^{max} \rightarrow \min_{t_1, t_2} \tag{7.45}$$

under the constraints

$$\max\left|\frac{d\tau_i}{dt}\right| \leq \dot{\tau}_i^{max} \tag{7.46}$$

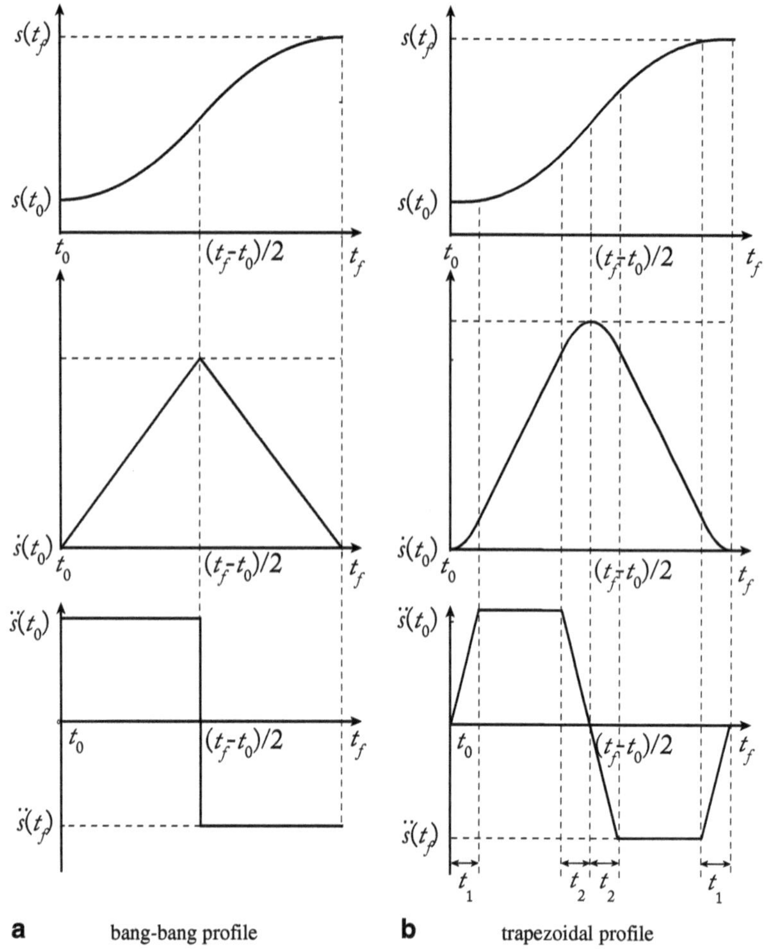

Fig. 7.19 Motion profiles used for the shaking force minimization

where τ_i is the input effort of the actuator i and $\dot{\tau}_i^{max}$ is the maximal admissible input effort variation for the actuator i. This problem is highly non linear, therefore it can be solved by numerical optimization methods. It should be mentioned that in the illustrative examples given in Sect. 7.3.2, the trapezoidal profile taking into account the actuators properties has been found using the optimisation function "fgoalattain" of Matlab.

7.3.1.3 Expression of the Manipulator Coordinates as a Function of the Mass Centre Parameters

Once the displacement of the manipulator *com* is defined, the second problem is to find the joint (or Cartesian) coordinates corresponding to this displacement. For this purpose, let us consider a manipulator composed of n links. The mass of the link i is denoted as m_i ($i = 1 \cdots n$) and the position of its *com* as $\mathbf{r}_{S_i}$. Once the joint coordinates $\mathbf{q}$ or Cartesian variables $\mathbf{x}$ are known, the values of $\mathbf{r}_{S_i}$ may easily be obtained using the manipulator kinematics relationships. As a result, the position of the manipulator *com S*, defined as

$$\mathbf{r}_S = \frac{1}{m_{tot}} \sum_{i=1}^{n} m_i \mathbf{r}_{S_i} \text{ where } m_{tot} = \sum_{i=1}^{n} m_i \tag{7.47}$$

may be expressed as a function of $\mathbf{x}$ or $\mathbf{q}$. However, in order to control the manipulator, the inverse problem should be solved, i.e. it is necessary to express variables $\mathbf{x}$ or $\mathbf{q}$ as a function of $\mathbf{r}_S$. Here, two cases should be distinguished:

1. $dim(\mathbf{r}_S) = dim(\mathbf{q})$, i.e. the manipulator has got as many actuators as controlled variables for the displacements $\mathbf{r}_S$ of the *com* (two variables for planar cases, three variables for spatial problems). In such case, the variables $\mathbf{x}$ or $\mathbf{q}$ can be directly expressed as a function of $dim(\mathbf{r}_S)$ using (7.47), i.e. $\mathbf{q} = g_1(\mathbf{r}_S)$,
2. $dim(\mathbf{r}_S) < dim(\mathbf{q})$, i.e. the manipulator has got more actuators than controlled variables. In such case, the problem is under-determined as there are more parameters in variables $\mathbf{x}$ or $\mathbf{q}$ than in $\mathbf{r}_S$. In order to solve it, let us consider that p_0 parameters of vector $\mathbf{q}_0$ (or $\mathbf{x}_0$) and p_f parameters of vector $\mathbf{q}_f$ (or $\mathbf{x}_f$) are fixed. In a first task, it is necessary to define the $m - p_0$ and $m - p_f$ other parameters of the initial and final manipulator configurations ($m = dim(\mathbf{q})$. The way to fix it is to find the manipulator initial and final configurations, taking into account the p_0 initial and p_f final fixed parameters, that will allow minimizing the norm of the vector $\mathbf{r}_{S_f} - \mathbf{r}_{S_0}$, i.e. the length of the displacement of the manipulator *com*. Then, the second task is to choose $m - k$ articular variables among the m possible of vector $\mathbf{q}$ ($k = dim(\mathbf{r}_S)$). These $m - k$ variables, denoted as $\mathbf{q}_{m-k}$, will be controlled using some classical displacement law given at (7.42) or can be used in order to minimize some other performance criteria, such as the shaking moments or some other interesting performance criterion (see Sect. 7.3.2). The k other variables, denoted as $\mathbf{q}_k$, should be expressed as a function of $\mathbf{r}_S$ and $\mathbf{q}_{m-k}$ using (7.47), i.e. $\mathbf{q}_k = g_2(\mathbf{r}_S, \mathbf{q}_{m-k})$.

In order to demonstrate the proposed balancing method, two illustrative examples are given in the following Section.

7.3.2 *Illustrative Examples*

7.3.2.1 The Planar 2R Serial Manipulator

Let us consider the shaking force minimization of a 2*R* serial manipulator (Fig. 7.20). This manipulator is controlled using two rotary actuators having two input parameters which are denoted as q_1 and q_2. For simulations the following parameters have been used:

Fig. 7.20 Schematics of the 2*R* serial manipulator

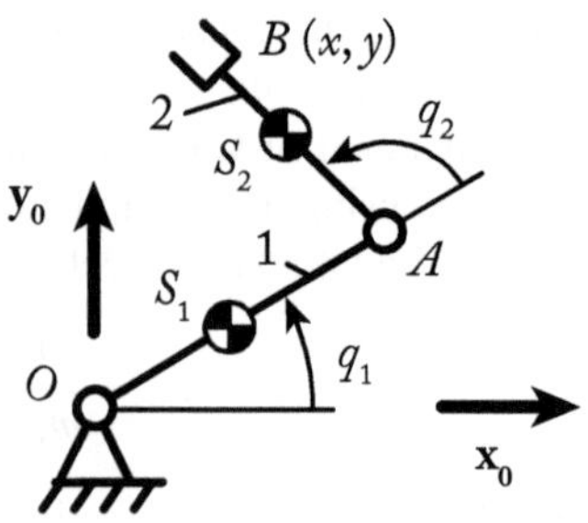

- $l_{OA} = 0.5$ m, $l_{OA} = 0.3$ m,
- $r_1 = 0.289$, where $l_{OS_1} = r_1 l_{OA}$ and $r_2 = 0.098$, where $l_{AS_2} = r_2 l_{AB}$.

The mass and inertia parameters are:

- $m_1 = 24.4$ kg and $m_2 = 8.3$ kg,
- $m_{tool} = 5$ kg, where m_{tool} is the payload,
- $I_{S_1} = 1.246$ kg m^2 and $I_{S_2} = 0.057$ kg m^2.

Let us now express the articulated joint positions $\mathbf{q}^T = [q_1 \; q_2]$ as a function of the position $\mathbf{r}_S$ of the manipulator *com*. From (7.47), we obtain:

$$\mathbf{r}_S = \begin{bmatrix} x_S \\ y_S \end{bmatrix} = \frac{m_1 r_1 l_{OA}}{m_{tot}} \begin{bmatrix} \cos q_1 \\ \sin q_1 \end{bmatrix} + \frac{m_2}{m_{tot}} \left(l_{OA} \begin{bmatrix} \cos q_1 \\ \sin q_1 \end{bmatrix} + r_2 l_{AB} \begin{bmatrix} \cos(q_1+q_2) \\ \sin(q_1+q_2) \end{bmatrix} \right)$$
$$+ \frac{m_{tool}}{m_{tot}} \left(l_{OA} \begin{bmatrix} \cos q_1 \\ \sin q_1 \end{bmatrix} + l_{AB} \begin{bmatrix} \cos(q_1+q_2) \\ \sin(q_1+q_2) \end{bmatrix} \right) \quad (7.48)$$

which leads to the following loop-closure equation:

$$\left(x_S - l_{eq_1} \cos q_1\right)^2 + \left(y_S - l_{eq_1} \sin q_1\right)^2 - l_{eq_2}^2 = 0 \quad (7.49)$$

where $l_{eq_1} = (m_1 r_1 + m_2 + m_{tool}) l_{OA}/m_{tot}$ and $l_{eq_2} = (m_2 r_2 + m_{tool}) l_{AB}/m_{tot}$.

Replacing $\cos q_1$ and $\sin q_1$ by $(1-t_1^2)/(1+t_1^2)$ and $2t_1/(1+t_1^2)$ ($t_1 = \tan(q_1/2)$), respectively, and developing (7.49), we obtain:

$$q_1 = 2\tan^{-1}\left(\frac{-b \pm \sqrt{b^2 - c^2 + a^2}}{c - a}\right) \quad (7.50)$$

where

$$a = -2 l_{eq_1} x_S, \; b = -2 l_{eq_1} y_S, \; c = x_S^2 + y_S^2 + l_{eq_1}^2 - l_{eq_2}^2 \quad (7.51)$$

In (7.50), the sign $\pm$ stands for the two possible working modes of the manipulator (for simulations, the working mode with the '+' sign is used). Once q_1 is known, q_2 may easily be found from (7.48):

$$q_2 = \tan^{-1}\left(\frac{y_S - l_{eq_1} \sin q_1}{x_S - l_{eq_1} \cos q_1}\right) - q_1 \quad (7.52)$$

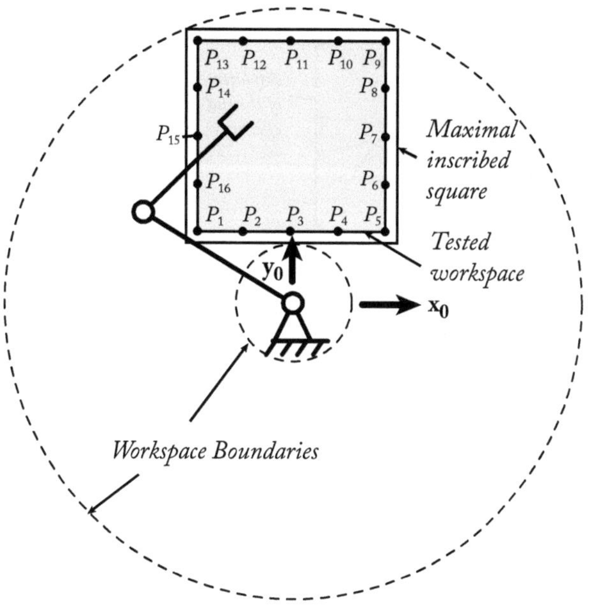

Fig. 7.21 The tested trajectories of the $2R$ serial manipulator

Let us now test the proposed approach. In order to show the efficiency of this optimal planning, several trajectories are tested. These trajectories are defined as follows. First, the maximal inscribed square inside of the workspace is found (Fig. 7.21). For this manipulator, it is a square of length 0.55 m, of which centre E is located at $x = 0$ m and $y = 0.475$ m. Then, in order to avoid problems due to the proximity of singular configuration, the tested zone is restricted to a square centred in E of edge length equal to 0.45 m (in grey on Fig. 7.21). Finally, we discretize each edge into four segments delimited by the points P_i $(i = 1 \cdots 16)$. The tested trajectories will be the segments P_1P_{13}, P_2P_{12}, P_3P_{11}, P_4P_{10}, P_5P_{19}, $P_{15}P_7$, $P_{14}P_8$ and $P_{13}P_9$. Each trajectory will have duration of 0.5 s and, for each trajectory, three different kinds of motion profiles are applied:

1. a fifth order polynomial profile is applied on the displacement of the manipulator end-effector,
2. a bang-bang acceleration profile is applied on the displacement of the manipulator *com*,
3. a trapezoidal acceleration variation is applied on the displacement of the manipulator centre of masses, taking into account that, for each actuator, the input effort variation is limited by 3.104 Nm/s.

The displacements of the end-effector and manipulator links centre of masses for the trajectory P_5P_9 are shown in Fig. 7.22. These trajectory parameters are implemented into Adams software and it is computed the variations of shaking forces. Figure 7.23 presents the shaking force transmitted by the manipulator for trajectory P_5P_9. The obtained results for the whole paths are summarized in Table 7.2. It is shown that the optimal trajectory planning (bang-bang profile) allows the reduction

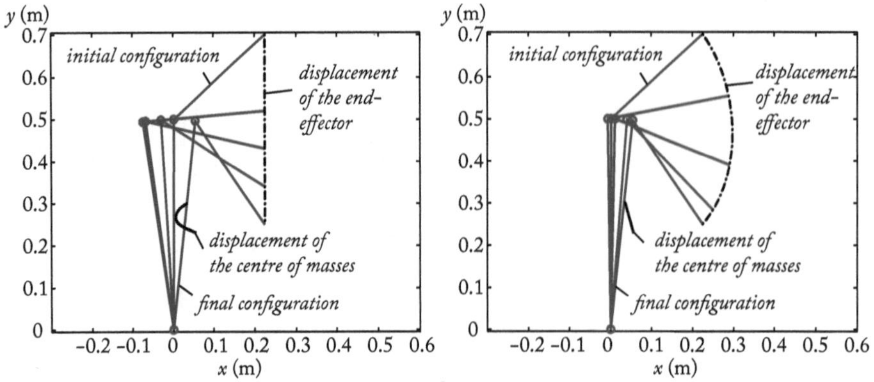

a for case 1 **b** for cases 2 and 3

Fig. 7.22 Manipulator end-effector displacements along the trajectory P_5P_9

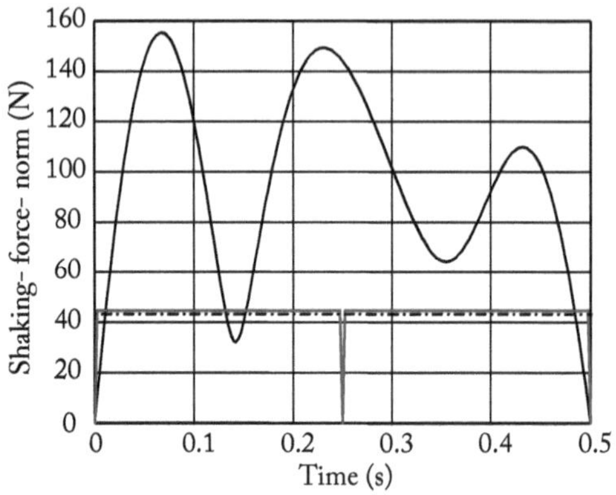

Fig. 7.23 Variations of the shaking force in the case of the trajectory P_5P_9: case 1 (*black full line*), case 2 (*black dashed line*) and case 3 (*grey full line*)

of the shaking force from 36 % up to 76.7 %. Moreover, it appears that for given actuator parameters, the minimizations obtained in the cases of the bang-bang and

Table 7.2 Maximal value of the shaking force norm for the tested trajectories of the 2*R* serial manipulator

		Followed path							
		P_1P_{13}	P_2P_{12}	P_3P_{11}	P_4P_{10}	P_5P_{19}	$P_{15}P_7$	$P_{14}P_8$	$P_{13}P_9$
$\max\left(\lVert \mathbf{f}^{sh} \rVert\right)$ (N)	Case 1	194.7	165.3	178.8	178.2	155.3	218.7	201.3	195.4
	Case 2	121.0	85.0	47.5	41.5	44.3	136.3	121.7	111.8
	Case 3	124.0	86.0	48.0	41.8	44.7	137.7	123.2	113.1
% of reduction	Case 2/1	37.8	48.6	73.4	76.7	71.5	37.8	39.5	42.8
	Case 3/1	36.3	48.0	73.2	76.5	71.2	37.0	38.9	42.1

trapezoidal profiles are very close (less than 1 %). It is due to the fact that the actuators can apply high input effort variations during a displacement. However, such a result depends on the actuator power capacity and it will be variable for each type of actuator. Obviously, the rate of reduction depends on the design parameters of the robot. For each system, it will be different

Let us now consider the second example.

7.3.2.2 The Planar 3R Serial Manipulator

This manipulator is controlled using three rotary actuators (Fig. 7.24), with three input parameters which are denoted as q_1, q_2 and q_3. The link parameters are the following:

- $l_{OA} = 0.5$ m, $l_{AB} = 0.3$ m, $l_{BC} = 0.1$ m,
- $r_1 = 0.289$, where $l_{OS_1} = r_1 l_{OA}$, $r_2 = 0.098$, where $l_{AS_2} = r_2 l_{AB}$, and $r_3 = 0.5$, where $l_{BS_3} = r_3 l_{BC}$,
- $m_1 = 24.4$ kg, $m_2 = 8.3$ kg and $m_3 = 2$ kg,
- $m_{tool} = 5$ kg, where m_{tool} is the payload,
- $I_{S_1} = 1.246$ kg m^2, $I_{S_2} = 0.057$ kg m^2 and $I_{S_3} = 0.025$ kg m^2.

In order to have the possibility to control the manipulator, let us express the relation between the articulated joint positions $\mathbf{q}^T = [q_1 \; q_2 \; q_3]$ and the position $\mathbf{r}_S$ of the manipulator *com*. From (7.47), we obtain:

$$\mathbf{r}_S = \begin{bmatrix} x_S \\ y_S \end{bmatrix} = \frac{m_1 r_1 l_{OA}}{m_{tot}} \begin{bmatrix} \cos q_1 \\ \sin q_1 \end{bmatrix} + \frac{m_2}{m_{tot}} \left(l_{OA} \begin{bmatrix} \cos q_1 \\ \sin q_1 \end{bmatrix} + r_2 l_{AB} \begin{bmatrix} \cos(q_1+q_2) \\ \sin(q_1+q_2) \end{bmatrix} \right)$$
$$+ \frac{m_3}{m_{tot}} \left(l_{OA} \begin{bmatrix} \cos q_1 \\ \sin q_1 \end{bmatrix} + l_{AB} \begin{bmatrix} \cos(q_1+q_2) \\ \sin(q_1+q_2) \end{bmatrix} + r_3 l_{BC} \begin{bmatrix} \cos\phi \\ \sin\phi \end{bmatrix} \right)$$
$$+ \frac{m_{tool}}{m_{tot}} \left(l_{OA} \begin{bmatrix} \cos q_1 \\ \sin q_1 \end{bmatrix} + l_{AB} \begin{bmatrix} \cos(q_1+q_2) \\ \sin(q_1+q_2) \end{bmatrix} + l_{BC} \begin{bmatrix} \cos\phi \\ \sin\phi \end{bmatrix} \right) \quad (7.53)$$

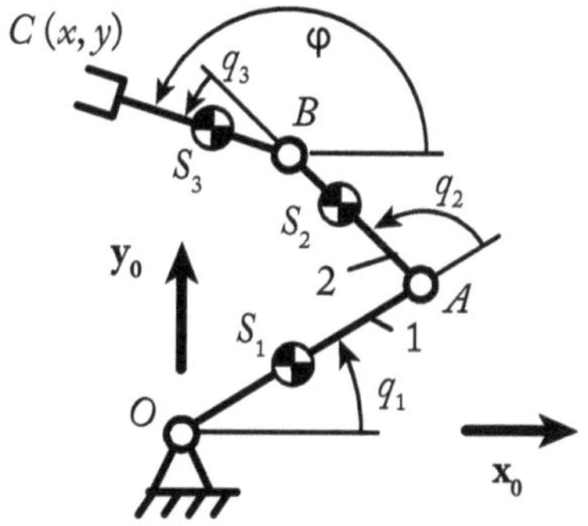

Fig. 7.24 Schematics of the 3*R* serial manipulator

where $\phi = q_1 + q_2 + q_3$ is the end-effector orientation.

In (7.53), there are three unknowns q_1, q_2, q_3 for two fixed parameters x_S and y_S. Therefore, as mentioned in Sect. 7.3.1, a way to solve this problem is to consider that one parameter, for example ϕ, is used to minimize some objective function. Then the expressions of q_1, q_2 and q_3 can be found as a function of x_S, y_S and ϕ. In the following of the paper, angle ϕ is used in order to minimize the shaking moment msh of the robot. Obviously, if necessary, it can be replaced by another criterion, such the energy, the torques, etc.

Equation (7.53) leads to the following loop-closure equation:

$$\left(x_S - l_{eq_3}\cos\phi - l_{eq_1}\cos q_1\right)^2 + \left(y_S - l_{eq_3}\sin\phi - l_{eq_1}\sin q_1\right)^2 - l_{eq2}^2 = 0 \quad (7.54)$$

where $l_{eq_1} = (m_1 r_1 + m_2 + m_3 + m_{tool}) l_{OA}/m_{tot}$, $l_{eq_2} = (m_2 r_2 + m_3 + m_{tool}) l_{AB}/m_{tot}$ and $l_{eq_3} = (m_3 r_3 + m_{tool}) l_{BC}/m_{tot}$.

Replacing $\cos q_1$ and $\sin q_1$ by $(1-t_1^2)/(1+t_1^2)$ and $2t_1/(1+t_1^2)$ $(t_1 = \tan(q_1/2))$, respectively, and developing (7.54), we obtain:

$$q_1 = 2\tan^{-1}\left(\frac{-b \pm \sqrt{b^2 - c^2 + a^2}}{c - a}\right) \quad (7.55)$$

where

$$\begin{aligned} a &= -2l_{eq_1}\left(x_S - l_{eq_3}\cos\phi\right) \\ b &= -2l_{eq_1}\left(y_S - l_{eq_3}\sin\phi\right) \\ c &= \left(x_S - l_{eq_3}\cos\phi\right)^2 + \left(y_S - l_{eq_3}\sin\phi\right)^2 + l_{eq_1}^2 - l_{eq2}^2 \end{aligned} \quad (7.56)$$

In expression (7.55), the sign $\pm$ stands for the two possible working modes of the manipulator (for simulations, the working mode with the '+' sign is used). Once q_1 is known, q_2 and q_3 may easily be found from (7.53):

$$q_2 = \tan^{-1}\left(\frac{y_S - l_{eq_3}\sin\phi - l_{eq_1}\sin q_1}{x_S - l_{eq_3}\cos\phi - l_{eq_1}\cos q_1}\right) - q_1 \quad (7.57)$$

$$q_3 = \phi - q_2 - q_1 \quad (7.58)$$

Let us now test the proposed approach with this manipulator. The tested trajectories are defined as follows. First, the maximal inscribed square inside of the workspace, for any end-effector orientation, is found (Fig. 7.25). For this manipulator, it is a square of length 0.375 m, of which centre E is located at $x = 0$ m and $y = 0.487$ m. Then, in order to avoid problems due to the proximity of singular configuration, the tested zone is restricted to a square centred in E of edge length equal to 0.3 m (in grey on Fig. 7.25). Finally, we discretize each edge into four segments delimited by the points P_i $(i = 1 \cdots 16)$. The tested trajectories will be

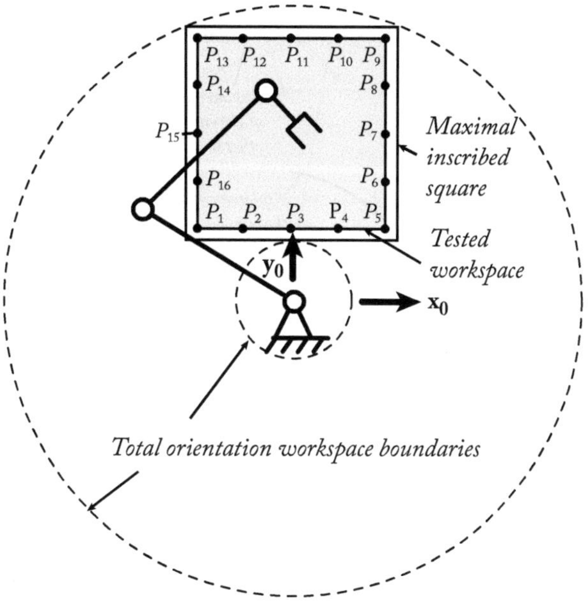

Fig. 7.25 The tested trajectories of the 3*R* serial manipulator

the segments P_1P_{13}, P_2P_{12}, P_3P_{11}, P_4P_{10}, P_5P_9, $P_{15}P_7$, $P_{14}P_8$ and $P_{13}P_9$. It should be noted that in this case there is an independent parameter ϕ, which can be defined from complementary condition describing the orientation of the end-effector. For numerical simulations, it is chosen to begin the tested trajectories with an end-effector orientation $\phi_0 = 0\,^\circ$ and to finish it at $\phi_f = 0\,^\circ$.

The simultaneous minimization of the shaking force and the shaking moment cannot be done without using an optimization algorithm in order to solve the following problem:

$$\max(M^{sh}) \rightarrow \min_{\phi} \tag{7.59}$$

under the constraints

$$\begin{cases} \phi(t_0) = \phi_0,\ \phi(t_f) = \phi_f \\ \dot{\phi}(t_0) = \dot{\phi}(t_f) = 0 \\ \ddot{\phi}(t_0) = \ddot{\phi}(t_f) = 0 \\ \mathbf{r}_S(t_0) = \mathbf{r}_{S_0},\ \mathbf{r}_S(t_f) = \mathbf{r}_{S_f} \end{cases} \tag{7.60}$$

Several motion profiles for ϕ can be tested. Here it is proposed to use polynomials. Our observations showed that the polynomial function that makes it possible to obtain optimal results is of degree 8.

Each trajectory will have duration of 0.5 s and, for each trajectory three different kinds of motion profile are applied:

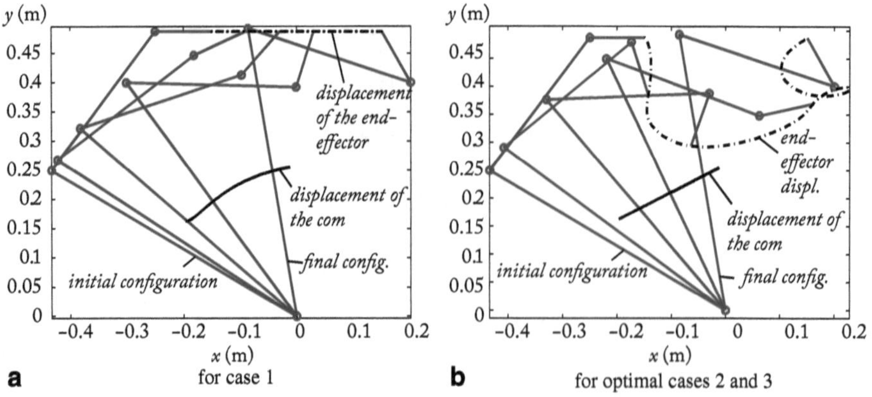

Fig. 7.26 Manipulator end-effector displacements along the trajectory $P_{15}P_7$

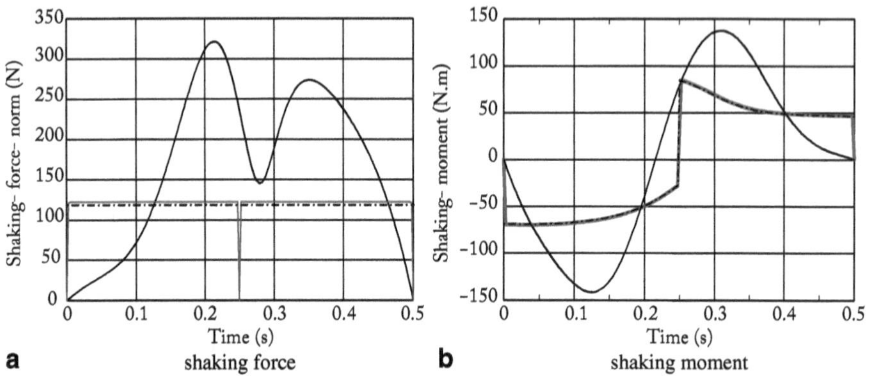

Fig. 7.27 Variations of the shaking force and shaking moment in the case of the trajectory $P_{15}P_7$: case 1 (*black full line*), case 2 (*black dashed line*) and case 3 (*grey full line*)

1. a fifth order polynomial profile is applied on the displacement (translation and rotation) of the manipulator end-effector,
2. a bang-bang profile is applied on the displacement of the manipulator *com* and the angle ϕ is optimized in order to minimize the shaking moment,
3. a trapeze acceleration profile is applied on the displacement of the manipulator *com*, taking into account that, for each actuator, the input effort variation is limited by 3.104 Nm/s; the trajectory for angle ϕ optimized in the previous case is used in order to compute the actuator displacements.

The displacements of the end-effector and manipulator links centre of masses for the trajectory $P_{15}P_7$ are shown in Fig. 7.26. Figure 7.27 presents the shaking force and shaking moment for the path $P_{15}P_7$. The obtained results for the whole paths are summarized in Tables 7.3 and 7.4. It is shown that the optimal trajectory planning (bang-bang profile) allows the reduction of the shaking forces from 48 %

Table 7.3 Maximal value of the shaking force norm for the tested trajectories of the 3*R* serial manipulator.

		Followed path							
		P_1P_{13}	P_2P_{12}	P_3P_{11}	P_4P_{10}	P_5P_{19}	$P_{15}P_7$	$P_{14}P_8$	$P_{13}P_9$
$\max\left(\left\|\mathbf{f}^{sh}\right\|\right)$	Case 1	158.7	144.9	133.6	119.6	103.1	321.3	266.3	238.7
(N)	Case 2	81.8	77.2	68.8	58.1	47.9	121.1	105.5	92.2
	Case 3	82.3	77.6	69.0	58.3	48.0	122.3	106.5	93.0
% of	Case 2/1	48.4	46.7	48.5	51.4	53.5	62.3	60.3	61.4
reduction	Case 3/1	48.1	46.4	48.3	51.2	53.4	61.9	60.0	61.0

Table 7.4 Maximal absolute value of the shaking moment for the tested trajectories of the 3*R* serial manipulator

		Followed path							
		P_1P_{13}	P_2P_{12}	P_3P_{11}	P_4P_{10}	P_5P_{19}	$P_{15}P_7$	$P_{14}P_8$	$P_{13}P_9$
$\max\left(M^{sh}\right)$	Case 1	70.0	63.2	58.3	50.7	42.3	154.0	130.8	119.6
(N.m)	Case 2	43.9	37.7	30.3	22.4	16.4	72.3	64.8	57.0
	Case 3	43.9	37.7	30.3	22.4	16.5	73.0	64.8	57.0
% of	Case 2/1	37.2	40.4	48.1	55.8	61.1	53.1	50.5	52.3
reduction	Case 3/1	37.2	40.4	48.1	55.8	61.0	52.6	50.5	52.3

up to 62.2 %. Moreover, with a simultaneous optimal control of angle ϕ, the shaking moment can be reduced from 37.2 % up to 61 %.

As previously mentioned, these results depend on the design parameters of the used robot. For another manipulator, they will be different. But, in any case the shaking force and moment shall be decreased.

7.3.2.3 Observations About Input Torques

The main drawback of the shaking force balancing by counterweights is the increase of the inertia of moving links caused by adding masses, and consequently, the increase of input torques. The advantage of the suggested balancing method is in the fact that the shaking forces are only reduced by optimal control of moving links, without adding counterweights. It results in the fact that the input torques are considerably lower than in the case of balancing by counterweights. To illustrate this advantage for examined 2*R* serial manipulator, three kinds of simulations have been carried out using dynamic simulation software:

1. unbalanced manipulator carrying out a straight line trajectory along P_5P_9 (Fig. 7.21) using a fifth order polynomial motion profile,

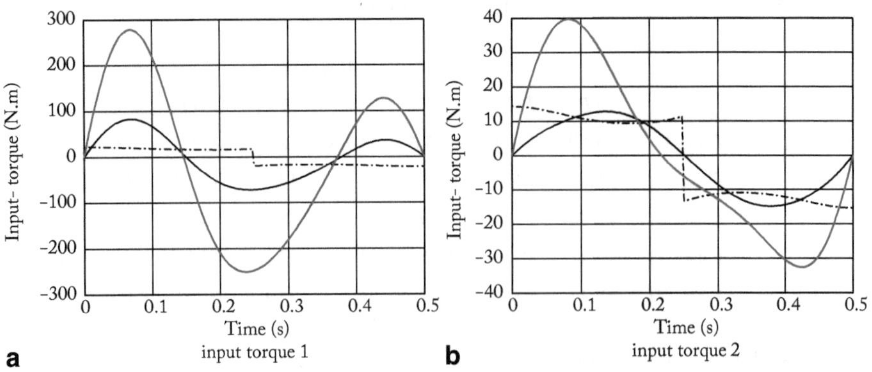

Fig. 7.28 Manipulator input torques for trajectory P_5P_9 corresponding to the three simulated models: (i) unbalanced manipulator carrying out a straight line trajectory of the end-effector using a fifth order polynomial motion profile (*black full line*); (ii) manipulator balanced by counterweights along the same trajectory (*grey full line*); (iii) manipulator controlled via optimal centre of mass displacement (*black dashed line*)

2. manipulator balanced by counterweights along the same trajectory[1],
3. manipulator controlled via optimal centre of mass displacement between points P_5 and P_9.

The obtained results are given in Fig. 7.28. The software simulations showed that in comparison with mass balanced manipulator a 92 % reduction in input torque is achieved, and, in comparison with unbalanced manipulator a 73 % reduction in input torque is achieved.

Finally, we would like to mention that the method proposed in this Section focused exclusively on the force balancing because it is carried out by optimisation of the trajectory of the manipulators *com*. However, as was shown above, it also allows the reduction of the shaking moment and the input torques. Such a result has been observed for many simulated manipulators. But it is not possible to pretend in any way that this will be true for any manipulator.

7.4 Balancing of Robot Manipulators via Optimal Motion Control

7.4.1 Dynamic Balancing of the SCARA Robot

7.4.1.1 Computation of the Shaking Force and Shaking Moment

The SCARA (**S**elective **C**ompliance **A**ssembly **R**obot **A**rm) is a high-performance robot with relatively simple structure, which is composed of three revolute joints

[1] The counterweights are located at 0.2m and 0.35m from the joints centres A and O, respectively.

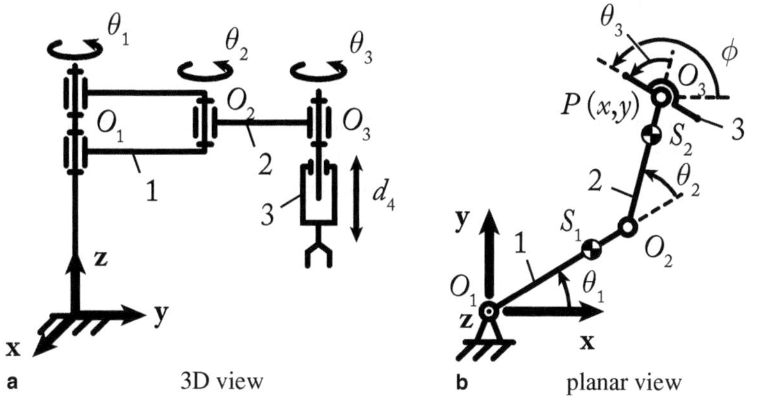

Fig. 7.29 Schematics of the SCARA robot developed at Yamanashi University

located at O_1, O_2 and O_3 (Fig. 7.29) making it possible to move and rotate the end-effector in the horizontal plane and one prismatic joint of length d_4 allowing its movement in the vertical direction. The robot was developed in the laboratory of Prof. Makino at Japon's Yamanashi University (Makino and Furuya 1982) and it is successfully applied in production industries. Today there are about 240 types of SCARA robots developed by 20 Companies.

It is well known that one of the fields of SCARA-type robots application is the fast pick-and-place manipulation. With increase of the manipulation speeds it becomes evident that the fast moving elements of mechanical systems bring about undesirable effects, such as vibration and noise.

In order to estimate the variable dynamic loads transmitted to the frame of a SCARA robot we carried out the simulations on the software ADAMS. For this purpose, a standard cycle of pick-and-place motion (25×300 mm in $t_f = 0.3$ s) defined by the following equations:

$$\begin{cases} x(t) = x_0 + s(t, t_f)(x_f - x_0) \\ y(t) = 0.2 \\ z(t) = \begin{cases} z_0 + s(t, t_f/2)(z_f - z_0) & t \in \left[0 \;\; t_f/2\right] \\ z_f + s(t - t_f/2, t_f/2)(z_0 - z_f) & t \in \left[t_f/2 \;\; t_f\right] \end{cases} \\ \phi(t) = \phi_0 + s_1(t, t_f)(\phi_f - \phi_0) \end{cases} \tag{7.61}$$

with

$$s_1(t, t_1) = \frac{10}{t_1^3}t^3 - \frac{15}{t_1^4}t^4 + \frac{6}{t_1^5}t^5 \tag{7.62}$$

was selected (for $x_0 = 0.15$ m, $x_f = -0.15$ m,$z_0 = 0$ m, $z_f = 0.025$ m and $\phi_0 = 83\,°$, $\phi_f = 157\,°$). The geometric and inertia parameters of the robot are the following:

- $l_{O_1O_2} = 0.125$ m, $l_{O_2O_3} = 0.225$ m, $l_{O_1S_1} = 0.0625$ m, $l_{O_2S_2} = 0.1125$ m,
- $m_1 = 1.2$ kg, $m_2 = 2$ kg, $m_3 = 0.5$ kg,
- $I_{S_1} = 0.014$ kg m^2, $I_{S_2} = 0.036$ kg m^2, $I_{S_3} = 0.0126$ kg m^2,

where $l_{O_iS_i}$ to the distance of the centre of masses S_i from the joint centre O_i, m_i is the mass of link i and I_{S_i} its axial moment of inertia.

Thus, its shaking force and moment computed at point O_1 are equal to:

$$\mathbf{f}^{sh} = \sum_{i=1}^{3} m_i \ddot{\mathbf{r}}_{S_i} \tag{7.63}$$

$$M_{O_1}^{sh} = \left(I_{S_1} + I_{S_2} + I_{S_3}\right)\ddot{\theta}_1 + \left(I_{S_2} + I_{S_3}\right)\ddot{\theta}_2 + I_{S_3}\ddot{\theta}_3 + \sum_{i=1}^{3} M_{O_1}(f_{xi}^{sh}, f_{yi}^{sh}) \tag{7.64}$$

where

$$\mathbf{r}_{S_1} = l_{O_1S_1}\begin{bmatrix} \cos\theta_1 \\ \sin\theta_1 \\ 0 \end{bmatrix} \tag{7.65}$$

$$\mathbf{r}_{S_2} = l_{O_1O_2}\begin{bmatrix} \cos\theta_1 \\ \sin\theta_1 \\ 0 \end{bmatrix} + l_{O_2S_2}\begin{bmatrix} \cos(\theta_1+\theta_2) \\ \sin(\theta_1+\theta_2) \\ 0 \end{bmatrix} \tag{7.66}$$

$$\mathbf{r}_{S_3} = l_{O_1O_2}\begin{bmatrix} \cos\theta_1 \\ \sin\theta_1 \\ 0 \end{bmatrix} + l_{O_2O_3}\begin{bmatrix} \cos(\theta_1+\theta_2) \\ \sin(\theta_1+\theta_2) \\ 0 \end{bmatrix} + \begin{bmatrix} 0 \\ 0 \\ d_4 \end{bmatrix} \tag{7.67}$$

$$\ddot{\mathbf{r}}_{S_1} = \frac{l_{O_1S_1}}{l_{O_1O_2}}\ddot{\mathbf{r}}_{O_2} \tag{7.68}$$

$$\ddot{\mathbf{r}}_{S_2} = \ddot{\mathbf{r}}_{O_2} + \frac{l_{O_2S_2}}{l_{O_2O_3}}\mathbf{a} \tag{7.69}$$

$$\ddot{\mathbf{r}}_{S_3} = \ddot{\mathbf{r}}_{O_2} + \mathbf{a} + \begin{bmatrix} 0 & 0 & m_3\ddot{d}_4 \end{bmatrix}^T \tag{7.70}$$

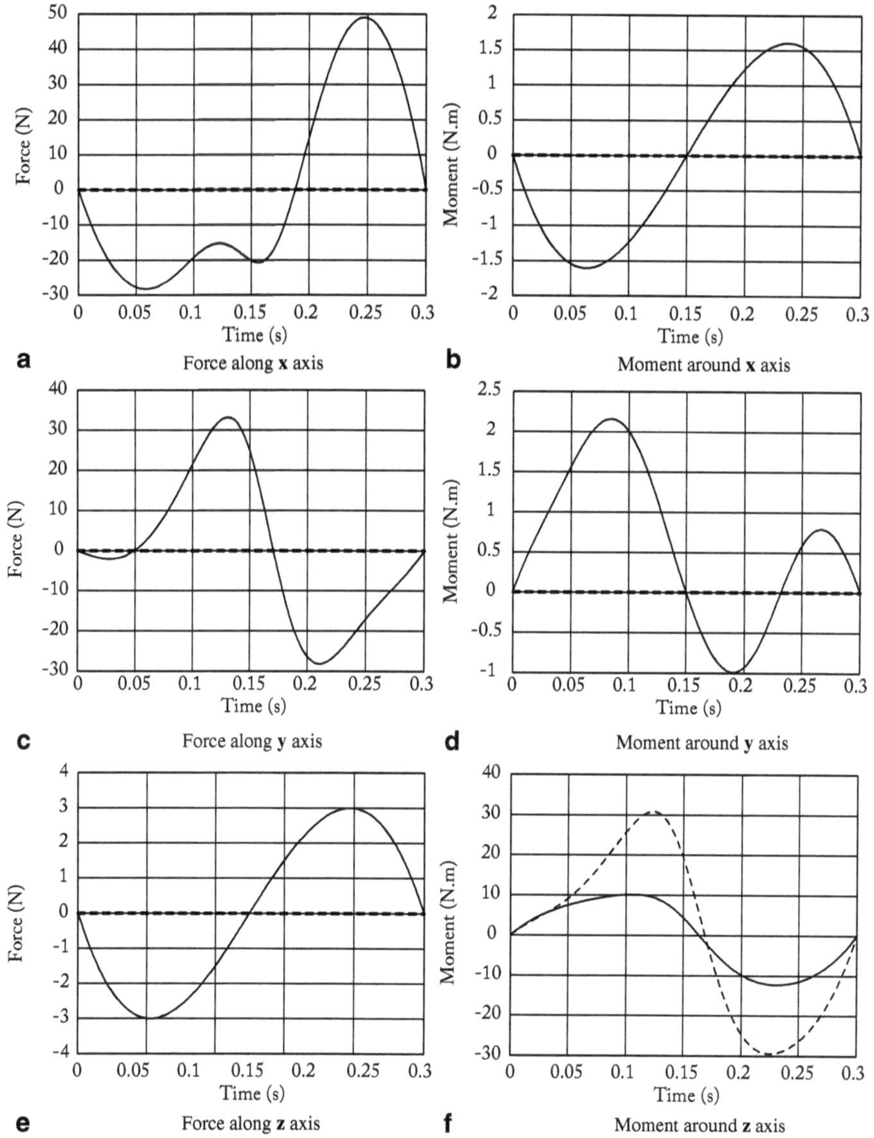

Fig. 7.30 Variations of the shaking force and shaking moment unbalanced (*full line*) and statically balanced (*dotted line*) SCARA robot

with

$$\ddot{\mathbf{r}}_{O_2} = l_{O_1O_2}\left(\ddot{\theta}_1\begin{bmatrix}-\sin\theta_1\\ \cos\theta_1\\ 0\end{bmatrix} - \dot{\theta}_1^2\begin{bmatrix}\cos\theta_1\\ \sin\theta_1\\ 0\end{bmatrix}\right) \tag{7.71}$$

and

$$\mathbf{a} = l_{O_2O_3}\left(\left(\ddot{\theta}_1 + \ddot{\theta}_2\right)\begin{bmatrix} -\sin(\theta_1+\theta_2) \\ \cos(\theta_1+\theta_2) \\ 0 \end{bmatrix} - \left(\dot{\theta}_1 + \dot{\theta}_2\right)^2 \begin{bmatrix} \cos(\theta_1+\theta_2) \\ \sin(\theta_1+\theta_2) \\ 0 \end{bmatrix}\right) \quad (7.72)$$

Figure 7.30 shows the shaking force and shaking moment variations of the unbalanced robot (full line) for the trajectory defined in (7.61).

7.4.1.2 Shaking Force and Shaking Moment Balancing

In order to achieve the dynamic balancing of the SCARA robot, we first have to ensure that it is force-balanced, i.e. statically balanced. With regard to the vertical motion along the **z** axis, we can cancel the inertia forces by counterweights executing similar but opposite movements to the end-effector motion, so that the *com* of link 3 becomes stationary and located in point O_3.

Figure 7.31 shows one of the design concepts carried out by pulley and cables. It should be noted that similar systems can be built by double slider-crank mechanisms, rhombic pantographs, and screw or cam systems.

Then, it is necessary to balance the shaking forces in the horizontal plane by making the total *com* of the moving links stationary (Fig. 7.30 dotted lines). For this purpose, the following conditions may be achieved:

$$\begin{aligned} m_3\overrightarrow{O_2O_3} &= -m_2\overrightarrow{O_2S_2} \\ m_1\overrightarrow{O_1S_3} &= -(m_2+m_3)\,\overrightarrow{O_1O_2} \end{aligned} \quad (7.73)$$

However the force balancing leads to the important increase of the moving masses of the robot, and as a result, its inertia couple (for the simulations, the following parameters for the balanced manipulator are used: $\overrightarrow{O_1S_1} = -0.936\,\overrightarrow{O_1O_2}$, $\overrightarrow{O_2S_2} = -0.210\,\overrightarrow{O_2O_3}$, $m_1 = 6.2$ kg, $m_2 = 4.8$ kg, $I_{S_1} = 0.175$ kg m^2, $I_{S_2} = 0.112$ kg m^2, $I_{S_3} = 0.0126$ kg m^2.). In Fig. 7.30 the shaking moment about the **z** axis is shown (dotted line). We can notice that it is increased by a factor 3.

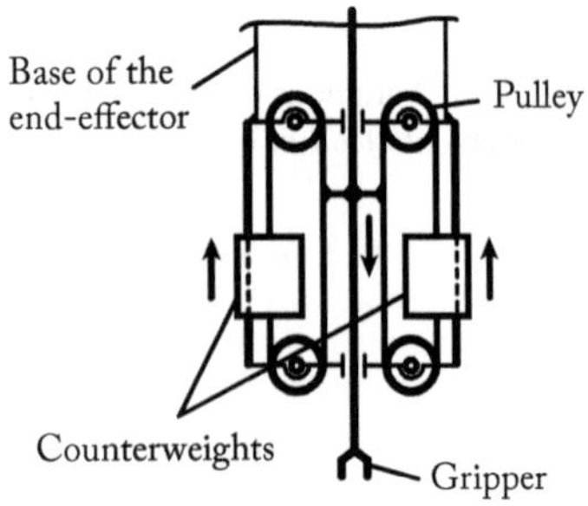

Fig. 7.31 Balancing of the vertical inertia forces by using a cable and pulley arrangement

Now that the shaking force balancing is achieved, we have to consider the cancellation of the shaking moment. The three angles of rotation of the SCARA robot (Fig. 7.29) are defined by the following relationships:

$$\theta_1 = 2\tan^{-1}\left(\frac{-l_{O_1O_2}y \pm \sqrt{l^2_{O_1O_2}\left(x^2+y^2\right) - \left(l^2_{O_1O_2} - l^2_{O_2O_3} + x^2 + y^2\right)^2}}{l^2_{O_1O_2} - l^2_{O_2O_3} + x^2 + y^2 - 2l_{O_1O_2}x}\right) \tag{7.74}$$

$$\theta_2 = \cos^{-1}\left(\frac{x - l_{O_1O_2}\cos\theta_1}{l_{O_2O_3}}\right) - \theta_1 \tag{7.75}$$

$$\theta_3 = \phi - \theta_1 - \theta_2 \tag{7.76}$$

where x, y and z are the position coordinates of the end-effector and ϕ its orientation. The sign '$\pm$' in (7.74) shows that for the same end-effector position and orientation, there are two possible configurations (working modes) of the robot. Inversely, the pose $[x \;\; y \;\; z \;\; \phi]^T$ of the end-effector is given by:

$$\begin{bmatrix} x \\ y \\ z \\ \phi \end{bmatrix} = l_{O_1O_2}\begin{bmatrix} \cos\theta_1 \\ \sin\theta_1 \\ 0 \\ 0 \end{bmatrix} + l_{O_2O_3}\begin{bmatrix} \cos\left(\theta_1+\theta_2\right) \\ \sin\left(\theta_1+\theta_2\right) \\ 0 \\ 0 \end{bmatrix} + \begin{bmatrix} 0 \\ 0 \\ d_4 \\ \theta_1+\theta_2+\theta_3 \end{bmatrix} \tag{7.77}$$

We can see from (7.77) that parameters x, y and z are defined only by the angles θ_1 and θ_2 and the stroke d_4 and do not depend on the angle θ_3. This property will be used for the shaking moment balancing.

After force balancing, the shaking moment is defined by the following relation:

$$M^{sh}_{O_1} = I_{S_1}\ddot{\theta}_1 + I_{S_2}\left(\ddot{\theta}_1 + \ddot{\theta}_2\right) + I_{S_3}\left(\ddot{\theta}_1 + \ddot{\theta}_2 + \ddot{\theta}_3\right) \tag{7.78}$$

As θ_3 is a parameter which defines the orientation of the end-effector between its initial and final positions but has not any influence to the prescribed Cartesian trajectory, it is decided to use it for cancelling the shaking moment defined in (7.78).

In order to have a shaking moment of the robot equal to zero for the given trajectory, the acceleration $\ddot{\theta}_3$ for the period $t_0 \le t \le t_f$, should vary by the following law:

$$\ddot{\theta}_3 = -\frac{I_{S_1}\ddot{\theta}_1 + I_{S_2}\left(\ddot{\theta}_1 + \ddot{\theta}_2\right)}{I_{S_3}} - \ddot{\theta}_1 - \ddot{\theta}_2 \tag{7.79}$$

The angular velocity $\dot{\theta}_3(t)$ and angular displacement $\theta_3(t)$ can be determined by simple integration of the obtained values of $\ddot{\theta}_3$.

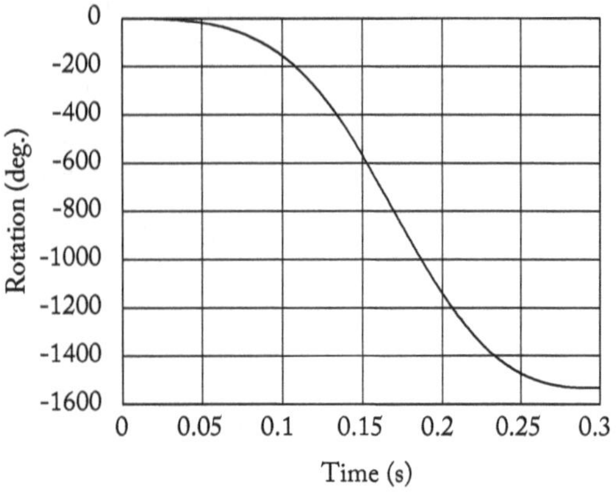

Fig. 7.32 Motion profile for the end-effector rotation ensuring the shaking moment balancing of the robot

For examined robot with parameters given above and taking into account that after shaking force balancing, the axial inertia moment of the end-effector is equal to 0.0126 kg m^2, we determine the law of rotation of the end-effector (Fig. 7.32).

It should be noted that the moment of inertia of the end-effector is relatively small compared to the moments of inertia of robot links. Thus, to ensure the balancing of the shaking moment it is necessary that it turns several times during the motion of the robot in the horizontal plane.

Let us now consider the balancing of the PAMINSA using the same approach.

7.4.2 Dynamic Balancing of a Position/Orientation Decoupled PAMINSA Robot

7.4.2.1 Computation of the Shaking Force and Shaking Moment

PAMINSA (PArallel Manipulator of the INSA) manipulators belong to a new family of partially decoupled parallel robots having between three and six *dof*, which has been presented in Sect. 6.2. The particularity of this family of manipulators is the decoupling of the displacements of the mobile platform in the horizontal plane from the translations along the vertical axis.

Figure 7.33a shows a CAD model of a PAMINSA architecture with six *dof*. The parallel structure of this manipulator has only three *dof* for displacements of the platform along the **x**, **y** and **z** axes. With regard to the orientation of the end-effector, a serial wrist with three *dof* is mounted on the platform. As in the previous architectures of PAMINSA manipulators, each leg of the PAMINSA is realized by a pantograph mechanism (Fig. 7.33b) with two input points A_i and B_i, and one output point C_i ($i = 1, 2, 3$). Each input point A_i is connected with the revolute actuator M_i by means of a prismatic guide mounted on a rotating link. This type of architecture allows the generation of translations in the horizontal plane by use of the rotating actuators M_1 and M_3 and the vertical translations by the linear actuator M_v. Thus, the displacements

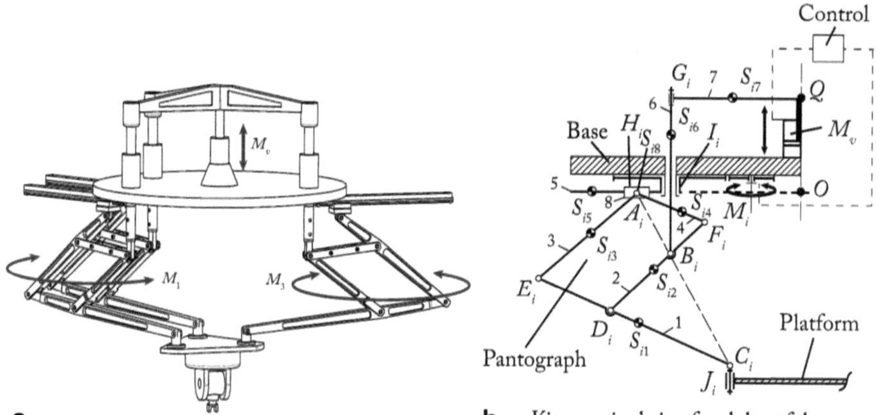

a 3D view **b** Kisnematic chain of each leg of the PAMINSA structure

Fig. 7.33 The hybrid PAMINSA manipulator

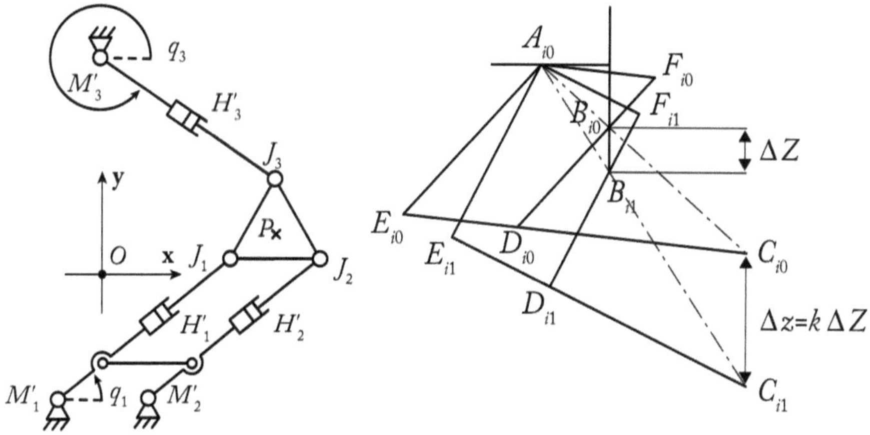

a Equivalent model for planar movements **b** Equivalent model for vertical translations

Fig. 7.34 Kinematic models for the manipulator under study

of point P of the platform in the horizontal plane Oxy (respectively the translations along the **x** and **y** axes) are independent of the vertical translations along **z**.

This implies that kinematic models controlling the displacement of the manipulator can be divided into two parts:

- one model for the displacements in the horizontal plane (Fig. 7.34a); this model is equivalent to a 3-*RPR* manipulator with two actuated revolute joints and two parallel legs. On this drawing, the displacement of pairs M_i' (or H_i') corresponds to the displacement of both passive pairs M_i (or H_i) and the pantograph linkage, and

Fig. 7.35 The 3-DOF serial wrist mounted on the platform

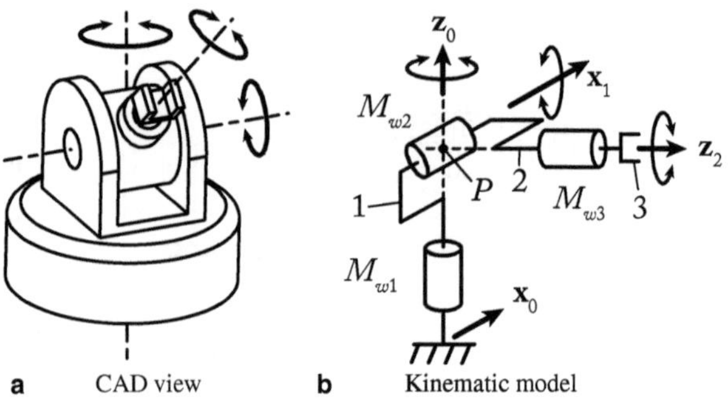

- one model for the translations along the vertical axis (Fig. 7.34b) equivalent to the model for the vertical translations of a pantograph linkage (k is the magnification factor of the pantograph linkage).

The PAMINSA manipulators were designed for high-load carrying applications, but could also be used in different applications, such as pick-and-place or micro-manipulation (as long as the magnification factor of the pantograph linkages does not scale up the displacements but, on the contrary, scales down the movement).

As depicted above, the parallel structure has only three translational *dof*. In order to ensure the three orientational *dof*, a serial wrist is added to the platform (Fig. 7.35). This wrist has three rotations which can be represented by expressing the direction cosines in terms of ZXZ Euler angles ϕ, θ and ψ (Khalil and Dombre 2002). This wrist will be used in order to balance the shaking moments during the displacements of the structure.

The positions of the *com* S_{ij} of the link ij ($j = 1, \dots 8$ and $i = 1, 2, 3$) are considered at the middle of each link. Moreover, it is assumed that the *com* S_{wi} of the i-th link of the link are located at point P.

Therefore, for this mechanism, the shaking force $\mathbf{f}^{sh}$ and shaking moment $\mathbf{m}_O^{sh}$ expressed in point O can be computed using the general equations:

$$\mathbf{f}^{sh} = m_p \ddot{\mathbf{r}}_P + \sum_{i=1}^{3} \sum_{j=1}^{8} m_{ij} \ddot{\mathbf{r}}_{S_{ij}} \tag{7.80}$$

$$\mathbf{m}_O^{sh} = m_p \mathbf{r}_p \times \ddot{\mathbf{r}}_P + \sum_{i=1}^{3} \sum_{j=1}^{8} \left(\mathbf{m}_{ij}^{rot} + m_{ij} \mathbf{r}_{S_{ij}} \times \ddot{\mathbf{r}}_{S_{ij}} \right) + \mathbf{m}_{wrist}^{rot} \tag{7.81}$$

with

$$\mathbf{m}_{ij}^{rot} = \dot{\mathbf{I}}_{ij} \omega_{ij} + \mathbf{I}_{ij} \dot{\omega}_{ij} \tag{7.82}$$

and

$$\mathbf{m}_{wrist}^{rot} = \sum_{i=1}^{3} \dot{\mathbf{I}}_{wi} \omega_{wi} + \mathbf{I}_{wi} \dot{\omega}_{wi} \tag{7.83}$$

where

- m_p is the mass of the platform with the wrist,
- $\ddot{\mathbf{r}}_P$ is the acceleration of point P of the end-effector,
- m_{ij} and $\mathbf{I}_{ij}$, represent the mass and the inertia matrix (expressed in the base frame) of the j-th limb of leg i, respectively,
- ω_{ij} is the vector of rotational velocities of the j-th limb of leg i,
- $\mathbf{I}_{wi}$ is the inertia matrix expressed in the base frame of the i-th link of the wrist,
- ω_{wi} is the vector of rotational velocities of the i-th link of the wrist.

Developing and simplifying (7.80), the expression of the shaking force $\mathbf{f}^{sh}$ can be rewritten under the form:

$$\mathbf{f}^{sh} = \sum_{i=1}^{3} \left(\begin{bmatrix} m_{C_{i1}} & 0 & 0 \\ 0 & m_{C_{i1}} & 0 \\ 0 & 0 & m_{C_{i2}} \end{bmatrix} \ddot{\mathbf{r}}_{C_i} + m_{E_i} \ddot{\mathbf{r}}_{E_i} + m_{S_{i5}} \ddot{\mathbf{r}}_{S_{i5}} \right) \tag{7.84}$$

where $\ddot{\mathbf{r}}_{C_i}$, $\ddot{\mathbf{r}}_{E_i}$, $\ddot{\mathbf{r}}_{S_{i5}}$ are the accelerations of points C_i, E_i and S_{i5}, respectively, and:

$$\begin{aligned} m_{C_{i1}} &= \frac{m_p}{3} + \frac{m_{i3} + m_{i4}}{2(1-k)} + \frac{m_{i1} + m_{i2}/k}{2} + \frac{m_{i2} + m_{i4} + 2m_{i8}}{2k(1-k)} \\ m_{C_{i2}} &= \frac{m_p}{3} + \frac{m_{i1}}{2} + \frac{2m_{i2} + m_{i4} + 2m_{i6} + 2m_{i7}}{2k} \\ m_{E_i} &= \frac{m_{i1} + m_{i3}}{2} + \frac{k-1}{2k} m_{i2} - \frac{m_{i2} + m_{i4} + 2m_{i8}}{2k} \\ m_{S_{i5}} &= m_{i5} \end{aligned} \tag{7.85}$$

where $k = l_{B_i F_i} / l_{D_i F_i}$ is the magnification factor of the pantograph linkage.

In order to estimate the variable dynamic loads transmitted to the base of PAMINSA, we carried out the simulations using ADAMS software. For this purpose, a standard cycle of pick-and-place motion (25 × 300 mm in 0.5 s) defined by the following equations:

$$\begin{cases} x(t) = 0.48 \\ y(t) = y_0 + s(t, t_f)(y_f - x_0) \\ z(t) = \begin{cases} z_0 + s(t, t_f/2)(z_f - z_0) & t \in [0 \ t_f/2] \\ z_f + s(t - t_f/2, t_f/2)(z_0 - z_f) & t \in [t_f/2 \ t_f] \end{cases} \end{cases} \tag{7.86}$$

where $s(t, t_f)$ is defined at (7.62), was selected (for $y_0 = 0.1$ m, $y_f = 0.2$ m and $z_0 = -0.65$ m, $z_f = -0.4$ mm). During the motion, the orientation of the wrist stays fixed. The geometric and inertia parameters of the robot are the following:

- positions of points M_i and J_i (in meters): $\mathbf{r}_{M_1} = -[0.1, 0.175, 0]^T$, $\mathbf{r}_{M_2} = [0.1, -0.175, 0]^T$, $\mathbf{r}_{J_3} = [0, 0.0.391, 0]^T$, $\mathbf{r}_{J_1} = \mathbf{r}_P - [0.1, 0.05, 0]^T$, $\mathbf{r}_{J_2} = \mathbf{r}_P + [0.1, -0.05, 0]^T$ and $\mathbf{r}_{J_3} = \mathbf{r}_P + [0, 0.11, 0]^T$

- magnification factor of the pantograph: $k = 3$,
- lengths of the links of the pantograph linkages: $l_{B_iG_i} = 0.442\text{m}$, $l_{A_iE_i} = l_{D_iF_i} = 0.42$ m, $l_{C_iE_i} = kl_{A_iF_i} = 0.63$ m, $l_{C_iJ_i} = 0.0275$ m, $l_{M_iS_{i6}} = 0.15$ m, $l_{M_iOi} = 0.308$ m,
- mass of the platform with the wrist: $m_p = 4.41$ kg,
- masses of the links of the pantograph linkages: $m_{i1} = 0.622$ kg, $m_{i2} = m_{i3} = 0.406$ kg, $m_{i4} = 0.107$ kg, $m_{i5} = 0.436$ kg, $m_{i6} = 0.921$ kg, $m_{i7} = 1.2$ kg, $m_{i8} = 0.305$ kg,
- terms of the inertia matrices of the links of the pantograph linkages: $I_{xx}^{(i1)} = 0.0012$ kg m^2, $I_{yy}^{(i1)} = I_{zz}^{(i1)} = 0.048$ kg m^2, $I_{xx}^{(i2)} = I_{xx}^{(i3)} = 0.0038$ kg m^2, $I_{yy}^{(i2)} = I_{yy}^{(i3)} = I_{zz}^{(i2)} = I_{zz}^{(i3)} = 0.02$ kg m^2, $I_{xx}^{(i4)} = 0.0008$ kg m^2, $I_{yy}^{(i4)} = I_{zz}^{(i4)} = 0.003$ kg m^2, $I_{xx}^{(i5)} = 0.0006$ kg m^2, $I_{yy}^{(i5)} = I_{zz}^{(i5)} = 0.02$ kg m^2, $I_{xx}^{(i6)} = 0.003$ kg m^2, $I_{yy}^{(i6)} = I_{zz}^{(i6)} = 0.071$ kg m^2,
- terms of the inertia matrices of the links of the wrist: $I_{xx}^{(w1)} = 0.019$ kg m^2, $I_{yy}^{(w1)} = I_{zz}^{(w1)} = 0.337$ kg m^2, $I_{xx}^{(w2)} = 0.039$ kg m^2, $I_{yy}^{(w2)} = I_{zz}^{(w2)} = 2.63$ kg m^2, $I_{xx}^{(w3)} = 0.088$ kg m^2, $I_{yy}^{(w3)} = I_{zz}^{(w3)} = 3.35$ kg m^2.

All the inertia matrices are expressed in local frames such that the local **x** axis is directed along:

- $\overrightarrow{C_iE_i}$, for link $i1$,
- $\overrightarrow{D_iF_i}$, for link $i2$,
- $\overrightarrow{A_iE_i}$, for link $i3$,
- $\overrightarrow{A_iF_i}$, for link $i4$,
- $\overrightarrow{A_iS_{5i}}$, for link $i5$,
- $\overrightarrow{B_iG_i}$, for link $i6$,
- $\overrightarrow{M_{w_1}M_{w_2}}$, for link $w1$,
- $\overrightarrow{M_{w_2}M_{w_3}}$, for link $w2$,
- $\overrightarrow{M_{w_2}M_{w_3}}$, for link $w3$.

Figure 7.36 shows the shaking force and moment variations of the unbalanced robot computed with ADAMS (in solid line).

7.4.2.2 Shaking Force and Shaking Moment Balancing

The shaking force can be balanced by cancelling the terms $m_{C_{i1}}$, $m_{C_{i2}}$, m_{E_i} and $m_{S_{i5}}$. This can be achieved by the redistribution of the *com* of the moving links, i.e.:

- for the cancellation of terms $m_{C_{i1}}$ and m_{E_i}, by the addition of counterweights of mass $m_{cw_{i1}}$ and $m_{cw_{i2}}$ positioned, by example, at points A_i and F_i respectively,
- for canceling the effects of term $m_{S_{i5}}$, by adding a counterweight $m_{cw_{i3}}$ on the link $i5$ positioned, by example, at point L_i such as $\mathbf{r}_{L_i} = \mathbf{r}_{O_i} + d\overrightarrow{O_iS_{i5}}$, $d < 0$,
- for canceling the effects of term $m_{C_{i2}}$, by adding counterweights undergoing similar but opposite movements to the end-effector motion (e.g. using the approach presented in Fig. 7.31).

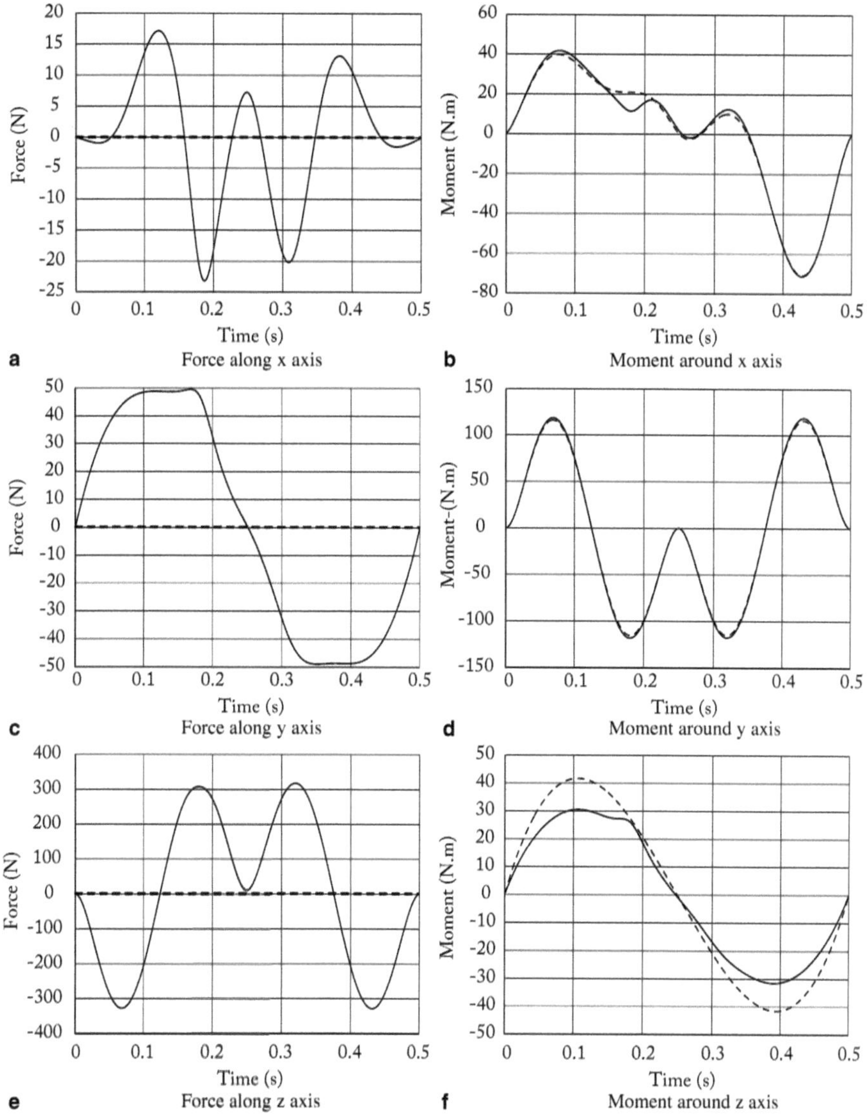

Fig. 7.36 Variations of the shaking force and shaking moment unbalanced (*full line*) and statically balanced (*dotted line*) PAMINSA robot

With the added counterweights and the supplementary system, (7.85) can be rewritten as:

$$m_{C_{i1}} = \frac{m_p}{3} + \frac{m_{i3} + m_{i4} + 2m_{cw_{i1}}}{2(1-k)} + \frac{m_{i1} + m_{i2}/k}{2} + \frac{m_{i2} + m_{i4} + 2m_{i8} + 2m_{cw_{i2}}}{2k(1-k)}$$

$$\begin{aligned} m_{C_{i2}} &= \frac{m_p}{3} + \frac{m_{i1}}{2} + \frac{2m_{i2} + m_{i4} + 2m_{i6} + 2m_{i7} + 2m_{cw_{i2}}}{2k} \\ m_{E_i} &= \frac{m_{i1} + m_{i3}}{2} + \frac{k-1}{2k} m_{i2} - \frac{m_{i2} + m_{i4} + 2m_{i8} + 2m_{cw_{i2}}}{2k} \\ m_{S_{i5}} &= m_{i5} + d m_{cw_{i3}} \end{aligned} \tag{7.87}$$

which leads to the expressions of the masses of the counterweights:

$$\begin{aligned} m_{cw_{i1}} &= (k-1)\left(\frac{m_p}{3} + \frac{m_{i3} + m_{i4}}{2(1-k)} + \frac{m_{i1} + m_{i2}/k}{2} + \frac{m_{i2} + m_{i4} + 2m_{i8} + 2m_{cw_{i2}}}{2k(1-k)}\right) \\ m_{cw_{i2}} &= k\left(\frac{m_{i1} + m_{i3}}{2} + \frac{k-1}{2k} m_{i2} - \frac{m_{i2} + m_{i4} + 2m_{i8}}{2k}\right) \\ m_{cw_{i3}} &= -m_{i5}/d \end{aligned} \tag{7.88}$$

The effects of counterweights on shaking force and shaking moment are shown in Fig. 7.36 (with a dotted line). For the manipulator under study, the masses of the counterweights are $m_{cw_{i1}} = 4.49$ kg, $m_{cw_{i2}} = 1.69$ kg, and $m_{cw_{i3}} = 0.436$ kg (taking into account that $d = -1$). Note that, although the added masses allow cancelling of the shaking force, there is an increase of about 35 % in the shaking moment around the **z** axis. Concerning the shaking moment around **x** and **y** axes, the values before and after mass balancing are similar.

After force balancing, the expression of the shaking moment becomes:

$$\begin{aligned} \mathbf{m}_O^{sh*} &= \mathbf{m}_O^{sh} + \sum_{i=1}^{3} \left(m_{cw_{i1}} \mathbf{r}_{A_i} \times \ddot{\mathbf{r}}_{A_i} + m_{cw_{i2}} \mathbf{r}_{F_i} \times \ddot{\mathbf{r}}_{F_i} + m_{cw_{i3}} \mathbf{r}_{L_i} \times \ddot{\mathbf{r}}_{L_i} \right) \\ &= m_p \mathbf{r}_p \times \ddot{\mathbf{r}}_P + \sum_{i=1}^{3} \sum_{j=1}^{8} \left(\mathbf{m}_{ij}^{rot} + m_{ij} \mathbf{r}_{S_{ij}} \times \ddot{\mathbf{r}}_{S_{ij}} \right) + \mathbf{m}_{wrist}^{rot} \\ &\quad + \sum_{i=1}^{3} \left(m_{cw_{i1}} \mathbf{r}_{A_i} \times \ddot{\mathbf{r}}_{A_i} + m_{cw_{i2}} \mathbf{r}_{F_i} \times \ddot{\mathbf{r}}_{F_i} + m_{cw_{i3}} \mathbf{r}_{L_i} \times \ddot{\mathbf{r}}_{L_i} \right) \end{aligned} \tag{7.89}$$

Here again, we have decided to adjust the rotations of the wrist in order to cancel the shaking moment. However, according to Eq. (7.89), which is nonlinear, the expressions of the rotations are difficult to compute. Therefore it is preferable to compute the values of the rotation angles starting from the expression of the angular momentum $\mathbf{h}_O$ of the mechanism. $\mathbf{h}_O$ can be written in the following form:

$$\mathbf{h}_O = \mathbf{h}_O^{pam} + \mathbf{h}_O^{wrist} \tag{7.90}$$

with

$$\mathbf{h}_O^{pam} = m_p \mathbf{r}_P \times \dot{\mathbf{r}}_P + \sum_{i=1}^{3} \sum_{j=1}^{8} \left(\mathbf{I}_{ij} \omega_{ij} + m_{ij} \mathbf{r}_{S_{ij}} \times \dot{\mathbf{r}}_{S_{ij}} \right) \tag{7.91}$$

$$+\sum_{i=1}^{3}\left(m_{cw_{i1}}\mathbf{r}_{A_i}\times\dot{\mathbf{r}}_{A_i}+m_{cw_{i2}}\mathbf{r}_{F_i}\times\dot{\mathbf{r}}_{F_i}+m_{cw_{i3}}\mathbf{r}_{L_i}\times\dot{\mathbf{r}}_{L_i}\right)$$

being the angular momentum of the PAMINSA and

$$\mathbf{h}_O^{wrist}=\sum_{i=1}^{3}\mathbf{I}_{wi}\omega_{wi} \tag{7.92}$$

being the angular momentum of the wrist.

In order to cancel the shaking moment of the mechanism, its angular moment $\mathbf{h}_O$ has to be constant. Here, we decide that $\mathbf{h}_O$ should be equal to zero. Thus, the following condition will be achieved:

$$\mathbf{h}_O^{wrist}=-\mathbf{h}_O^{pam} \tag{7.93}$$

Simplifying the expression of $\mathbf{h}_O^{wrist}$, one can rewrite this term as a linear combination of the angular velocities of the actuators of the wrist:

$$\mathbf{h}_O^{wrist}=\mathbf{N}\begin{bmatrix}\dot{\theta}\\ \dot{\phi}\\ \dot{\psi}\end{bmatrix} \tag{7.94}$$

with

$$\mathbf{N}=\begin{bmatrix}\mathbf{0}_{2\times2} & \mathbf{0}_{2\times1}\\ \mathbf{0}_{1\times2} & 1\end{bmatrix}\mathbf{I}_{w1}+(\mathbf{I}_{w2}+\mathbf{I}_{w3}\mathbf{E}_3)\begin{bmatrix}1&0&0\\0&0&0\\0&0&1\end{bmatrix}+\mathbf{I}_{w3}\begin{bmatrix}\mathbf{0}_{1\times2}&0\\ \mathbf{0}_{1\times2}&1\\ \mathbf{0}_{1\times2}&0\end{bmatrix} \tag{7.95}$$

where $\mathbf{0}_{i\times j}$ is a zero matrix of dimension $(i\times j)$ and $\mathbf{E}_3$ is a rotation matrix:

$$\mathbf{E}_3=\mathbf{Rot}\,(\phi,\mathbf{z})\,\mathbf{Rot}\,(\theta,\mathbf{x}) \tag{7.96}$$

with

$$\mathbf{Rot}\,(\phi,\mathbf{z})=\begin{bmatrix}\cos\phi & -\sin\phi & 0\\ \sin\phi & \cos\phi & 0\\ 0&0&1\end{bmatrix} \tag{7.97}$$

and

$$\mathbf{Rot}\,(\theta,\mathbf{x})=\begin{bmatrix}1&0&0\\ 0&\cos\theta & -\sin\theta\\ 0&\sin\theta&\cos\theta\end{bmatrix} \tag{7.98}$$

Thus, by introducing (7.94) into (7.93), it is possible to determine the trajectory of the wrist, which can cancel the angular momentum, and as a result, the shaking moment, by:

$$\begin{bmatrix} \dot{\theta} \\ \dot{\phi} \\ \dot{\psi} \end{bmatrix} = -\mathbf{N}^{-1}\mathbf{h}_O^{pam} \tag{7.99}$$

Taking into account that matrix **N** depends on the angles of rotation of the wrist, the value of θ, ϕ and ψ at time t can be found by a numerical integration:

$$\begin{bmatrix} \theta(t) \\ \phi(t) \\ \psi(t) \end{bmatrix} = \begin{bmatrix} \theta(t-\delta t) \\ \phi(t-\delta t) \\ \psi(t-\delta t) \end{bmatrix} + \begin{bmatrix} \dot{\theta}(t-\delta t) \\ \dot{\phi}(t-\delta t) \\ \dot{\psi}(t-\delta t) \end{bmatrix} \delta t \tag{7.100}$$

where δt corresponds to the integration step, and the values of the angles and angular velocities at time $t - \delta t$ are considered known.

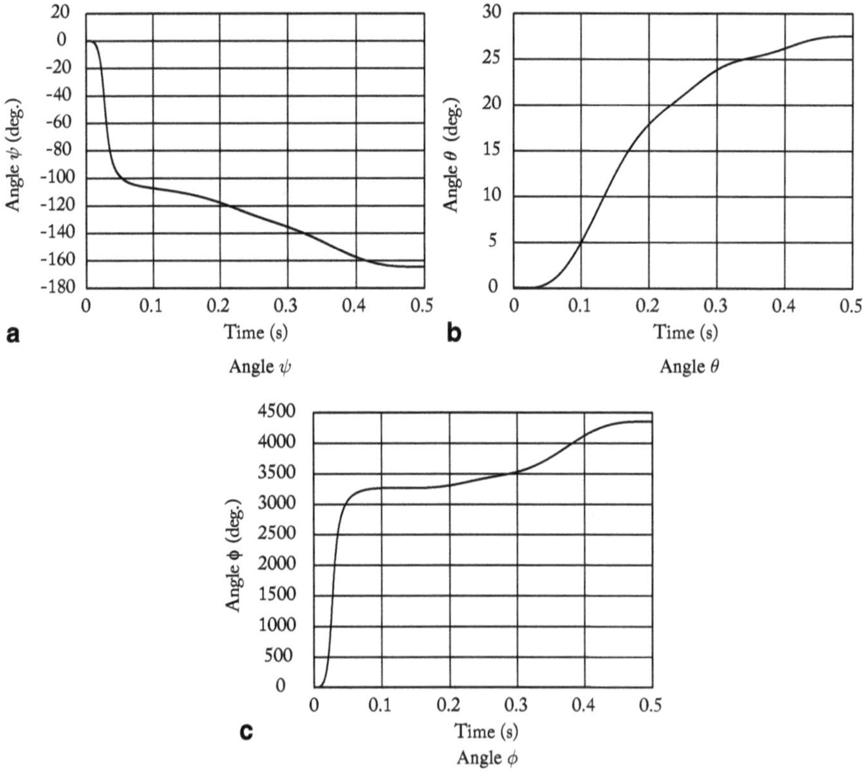

Fig. 7.37 Rotations of the wrist used for canceling the shaking moment

For the examined manipulator, we determine the required rotations of the end-effector, which makes it possible to cancel shaking moment (Fig. 7.37).

It should be noticed that the moment of inertia of link 3 of the wrist is relatively small compared to the moments of inertia of the robot links. Thus, to ensure the balancing of the shaking moment, it is necessary that it makes several full revolutions during the motion of the robot in the horizontal plane.

Finally, it should be mentioned that in practice there are small deviations from the ideal zero reaction. Sources of such deviations include manufacturing errors, friction and clearances in the joints and drive actuators, as well as manipulator unbalancing due to a payload. However, in all of these cases, the proposed balancing method will eliminate the most significant base reactions due to the accelerating links of the robot.

Chapter 8
Gravitational Force Balancing of Robotic Systems

Abstract This Chapter deals with the optimal balancing of gravitational forces. In Sect. 8.1, the balancing methods of manipulator mechanisms with the reduced number of springs, which are based on the copying properties of the pantograph mechanism, are presented. Then, in Sect. 8.2, a newly designed parallel robot for medical 3D-ultrasound imaging is considered for an optimum static balancing. The optimum solutions reduces substantially the effect of gravity with simple mechanical system. The efficiency of the suggested solutions is illustrated by numerical simulations.

Section 8.3 deals with an analytically tractable solution for the gravity balancing considering the spring mass. For this purpose, the relationship between the stiffness coefficient of the spring and its mass is provided. Then this relationship is introduced into the balancing equation and spring elastic force is determined taking into account its mass. For zero-free length springs, the stiffness coefficient of the springs is determined from a quadratic equation and for non-zero-free length springs from a cubic equation. In this way, an exact balancing of gravitational forces is achieved, which allows improving the balancing accuracy of robotic systems. The efficiency of the suggested approach is illustrated by numerical examples. An application to the balancing of the leg orthosis for robotic rehabilitation is also presented.

The last Section (Sect. 8.4) improves the known design concepts permitting the dynamic decoupling of serial manipulators with an optimal balancing schemes permitting relatively small increase in the total mass of the moving links.

8.1 Balancing of Pantograph Mechanisms

In this Section are presented the balancing methods of manipulator mechanisms with the reduced number of springs. Let us consider the balancing of the pantograph mechanism by a spring.

Figure 8.1 shows a pantograph mechanism with symmetric links. To obtain the conditions of balancing, we will use the method of static substitution of distributed masses by concentrated point masses: if the mass of the link 2 is statically replaced by concentrated masses m_{A2} and m_{O2} located at the centres of the hinges O and A, the mass m_3 by the concentrated masses m_{O3} and m_{C3} located at the centres of the hinges O and C, all masses of the pantograph mechanism will be substituted by

V. Arakelian, S. Briot, *Balancing of Linkages and Robot Manipulators*,
Mechanisms and Machine Science 27, DOI 10.1007/978-3-319-12490-2_8

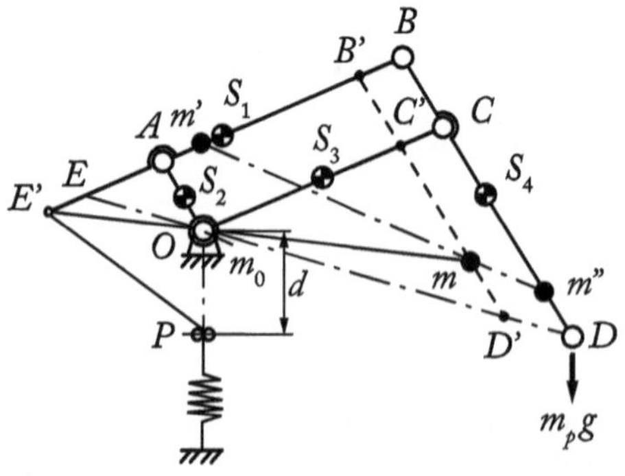

Fig. 8.1 Balancing of the pantograph mechanism

an equivalent mass system comprising three concentrated masses m_O, m' and m'', located at the points O, S' and S'' respectively, and determined by the expressions:

$$m_O = m_2 l_{AS_2}/l_{OA} + m_3 l_{CS_3}/l_{OC} \tag{8.1}$$

$$m' = m_1 + m_2 l_{OS_2}/l_{OA} \tag{8.2}$$

$$m'' = m_3 l_{OS_3}/l_{OC} + m_4 + m_p \tag{8.3}$$

where, m_i are the masses of the corresponding links ($i = 1 \ldots 4$); m_p is the payload; l_{OA} is the distances between the centres of the joints O and A; l_{OC} is the distance between the centres of the joints O and C; l_{AS_2} is the distance between the centre of the joint A and the centre of mass S_2 of link 2; l_{CS_3} is the distance between the centre of the joint C and the centre of mass S_3 of link 3; l_{CS_2} is the distance between the centre of the joint C and the centre of mass S_2 of link 2; l_{OS_3} is the distance between the centre of the joint O and the centre of mass S_3 of link 3.

Thus, the balancing problem of the pantograph mechanism $OEABCD$ is reduced to the balancing problem of the concentrated mass located on the link $B'D'$ of a certain virtual pantograph. The balancing of this concentrated mass is provided by a spring having the following parameters:

$$k\lambda d = mg \tag{8.4}$$

where, k is the stiffness coefficient of the spring; $d = l_{OP}$ (see Fig. 8.1); $\lambda = l_{OE'}/l_{OM}$ is the scale factor of the virtual pantograph and g is the gravitational acceleration.

The mechanisms with closed kinematic chains can also be balanced by a similar way. By example, the balancing of the Chebyshev linkage can be achieved by adding a dyad $B'C'D$ wearing a spring (Fig. 8.2) and forming with initial linkage a pantograph.

In this case, all link masses of the mechanism will be substituted by an equivalent mass system comprising three concentrated masses m_O, m_C and m, located at points O, C and S respectively, and determined by the expressions:

$$m_O = m_1 l_{AS_1}/l_{OA} \tag{8.5}$$

$$m_C = m_3 l_{BS_3}/l_{BC} \tag{8.6}$$

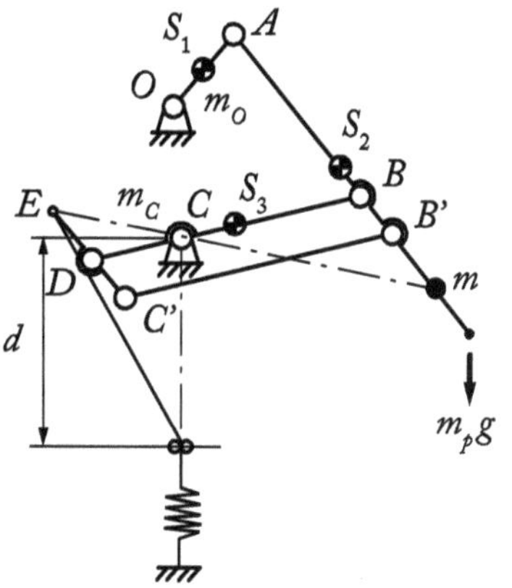

Fig. 8.2 Balancing of the Chebyshev linkage by a spring

$$m = m_1 l_{OS_1}/l_{OA} + m_2 + m_3 l_{CS_3}/l_{BC} + m_p \tag{8.7}$$

where, m_i are the masses of the corresponding links ($i = 1 \ldots 3$); m_p is the payload; l_{OA} is the distance between the centres of the joints O and A; l_{BC} is the distance between the centres of the joints B and C; l_{AS_1} is the distance between the centre of the joint A and the centre of mass S_1 of link 1; l_{BS_3} is the distance between the centre of the joint B and the centre of mass S_3 of link 3; l_{OS_1} is the distance between the centre of the joint O and the centre of mass S_1 of link 1; l_{CS_3} is the distance between the centre of the joint C and the centre of mass S_3 of link 3.

Thus, the balancing of the Chebyshev linkage is reduced to the balancing problem of the concentred mass m, which was examined above.

8.2 Optimal Balancing of the Parallel Robot for Medical 3D-ultrasound Imagining

This section deals with the optimum static balancing of the parallel robot developed for medical 3D-ultrasound imaging. Such a robot has been developed at École de technologie supérieure in Montréal (Canada) (CAD model showed in Fig. 8.3). It consists of two five-bar planar mechanisms, which are connected to the tool holder by spherical and universal joints. The four rotating actuators of the two five-bar mechanisms are mounted on a single frame. A fifth rotating actuator is attached directly to the tool holder allowing large amplitude turn. It should be noted that the tool holder has a passive translation making possible the necessary robot displacements. The last actuator is linear and it allows horizontal displacements of the robot. Thus, such architecture (Lessard et al. 2006) has six degrees of freedom and it can move in a position workspace represented by a parallelepiped of dimensions $\delta x = 1.5$ m, $\delta y = 0.5$ m and $\delta z = 0.5$ m for any orientations

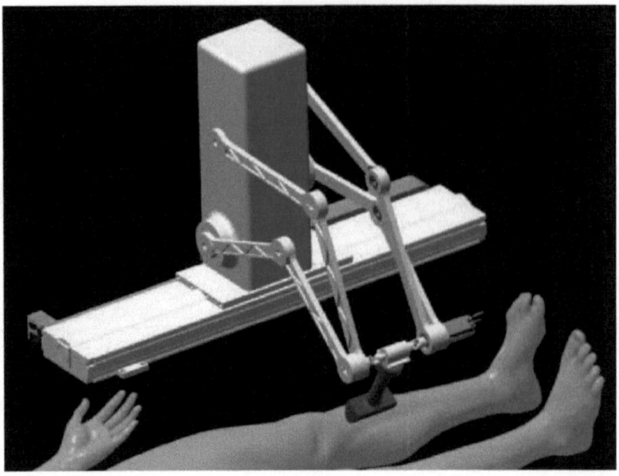

Fig. 8.3 CAD model of the prototype of parallel robot for medical 3D-ultrasound imaging

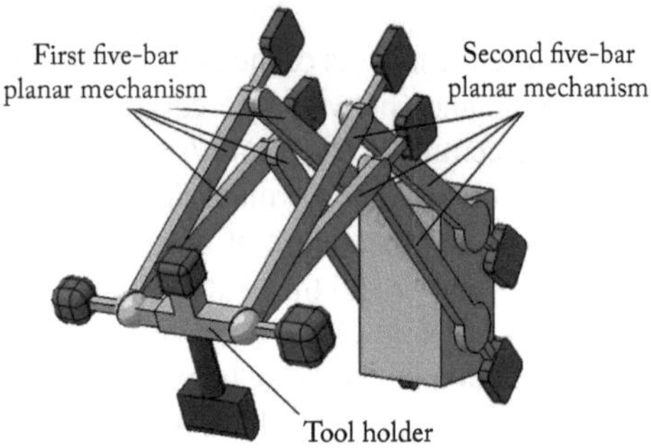

Fig. 8.4 Complete balancing by counterweights

- around **x** axis from −75 to 90°,
- around **y** axis from −50 to 35°, and
- around **z** axis from −45 to 45°.

A part of the research program devoted to the design of this parallel robot for medical 3D-ultrasound imaging was the minimization of the input torques. We have considered several balancing approaches for this robot that allows a significant reduction of the input torques by means of simpler design solutions.

8.2.1 Complete Static Balancing

Complete static equilibrium can be carried out by counterweights mounted on the moving links. Figure 8.4 illustrates the complete static balancing of the Parallel Robot for Medical 3D-Ultrasound. However, the added counterweights increase the total mass of the robot and the load on each joint. With regard to the design of the

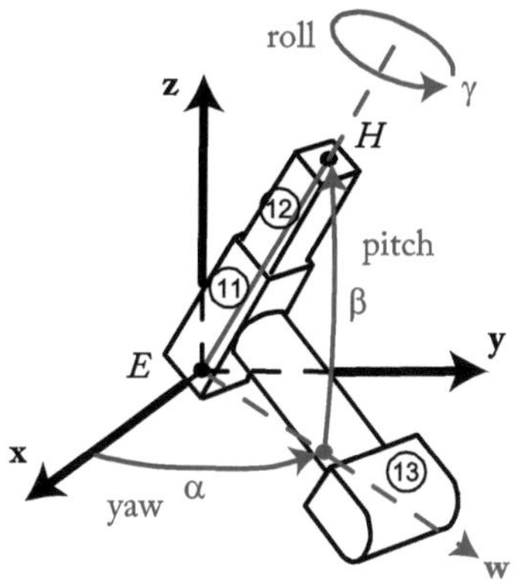

Fig. 8.5 Euler angles of the tool

present robot with counterweights, the last motor for the roll angle would be difficult to assembly with a counterweight because there is a universal joint.

In the case of a complete balancing with springs, it is necessary to add auxiliary mechanisms (parallelograms, etc.), which lead to a complicated robot architecture.

So the mechanical system after complete balancing becomes either very heavy or a complicated assembly. That's why we have developed a partial balancing approach. It is obvious that in this case the elimination of the loads on the actuators is partial, but on another hand the design solution is very simple.

8.2.2 *Input Torques*

The input torque of the ith actuator can be expressed as:

$$\tau_i = \sum_{j=1}^{3} \mathbf{j}_{ij}^T \mathbf{g}_j, \text{ for } j = 1, \ldots, 16 \tag{8.8}$$

where $\mathbf{j}_{ij}$ is the Jacobian matrix that express the relation between the velocity along the vertical axis of the *com* S_j of the link j and the actuated variables q_i, $\mathbf{g}_j$ are the gravity forces applied on the link j (Figs. 8.5, 8.6 and 8.7).

The structure of this robot has some particularities, which allows a considerable simplification of torques determination. The robot architecture can be divided in two parts because the tool holder represents an Assur group (Leinonen 1991) with two links and three joints (spherical, universal and revolute pairs) dividing the robot into two parts. Thus the tool holder can be disconnected from the robot structure to be examined separately.

From this point forward, the tool is considered statically balanced around the sixth motor, i.e. the tool's *com* coincides with the motor rotation axis. Such an approach simplifies reaction forces and input torques calculations.

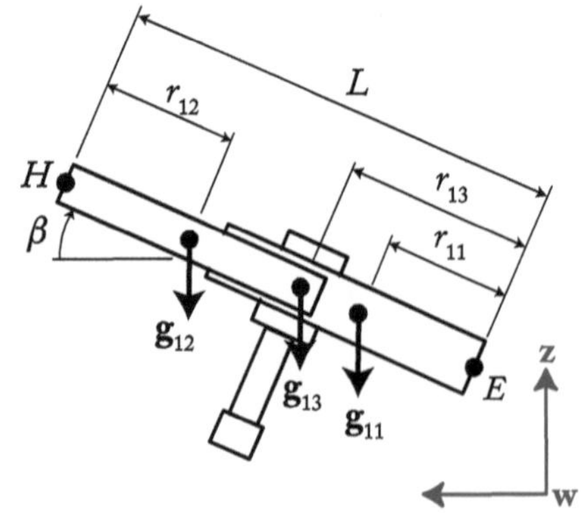

Fig. 8.6 Schematic of the tool assembly

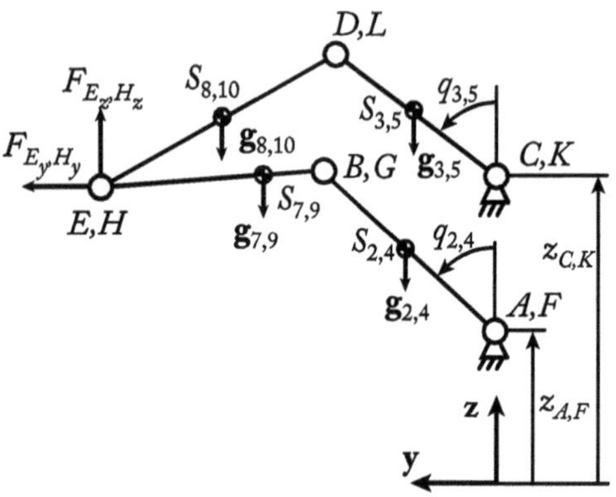

Fig. 8.7 Schematics of the two five-bars mechanisms with applied reaction forces

Thus the reaction forces $\mathbf{f}_E$ and $\mathbf{f}_H$ (respectively applied by the tool holder at points E and H) are function of the yaw and pitch angles only, the roll angle being equal to the sixth motor rotation angle. This motor input torque is independent of the other motors.

The length L of the passive linear joint is calculated as a function of the orientation angles:

$$L = L_f \frac{\cos \beta}{\cos \alpha} \tag{8.9}$$

where L_f is the distance between the two planar five bar mechanisms, i.e. $L_f = |x_E - x_H|$.

Thus, the vertical reaction forces on point H and E can be determined from the static equilibrium equations:

$$F_{H_z} = \frac{(m_{12}(L - r_{12}) + m_{11}r_{11} + m_{13}r_{13})\, g}{L} \tag{8.10}$$

$$F_{E_z} = (m_6 + m_{11} + m_{12} + m_{13})\, g - F_{H_z} \tag{8.11}$$

where m_i and r_i are respectively the mass and the position of the *com* of the i-th link and g is the gravity; m_6 is the mass of actuator 6.

The horizontal reaction forces are projected from axis **w** to axis **y**:

$$F_{H_y} = \left(m_{12}g - F_{H_z}\right)\tan\beta\sin\alpha \tag{8.12}$$

$$F_{E_y} = -F_{H_y} \tag{8.13}$$

Thus, it is possible to deduce the potential energy V of the first five-bar mechanism (Fig. 8.7):

$$V = (m_2z_2 + m_3z_3 + m_7z_7 + m_8z_8)\,g + F_{E_y}y_E + F_{E_z}z_E \tag{8.14}$$

with

$$z_2 = z_A + r_2\cos q_2 \tag{8.15}$$

$$z_3 = z_C + r_3\cos q_3 \tag{8.16}$$

$$z_7 = (l_{AB}\cos q_2 + z_A - z_E)\frac{r_7}{l_{BE}} + z_E \tag{8.17}$$

$$z_8 = (l_{CD}\cos q_3 + z_C - z_E)\frac{r_8}{l_{DE}} + z_E \tag{8.18}$$

where z_i is the vertical coordinate of the *com* S_i of the i-th link and $r_2 = l_{AS_2}$, $r_3 = l_{CS_3}$, $r_7 = l_{ES_7}$, $r_8 = l_{ES_8}$.

Then the input torques are determined from

$$\tau = \begin{bmatrix}\tau_2\\ \tau_3\end{bmatrix} = \frac{\partial V}{\partial \mathbf{q}} + \mathbf{J}_E^T\frac{\partial V}{\partial \mathbf{x}_E} \tag{8.19}$$

where $\mathbf{q}^T = [q_2\ q_3]$ is the motor articulation position vector, $\mathbf{x}_E^T = [x_E\ z_E]$ is the end-effector position and $\mathbf{J}_E^T$ is the Jacobian matrix that relates the end-effector velocities to the motor velocities, i.e.

$$\mathbf{J}_E = \left[\frac{\partial \mathbf{x}_E}{\partial \mathbf{q}}\right] \tag{8.20}$$

The torques τ_4 and τ_5 of the actuators 4 and 5 of the second five-bar mechanism are determined by a similar way.

8.2.3 Minimization of the Root-mean-square Values of the Input Torques

The input torques minimization is carried out by each actuator torque root-mean-square values minimization for all the workspace. The workspace used in the calculation is only the **yz** plane because the linear motorized axis is not subject to gravity balancing. Three solutions are considered for optimum balancing.

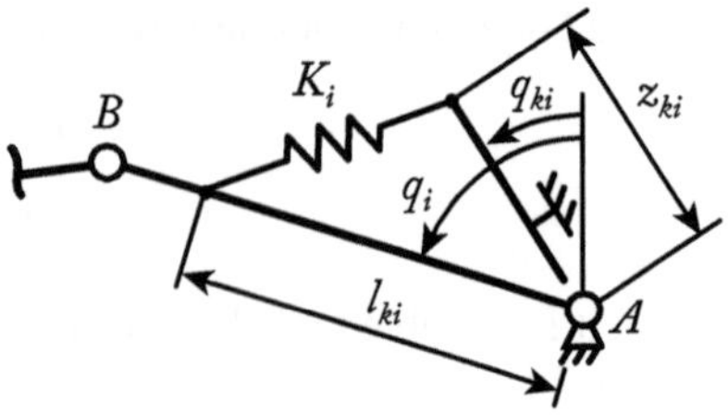

Fig. 8.8 Linear spring balancing

8.2.3.1 Tension/Compression Spring Equilibrium

Firstly, one solution is an equilibrium mechanical system composed of zero free length springs (tension or compression) attached on each motorized arms. The fixed end of each spring is positioned optimally to release the actuator.

The spring applied torque's potential energy is:

$$V_{k_i} = \frac{K_i}{2}\left(\left(z_{k_i}\cos q_{k_i} - l_{k_i}\cos q_i\right)^2 + \left(z_{k_i}\sin q_{k_i} - l_{k_i}\sin q_i\right)^2\right) \tag{8.21}$$

where K_i is the stiffness coefficient of the spring, which is attached on a fix arm of length z_{k_i} and orientation q_{k_i} (Fig. 8.8), q_i is the angular position of the i-th actuator. The other end is linked on the motorized arm at length l_{k_i}.

By differentiating Eq. (8.21) we determine the input torque of the i-th due to the spring:

$$\tau_{k_i} = \frac{\partial V_{k_i}}{\partial q_i} = K_i l_{k_i} z_{k_i} \sin\left(q_i - q_{k_i}\right) = C_i \sin\left(q_i - q_{k_i}\right) \tag{8.22}$$

Two parameters have to be optimized for each motor:

- the constant C_i and
- the angular position of the fix arm q_{k_i}.

The torque root-mean-square value minimization leads to the following condition:

$$\Delta_i = \sum_{WS}\left(\tau_i + \tau_{k_i}\right)^2 \rightarrow \min_{C_i, q_{k_i}} \text{ for } i = 1,\ldots,5 \tag{8.23}$$

where $WS :=$ Workspace.

For this purpose, we shall achieve the conditions:

$$\frac{\partial \Delta_i}{\partial C_i} = 0 \text{ and } \frac{\partial \Delta_i}{\partial q_{k_i}} = 0 \tag{8.24}$$

from which we determine:

$$q_{k_i} = 2\tan^{-1} t_{m_i} \text{ for } m = 1,\ldots,6 \tag{8.25}$$

$$C_i = \frac{-\sum_{WS}\tau_i \sin\left(q_i - q_{k_i}\right)}{\sum_{WS}\sin^2\left(q_i - q_{k_i}\right)} \tag{8.26}$$

where t_{m_i} are the roots of a polynomial $p_i(t)$ given by:

$$p_i(t) = \sum_{j=0}^{6} a_j t^j \tag{8.27}$$

with

$$\begin{aligned} &a_0 = b_5,\ a_1 = 2b_7,\ a_2 = -3b_5 + 4b_8,\ a_3 = -4b_7 + 8b_6, \\ &a_4 = -a_2,\ a_5 = a_1,\ a_6 = -a_0 \end{aligned} \tag{8.28}$$

and

$$b_0 = -\sum_{WS} \tau_i \sin q_i,\ b_1 = -\sum_{WS} \tau_i \cos q_i \tag{8.29}$$

$$b_2 = \sum_{WS} \sin^2 q_i,\ b_3 = \sum_{WS} \cos^2 q_i \tag{8.30}$$

$$b_4 = -2\sum_{WS} \sin q_i \cos q_i \tag{8.31}$$

$$b_5 = \sum_{WS} \tau_i b_2 \cos q_i + b_0 \sin q_i \cos q_i \tag{8.32}$$

$$b_6 = \sum_{WS} \tau_i b_3 \sin q_i - b_1 \sin q_i \cos q_i \tag{8.33}$$

$$b_7 = \sum_{WS} \tau_i \left(b_4 \cos q_i + b_2 \sin q_i\right) + b_0 \left(\sin^2 q_i - \cos^2 q_i\right) + b_1 \sin q_i \cos q_i \tag{8.34}$$

$$b_8 = \sum_{WS} \tau_i \left(b_3 \cos q_i + b_4 \sin q_i\right) + b_1 \left(\sin^2 q_i - \cos^2 q_i\right) - b_0 \sin q_i \cos q_i \tag{8.35}$$

8.2.3.2 Torsion Spring Equilibrium

Secondly, static balance can be achieved using torsion spring mounted on each actuator axes (Fig. 8.9). The spring torque is linear to the bending angle:

$$\tau_{k_i} = K_i \left(q_{k_i} - q_i\right) \tag{8.36}$$

The minimization conditions are similar to the previous ones:

$$\Delta_i = \sum_{WS} \left(\tau_i + \tau_{k_i}\right)^2 \rightarrow \min_{K_i, q_{k_i}} \text{ for } i = 1, \ldots, 5 \tag{8.37}$$

$$\frac{\partial \Delta_i}{\partial K_i} = 0 \text{ and } \frac{\partial \Delta_i}{\partial q_{k_i}} = 0 \tag{8.38}$$

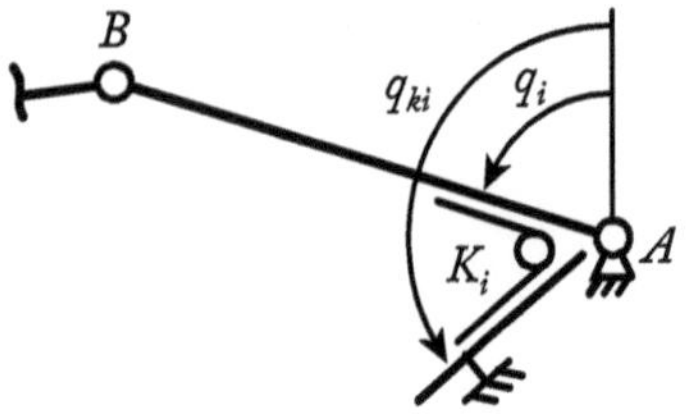

Fig. 8.9 Torsion spring balancing

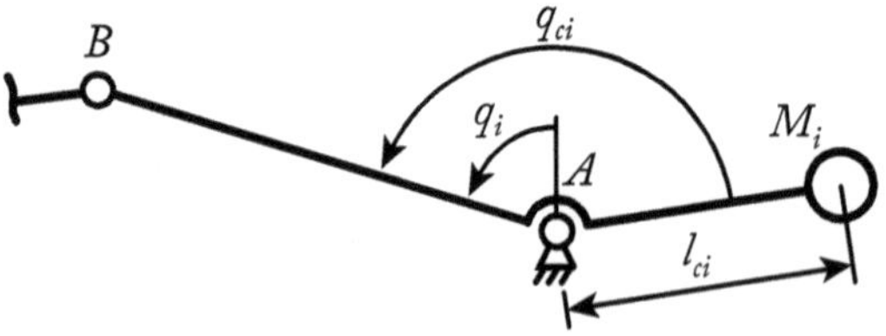

Fig. 8.10 Counterweight balancing

from which we determine:

$$q_{k_i} = \frac{\sum_{WS} \tau_i q_i \sum_{WS} q_i - \sum_{WS} \tau_i \sum_{WS} q_i^2}{N \sum_{WS} \tau_i q_i - \sum_{WS} \tau_i \sum_{WS} q_i} \tag{8.39}$$

$$K_i = -\frac{\sum_{WS} \tau_i \left(q_i - q_{k_i}\right)}{\sum_{WS} \left(q_i - q_{k_i}\right)^2} \tag{8.40}$$

where N is the number of calculated positions in the workspace.

8.2.3.3 Counterweight Equilibrium

Lastly, a balancing approach is carried out by adding a counterweight of mass m_{cw_i} on each motorized axis (Fig. 8.10).

The counterweight m_{cw_i} is placed at a certain angle so that the maximum torque applied is at a specific desired motor angle. The optimum counterweight system is then function of the mass m_{cw_i} and position q_{c_i} for a given length l_{c_i}. The added torque due to the counterweight is equal to:

$$\tau_{c_i} = -m_{cw_i} l_{c_i} \sin\left(q_i - q_{c_i}\right) = -C_i \sin\left(q_i - q_{c_i}\right) \tag{8.41}$$

Determination of parameters C_i and q_{c_i} is equivalent to the determination of parameters C_i and q_{k_i} of Sect. 8.2.3.1.

Table 8.1 Tension/compression spring optimum configurations

Motor	Spring parameters		*rms* red. (%)	Max. torque red. (%)
2	$q_{k_2} = 1.7°$	$C_2 = 0.104$ N.m	57.6	34.7
3	$q_{k_3} = -42.7°$	$C_3 = 0.263$ N.m	91.2	45.6
4	$q_{k_4} = 3.4°$	$C_4 = 0.081$ N.m	50.1	33.1
5	$q_{k_5} = -43.0°$	$C_5 = 0.187$ N.m	90.2	43.1

Table 8.2 Torsion spring optimum configurations

Motor	Spring parameters		*rms* red. (%)	Max. torque red. (%)
2	$q_{k_2} = 7.7°$	$K_2 = -4.55$ N.m/rad	56.9	45.4
3	$q_{k_3} = 180°$	$K_3 = 5.29$ N.m/rad	90.4	45.2
4	$q_{k_4} = 6.2°$	$K_4 = -3.33$ N.m/rad	50.0	46.7
5	$q_{k_5} = 180°$	$K_5 = 3.74$ N.m/rad	89.2	34.3

Table 8.3 Counterweight optimum configurations

	Counterweight parameters		*rms* red. (%)	Max. torque red. (%)
2	$q_{c_2} = -178.4°$	$C_2 = 4.63$N.m	57.6	34.7
3	$q_{c_3} = -221.4°$	$C_3 = 11.63$N.m	91.5	46.4
4	$q_{c_4} = -176.9°$	$C_4 = 3.62$N.m	50.1	33.1
5	$q_{c_5} = -224.2°$	$C_5 = 8.36$N.m	90.2	44.1

8.2.4 Results

The proposed robot (Fig. 8.3) with the following geometrical parameters and mass distribution:

- $l_{AB} = l_{FG} = 0.5$ m, $l_{CD} = l_{KL} = 0.7$ m, , $l_{BE} = l_{DE} = l_{GH} = l_{LH} = 0.6$ m, $L_f = 0.1$ m,
- $r_2 = r_4 = 0.2094$ m, $r_3 = r_5 = 0.3046$ m, $r_7 = r_8 = r_9 = r_{10} = 0.3$ m, $r_{11} = 0.0355$ m, $r_{12} = 0.0315$ m,
- $z_A = z_F = 0.13$ m, $z_C = z_K = 0.28$ m,
- $m_2 = m_4 = 1.235$ kg, $m_3 = m_5 = 1.549$ kg, $m_6 = 0.331$ kg, $m_7 = m_8 = m_9 = m_{10} = 0.536$ kg, $m_{11} = 0.107$ kg, $m_{12} = 0.083$ kg, $m_{13} = 0.111$ kg.

was used for numerical simulation. The three mechanical solutions were tested. For comparison basis, the *rms* and maximum motor torques are given in Tables 8.1–8.3. In Table 8.2, negative values of the spring constant K_i is equivalent to say that springs must be mounted in the other sense.

The minimization of the root-mean-square sum of static torques was reduced up to 50–91.5 %. In more practical terms, the required maximum motor torque was reduced up to 34.3–46.7 %.

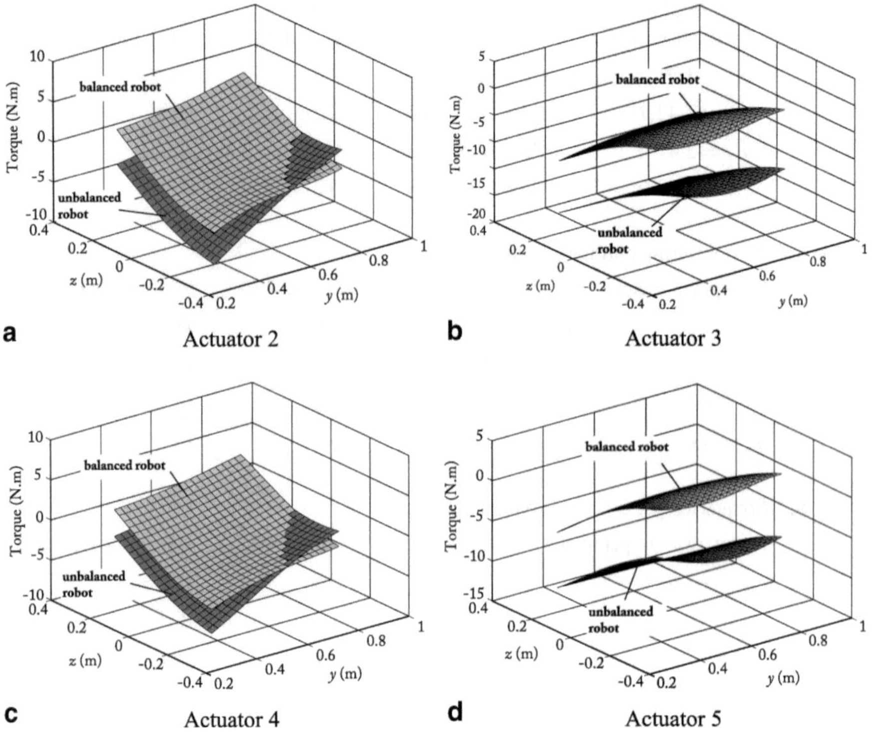

Fig. 8.11 Variation of the actuator torques for zero orientation of the tool

Figure 8.11 shows input torques variations for unbalanced and optimum balanced robots with extension springs (the simulation was carried out in a static mode of operation).

8.3 Improvement of Balancing Accuracy of Robot-manipulators Taking into Account the Spring Mass

8.3.1 Improvement of Balancing Accuracy by Taking into Account the Spring Mass

The balancing of the gravitational forces of a link 1, which rotates around a horizontal axis, is schematically shown in Fig. 8.12. In this scheme, for weigh balancing a helical spring 2, jointed between a point A of the link and a fixed point B, is used. The movable coordinate axis system attached to link 1 was chosen so that the point A is upon the $\mathbf{x}_1$ axis.

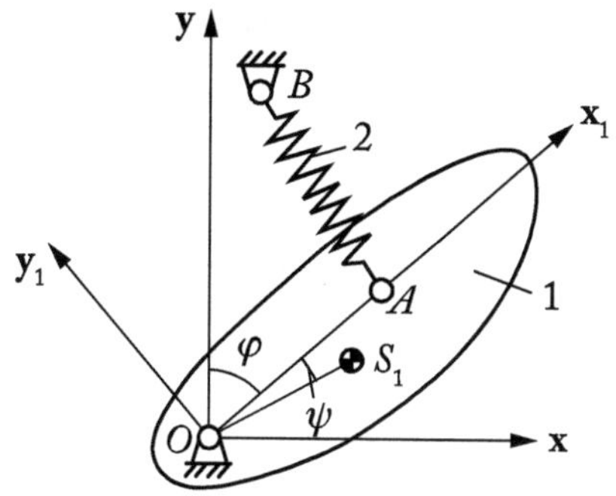

Fig. 8.12 Balancing of a rotating link

The unbalanced moment can be expressed as follows:

$$M_u = M_g + M_b \tag{8.42}$$

where M_g is the moment of the gravitational forces, M_b is the balancing moment of the elastic force of the spring.

For the cancellation of the unbalanced moment it is necessary to achieve the following condition:

$$[m_1 s_1 \sin(\varphi + \psi) + m_{2A} r \sin\varphi]\, g + F_s \frac{x_B y_A - x_A y_B}{l} = 0 \tag{8.43}$$

where m_1 is the mass of the rotating link, $s_1 = l_{OS_1}$ is the distance of gravity center S_1 from axis O, φ is the angle between **y**-axis and $\mathbf{x}_1$-axis, ψ is the angle between the axis $\mathbf{x}_1$ and OS_1, m_{2A} is the concentrated point mass of the spring situated at the point$A$ ($m_{2A} = m_2 s_2/l$), m_2 is the mass of spring, $s_2 = l_{BS_2}$ is the distance of gravity center S_2 of the spring from point B, $l = l_{AB}$ the length of the spring at current angle φ, $F_s = F_0 + k(l - l_0)$ is the elastic force of the helical spring, l_0 is the initial length of the spring, F_0 is the initial force of the spring (the initial force is the internal force that holds the coils tightly together), k is the stiffness coefficient of the spring, $x_A = r \sin\varphi$, $y_A = r\cos\varphi$, x_B, y_B are the coordinates of the points A and B in the fixed coordinate axis system xOy, $r = l_{OA}$ is the distance of point A from axis O.

The analysis of Eq. (8.42) shows that there are two solutions: (i) complete balancing when the zero free length spring is applied, (ii) partial balancing when the non-zero free length spring is applied.

8.3.1.1 Zero Free Length Springs

The balancing of rotating link is favored by the use of zero free length springs, which is distinguished by the relationship $F_0 = kl_0$ or $F_0 = l_0 = 0$.

In this case, when $\psi = 0$ and $x_B = 0$, Eq. (8.42) leads to the following condition:

$$k = \frac{g}{y_B r}\left(m_1 s_1 + m_2 \frac{r s_2}{l}\right) \tag{8.44}$$

Considering that the extension of the spring is regular and $s_2 = l/2$, Eq. (8.44) can be written as:

$$k = \frac{g}{y_B r}(m_1 s_1 + 0.5 m_2 r) \tag{8.45}$$

The stiffness coefficient and the mass of the spring also depend on its geometric and material parameters (Wahl 1963):

$$k = \frac{Gd^4}{8D^3 n} \tag{8.46}$$

and

$$m_2 = \frac{\rho L_W \pi d^2}{4} \tag{8.47}$$

where G is the Shear modulus calculated from the material's elastic modulus E and Poisson ratio ν ($G = E/(2(1+\nu))$), d is the diameter of the wire that is wound into a helix, D is the mean diameter of the helix, n is the number of active coils, ρ is the material mass density, L_W is the length of wire.

From Eqs. (8.46) and (8.47) we can obtain the relationship between the mass of the spring and its stiffness coefficient:

$$m_2 = \rho \pi L_W \sqrt{\frac{D^3 nk}{2G}} \tag{8.48}$$

On substituting this expression in (8.45), we obtain aquadratic equation from which we determine the stiffness coefficient:

$$k = \left(-\frac{q}{2} \pm \sqrt{\frac{q^2}{4} - u} \right)^2 \tag{8.49}$$

where

$$q = -\frac{\rho g \pi L_W}{2 y_B} \sqrt{\frac{D^3 n}{2G}} \tag{8.50}$$

and

$$u = -\frac{m_1 s_1 g}{r y_B} \tag{8.51}$$

taking into account that

$$k = -\frac{q}{2} \pm \sqrt{\frac{q^2}{4} - u} > 0 \tag{8.52}$$

Now let us consider the balancing by non-zero length springs.

8.3.1.2 Non-zero Free Length Springs

There are two cases of the force-length characteristic of non-zero free length springs: with initial zero force ($F_0 = 0$) and non-zero initial force (taking into account that $F_0 \neq kl_0$).

In this case the complete balancing of a rotating link is impossible. Thus, an approximate solution may be applied.

For this purpose, we propose to minimize the *rms* value of unbalanced moment:

$$rms = \sqrt{\sum_{i=1}^{N} (M_{gi} + M_{bi})^2 / N} \tag{8.53}$$

where N is the number of calculated positions of rotating link.

For the minimization of the *rms*, it is necessary to minimize the sum:

$$\Delta = \sum_{i=1}^{N} (M_{gi} + M_{bi})^2 \to \min_{k} \tag{8.54}$$

or

$$\Delta = \sum_{i=1}^{N} (C_1 + C_2\sqrt{k} + C_3 k)^2 \to \min_{k} \tag{8.55}$$

where

$$C_1 = \sum_{i=1}^{N} m_1 s_1 g \sin(\varphi_i + \psi) \tag{8.56}$$

$$C_2 = 0.5 \sum_{i=1}^{N} \left(0.5 \rho \pi g r L_W \sqrt{\frac{D^3 n}{2G}} \right) \sin \varphi_i \tag{8.57}$$

$$C_3 = \sum_{i=1}^{N} \left(\sqrt{(x_B - r \sin \varphi_i)^2 + (y_B - r \cos \varphi_i)^2} - l_0 \right) \times \frac{x_B r \cos \varphi_i - y_B r \sin \varphi_i}{\sqrt{(x_B - r \sin \varphi_i)^2 + (y_B - r \cos \varphi_i)^2}} \tag{8.58}$$

For this purpose, we shall achieve the condition:

$$\partial \Delta / \partial k = 0 \tag{8.59}$$

from which we obtain the following cubic equation:

$$z^3 + az^2 + bz + c = 0 \tag{8.60}$$

where

$$a = \frac{3C_2}{2C_3} \quad (8.61)$$

$$b = \frac{C_2^2 + 2C_1C_3}{2C_3^2} \quad (8.62)$$

$$c = \frac{C_1C_2}{2C_3^2} \quad (8.63)$$

$$z = \sqrt{k} \quad (8.64)$$

The solution of Eq. (8.60) with real coefficient can be expressed in algebraic form by means of Viette-Cordano method.

For determination of roots, first of all, we shall calculate:

$$Q = (a^2 - 3b)/9 \quad (8.65)$$

$$R = (2a^3 - 9ab + 27c)/54 \quad (8.66)$$

When $R^2 < Q^3$, cubic equation has three real roots, determined by the following expressions:

$$z_1 = -2\sqrt{Q}\cos t - a/3 \quad (8.67)$$

$$z_2 = -2\sqrt{Q}\cos(t + 2\pi/3) - a/3 \quad (8.68)$$

$$z_3 = -2\sqrt{Q}\cos(t - 2\pi/3) - a/3 \quad (8.69)$$

$$t = \cos^{-1}\left(R/\sqrt{Q^3}\right)/3 \quad (8.70)$$

When $R^2 \geq Q^3$, general cubic equation case has one real root and two real roots for confluent case.

For determination the complex roots, it is necessary to calculate:

$$A = -sign(R)\sqrt[3]{|R| + \sqrt{R^2 - Q^3}} \quad (8.71)$$

$$B = Q/A \text{ if } A \neq 0 \text{ and } B = 0 \text{ if } A = 0 \quad (8.72)$$

The real root is then:

$$z_1 = A - B - a/3 \quad (8.73)$$

In the case, when $A = B$, the complex roots become the real roots:

$$z_2 = -A - a/3 \quad (8.74)$$

After determination of z, we determine the stiffness of the spring (Eq. (8.64)) taking into account that $z > 0$.

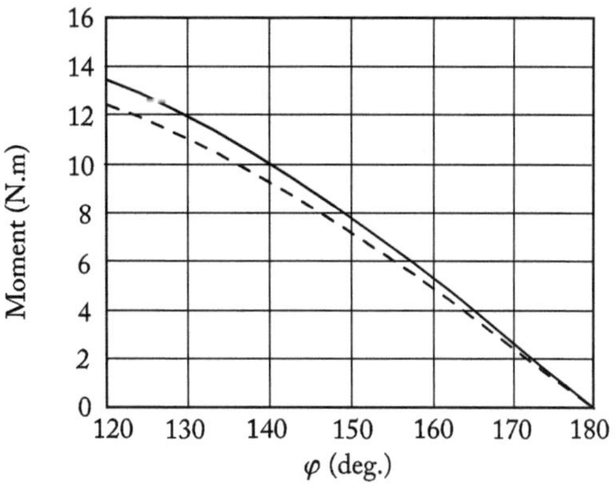

Fig. 8.13 Balancing moments for two examined cases: first model with weightless spring (*dashed line*) and second model with spring mass (*full line*)

8.3.2 *Numerical Examples and Error Analysis*

For illustration of the suggested approach let us consider numerical examples. Numerical simulations were carried out for the balancing of the rotating link with following parameters: $m_1 = 8$ kg; $s_1 = 0.183$ m; $\psi = 0°$ and $\varphi \in [2\pi/3, \pi]$. The parameters of the fixed points of the spring are the following: $y_B = 0.16$ m and $r = 0.3$ m.

Firstly, this rotating link will be balanced by zero free length spring and then by non-zero free length spring.

8.3.2.1 Balancing by Zero Free Length Spring

The simple model of link balancing with weightless spring leads to the following parameters: $k = 299$ N/m; $F_0 = 121$ N.

Now let us balance the same link with zero free length spring taking into account its mass. The parameters which characterize the selected spring are the following: $G = 81,000$ N/mm^2, $\rho = 7800$ kg/m^3, $D = 0.04$ m, $n = 80$ and $L_W = 10.38$ m.

When the spring mass is included in the balancing equation we obtain $k = 324.5$ N/m and $F_0 = 131$ N. The mass of the spring is 0.8 kg.

It should be noted that the parameters of the spring was selected, taking into account the admissible maximum extension and the spring index ($4 \leq D/d \leq 15$).

Figure 8.13 shows the balancing moments for two examined cases. The error caused by neglect of the spring mass is 8.3 %.

8.3.2.2 Balancing by Non- zero Free Length Spring

Now we consider the same problem with $F_0 = 0$ N and $l_0 = 0.3$ N. The simple model of link balancing with weightless spring leads to $k = 990$ N/m. The spring with such stiffness and the selected geometric parameters (with $L_W = 8$ m) has a

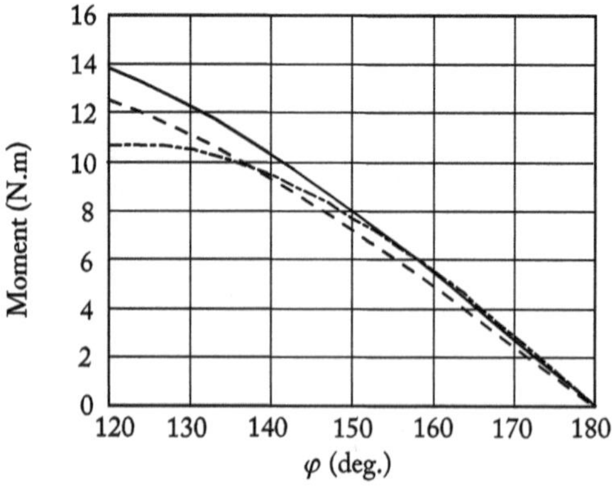

Fig. 8.14 Moment of the gravitational forces and the balancing moment of the spring elastic force: moment of the gravitational forces for first model with weightless spring (*dashed line*), moment of the gravitational forces for second model taking into account the spring mass (*full line*) and balancing moment with $k = 990$ N/m (*dash-dotted line*)

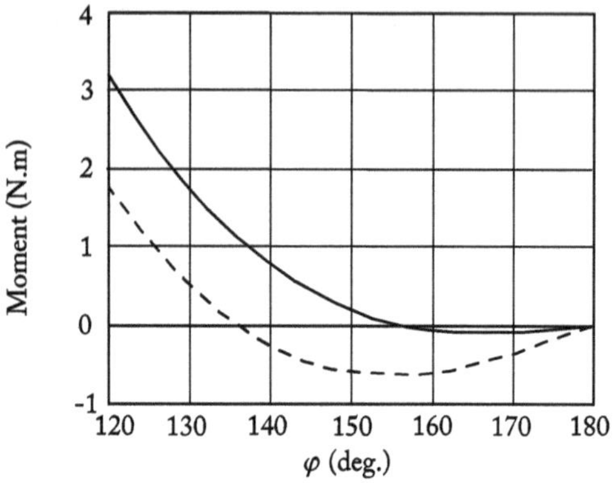

Fig. 8.15 Theoretical values of the unbalanced moment (*dashed line*)and the effective values of this moment with spring mass (*full line*)

weight of 1.1 kg. This weight, which is neglected in the simple model add some unbalanced moment (Fig. 8.14).

Figure 8.15 shows the theoretical values of the unbalanced moment and the effective values of this moment with spring mass.

The obtained results showed that the maximum value of the theoretical unbalanced moment is 1.81 N.m and the maximum value of the effective value (with spring mass) is 3.2 N.m.

Let us balance now the same link taking into account the spring mass. With the same parameters selected for our first example, from (8.56–8.58) we obtain $C_1 = 88.454$, $C_2 = 0.316$, $C_3 = -0.089$ and the following cubic equation:

$$z^3 - 5.3z^2 - 989.7z + 1748.8 = 0 \tag{8.75}$$

with $Q = 333$ and $R = -5.3$.

Thus $R^2 < Q^3$, we have three real roots determined by (8.67–8.69): $z_1 = 1.76$, $z_2 = 33.38$ and $z_3 = -29.84$. Taking $z = z_2$, we determine the stiffness of the spring $k = 1114$ N/m.

The spring with such a stiffness and the selected geometric parameters has a weight of 1.16 kg.

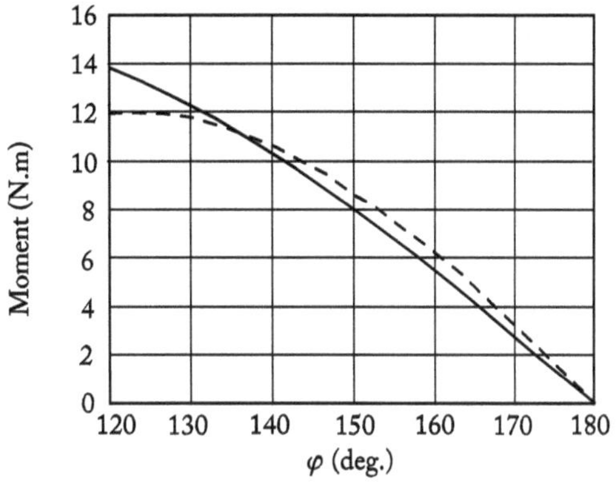

Fig. 8.16 Moment of the gravitational forces of the rotating link taking into account the spring mass (*full line*) and the balancing moment of the spring with $k = 1114$ N/m (*dashed line*)

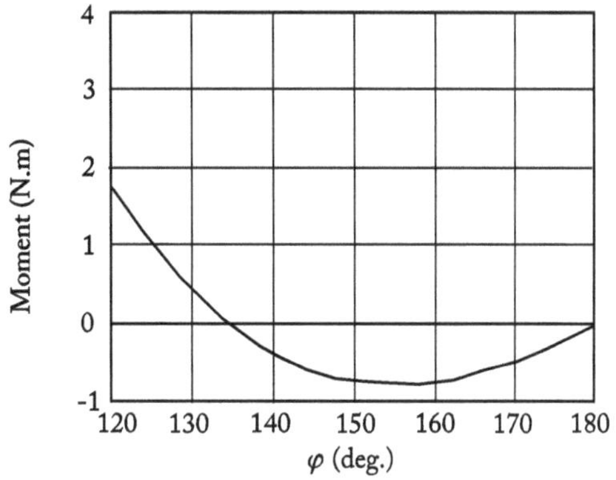

Fig. 8.17 Effective values of the unbalanced moment considering the spring mass

Figure 8.16 shows the effective unbalanced moment due to the weights of the rotating link and spring, as well as the balancing moment of the spring with $k = 1114$ N/m.

Figure 8.17 shows the effective values of the unbalanced moment when the mass of the spring was taken into account in the balancing equation.

Thus, in this case, the maximum value of the effective unbalanced moment is 1.86 N.m, which shows that significant improvement in balancing performance can be achieved through the use of the suggested approach.

It should be noted that the minimization of the function (8.54) was carried out by only one parameter: the spring stiffness. However this approach can be further optimized for spring connection points and other parameters of spring such as free length, i.e.

$$\Delta = \sum_{i=1}^{N} (M_{gi} + M_{bi})^2 \rightarrow \min_{k, x_B, y_B, r, \psi, l_0} \tag{8.76}$$

The unknowns might be determined from the system of six equations obtained from the following conditions: $\partial \Delta / \partial \chi = 0$, where $\chi = k, x_B, y_B, r, \psi, l_0$. In this

Fig. 8.18 Balancing device for the leg of a walking person

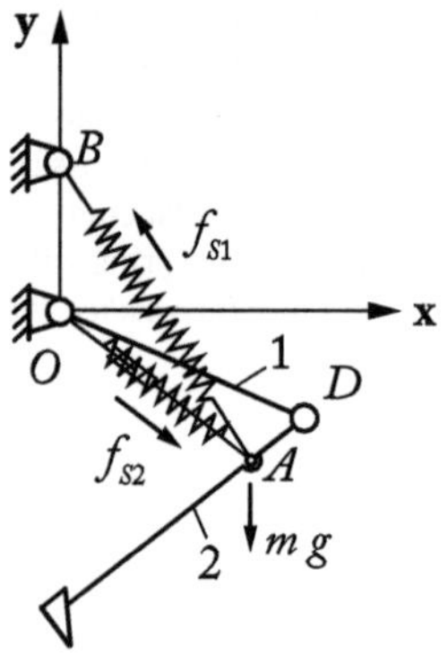

case, the solution cannot be achieved by analytical methods and the unknowns can be only determined by numerical investigation.

8.3.3 Application to the Balancing of Leg Orthosis for Rehabilitation Devices

Let us consider the design of a rehabilitation device, which can support the weight of leg during walking (Fig. 8.18). It is obvious that especially for the rehabilitation devices, it is hoped that the balancing will be perfect.

The following mass distribution and geometric parameters are considered for the leg (Agrawal and Agrawal 2005): $l_1 = 0.4322$ m, $l_2 = 0.421$ m, $m_1 = m_{tight} = 7.39$ kg, $m_{shank} = 3.11$ kg, $m_{foot} = 0.97$ kg, $m_2 = m_{shank} + m_{foot} = 4.08$ kg, $l_{OS_1} = 0.41l_1$ and $l_{DS_1} = 0.44l_2$. It should be noted that for two links design example, m_2 consists of the mass of the shank plus the foot.

Let us consider the exact balancing of the leg with zero free length spring. For this purpose let us substitute mass m_1 of link 1 by two concentrated masses m_{1O} and m_{1D} situated at the centers of joints O and D. Then we determine the common center of mass of the link 2 with concentrated mass m_{1D}:

$$s_M = \frac{m_2 l_{DS_2}}{m} \tag{8.77}$$

with

$$m = m_2 + m_1 l_{OS_1}/l_1 \tag{8.78}$$

where s_M is the distance of the mass m from the center of joint D, l_{DS_2} is the distance of the center of S_2 from the center of joint D.

Thus, the masses of moving links are replaced by two masses: m_{1O}, which is fixed and m situated at the point s_M.

Now we connect a zero free length extension spring with the body at the point B and with the link 2 at the point A, and a compression spring with the body at the point O and with the link 2 at the point A. Please note that the point A coincides with the center of masses s_M.

Let us balance this system by considering the potential energy. The potential energy of the system can be written as:

$$V = V_g + V_{S_1} + V_{S_2} \tag{8.79}$$

$$V_g = -\left[m + 0.5(m_{S_1} + m_{S_2})\right] g l_{OA} \sin\beta \tag{8.80}$$

$$V_{S_1} = 0.5 k_{S1} l_{AB}^2 \tag{8.81}$$

$$V_{S_2} = 0.5 k_{S2} l_{OA}^2 \tag{8.82}$$

where $\beta = \angle XOA$, m_{S_1} and m_{S_2} are the masses of the springs, l_{OA} and l_{AB} are the distances between the corresponding points, $k_{S1} = k$ and $k_{S2} = -k$. Note please that k is the stiffness coefficient of springs, which is the same for both springs.

On substituting

$$l_{AB}^2 = l_{OB}^2 + l_{OA}^2 + 2 l_{OB} l_{OA} \sin\beta \tag{8.83}$$

in Eq. (8.81), and thenafter substituting Eqs. (8.80–8.82) in (8.79), we get

$$V = \left[k l_{OB} - mg - 0.5g(m_{S_1} + m_{S_2})\right] l_{OA} \sin\beta + 0.5 k l_{OB}^2 \tag{8.84}$$

Thus, the potential energy becomes constant when the coefficient of $l_{OA} \sin\beta$ is zero, i.e.

$$k l_{OB} - mg - 0.5g(m_{S_1} + m_{S_2}) = 0 \tag{8.85}$$

This expression taking into account Eq. (8.48) can be rewritten as:

$$mg + 0.5g\pi\sqrt{k}\left(\rho_1 L_{W1}\sqrt{\frac{D_1^3 n_1}{G_1}} + \rho_2 L_{W2}\sqrt{\frac{D_2^3 n_2}{G_2}}\right) - k l_{OB} = 0 \tag{8.86}$$

from which we can determine the stiffness of springs.

Thus, by using two springs with parameters $G_1 = G_2 = 81{,}000$ N/mm^2, $D_1 = D_2 = 0.4$ m, $\rho_1 = \rho_2 = 7800$ kg/m^3, $n_1 = 66$, $L_{W1} = 8.6$ m, $n_2 = 62$, $L_{W2} = 7.8$ m, we obtain $k_{S_1} = 522$ N/m and $k_{S_2} = -522$ N/m and the potential energy of the system becomes constant for all possible configurations, i.e. zero torques are required. In this way, an exact balancing of gravitational forces is achieved, which allows improving the balancing accuracy of rehabilitation devices.

8.4 Optimal Balancing of Serial Manipulators with Decoupled Dynamics

This Section improves the known design concept described permitting the dynamic decoupling of serial manipulators with an optimal balancing schemes which allows relatively small increase in the total mass of the moving links. It is also discussed the dynamic decoupling of 3-*dof* spatial manipulator.

8.4.1 Complexity and the Nonlinearity of Robot Arm Dynamics: Basic Notions

Let us consider a robot arm composed of n links. According to Lagrangian dynamics, the equations of motion can be written as

$$\tau_i = \frac{d}{dt}\left(\frac{\partial L}{\partial \dot{\theta}_i}\right) - \frac{\partial L}{\partial \theta_i} \tag{8.87}$$

where τ_i is the generalized torque (or force); θ_i is the generalized coordinate; $L = T - V$ is the Lagrangian; T is the kinetic energy and V is the potential energy.

The kinetic energy of segment i can be expressed as

$$T_i = 0.5 m_i \mathbf{v}_i^T \mathbf{v}_i + 0.5 \omega_i^T \mathbf{I}_i \omega_i \tag{8.88}$$

in which m_i is the mass of the link, $\mathbf{I}_i$ is the inertia matrix, and $\mathbf{v}_i$ and ω_i are the translational and angular velocities at the *com* respectively, which can be expressed as

$$\mathbf{v}_i = \sum_{j=1}^{i} f_{ij}^{(v)}(\theta_1, \ldots, \theta_i)\dot{\theta}_i \tag{8.89}$$

$$\omega_i = \sum_{j=1}^{i} f_{ij}^{(\omega)}(\theta_1, \ldots, \theta_i)\dot{\theta}_i \tag{8.90}$$

When Eqs. (8.89) and (8.90) are substituted into Eq. (8.88), we obtain

$$T = \sum_{i=1}^{n} T_i = \sum_{i=1}^{n}\sum_{j=1}^{i}\sum_{k=1}^{i} A_{ijk}(\theta_1, \ldots, \theta_i)\dot{\theta}_j\dot{\theta}_k \tag{8.91}$$

The potential energy is the function of the generalized coordinates θ only, which can be written as

$$V = \sum_{i=1}^{n} f_i^{(p)}(\theta_1, \ldots, \theta_i) \tag{8.92}$$

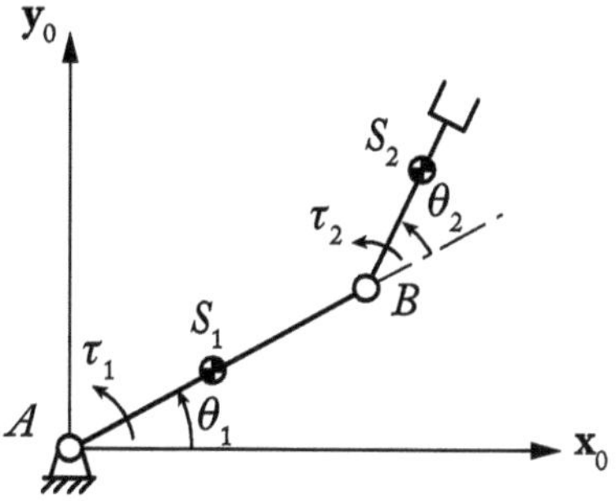

Fig. 8.19 A 2-*dof* planar serial manipulator

By putting Eqs. (8.91) and (8.92) into Eq. (8.87), we get

$$\tau_i = \sum_{j=1}^{n}\sum_{k=1}^{j} A_{ijk}(\theta_1,\ldots,\theta_j)\ddot{\theta}_k + \frac{1}{2}\sum_{j=1}^{n}\dot{\theta}_j\sum_{k=1}^{n}\left(\frac{\partial A_{ij}}{\partial\theta_k} + \frac{\partial A_{ik}}{\partial\theta_j} - \frac{\partial A_{jk}}{\partial\theta_i}\right)\dot{\theta}_k - \frac{f_i^{(p)}(\theta_1,\ldots,\theta_i)}{\partial\theta_i} \tag{8.93}$$

Thus, we can see that there is nonlinearity in the manipulator dynamics and our goal to develop an improved design concept permitting the dynamic decoupling of serial manipulators.

To understand better the suggested design concept let us first consider the dynamic equations for a planar manipulator with two degrees of freedom shown in Fig. 8.19

In this case Eq. (8.93) can be written as

$$\begin{bmatrix}\tau_1\\ \tau_2\end{bmatrix} = \begin{bmatrix}D_{11} & D_{12}\\ D_{21} & D_{22}\end{bmatrix}\begin{bmatrix}\ddot{\theta}_1\\ \ddot{\theta}_2\end{bmatrix} + \begin{bmatrix}D_{111} & D_{122}\\ D_{211} & D_{222}\end{bmatrix}\begin{bmatrix}\dot{\theta}_1^2\\ \dot{\theta}_2^2\end{bmatrix} + \begin{bmatrix}D_{112} & D_{121}\\ D_{212} & D_{221}\end{bmatrix}\begin{bmatrix}\dot{\theta}_1 & \dot{\theta}_2\\ \dot{\theta}_1 & \dot{\theta}_2\end{bmatrix} + \begin{bmatrix}D_1\\ D_2\end{bmatrix} \tag{8.94}$$

with

$$D_{11} = m_1 l_{AS_1}^2 + m_2 l_1^2 + m_2 l_{BS_2}^2 + 2m_2 l_1 l_{BS_2}\cos\theta_2 + I_{S_1} + I_{S_2} \tag{8.95}$$

$$D_{12} = D_{21} = m_2 l_{BS_2}^2 + m_2 l_1 l_{BS_2}\cos\theta_2 + I_{S_2} \tag{8.96}$$

$$D_{22} = m_2 l_{BS_2}^2 + I_{S_2} \tag{8.97}$$

$$D_{111} = D_{222} = D_{212} = D_{221} = 0 \tag{8.98}$$

$$D_{122} = -D_{211} = D_{112} = D_{121} = -m_2 l_1 l_{BS_2}\sin\theta_2 \tag{8.99}$$

$$D_1 = (m_1 l_{AS_1} + m_2 l_1)g\cos\theta_1 + m_2 g l_{BS_2}\cos(\theta_1+\theta_2) \tag{8.100}$$

$$D_2 = m_2 g l_{BS_2}\cos(\theta_1+\theta_2) \tag{8.101}$$

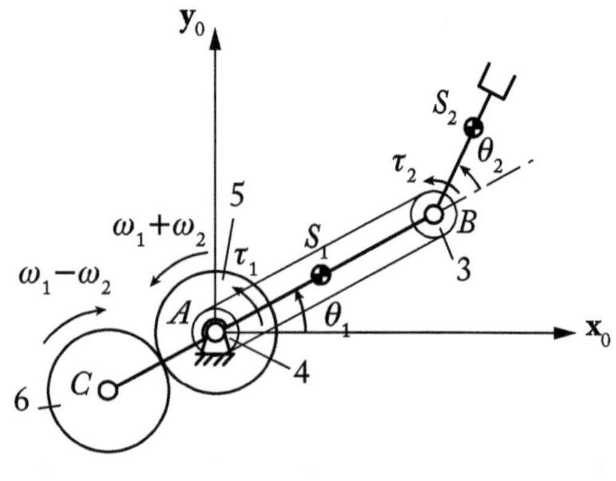

Fig. 8.20 Design of the decoupled 2-*dof* planar serial manipulator

where l_1 and l_2 are the lengths of links 1 an 2; θ_1 is the angular displacement of link 1 relative to the base; θ_2 is the angular displacement of link 2 relative to link 1; $\dot{\theta}_1$ is the angular velocity of link 1 relative to the base; $\dot{\theta}_2$ is the angular velocity of link 2 relative to link 1; m_1 and m_2 are the masses of links 1 and 2; l_{AS_1} is the distance between the centre of mass S_1 of link 1 and joint centre A; l_{BS_2} is the distance between the centre of mass S_2 of link 2 and joint centre B, I_{S_1} is the axial moment of inertia of link 1 relative to the centre of mass S_1 of link 1; I_{S_2} is the axial moment of inertia of link 2 relative to the centre of mass S_2 of link 2; g is the gravitational acceleration.

There are two intrinsic problems involved in Eq. (8.94) regarding the manipulator dynamics: complexity and nonlinearity. The proposed design methodology described bellow will eliminate both the complexity and the nonlinearity. Notice that the suggested method is more efficient from point of view of the mass increasing of the manipulator.

8.4.2 Design of Decoupled 2-dof Planar Serial Manipulator

The proposed 2-*dof* manipulator illustrated in Fig. 8.20 differs from the traditional scheme by the fact that two counter rotating links 5 and 6 are mounted on actuated element 1 and linked kinematically with second actuated element 2 through a belt transmission. It should be noted that a similar architecture with relocated counterweights has also been proposed in (Hess-Coelho et al. 2004).

Let us rewrite the Lagrangian of the manipulator with the added links:

$$L = T - V = \sum_{i=1}^{6} T_i - \sum_{i=1}^{6} V_i$$

$$= 0.5(m_1 \mathbf{v}_{S_1}^T \mathbf{v}_{S_1} + m_2 \mathbf{v}_{S_2}^T \mathbf{v}_{S_2} + m_3 \mathbf{v}_B^T \mathbf{v}_B + m_6 \mathbf{v}_C^T \mathbf{v}_C)$$

$$+ 0.5(I_{S_1}\dot{\theta}_1^2 + (I_{S_2} + I_{S_3} + I_{S_4} + I_{S_5})(\dot{\theta}_1 + \dot{\theta}_2)^2 + I_{S_6}(\dot{\theta}_1 - \dot{\theta}_2)^2) - \sum_{i=1}^{6} V_i$$

$$= 0.5((k_1 + 2k_2 \cos\theta_2)\dot{\theta}_1^2 + k_3\dot{\theta}_2^2 + (k_4 + 2k_2 \cos\theta_2)\dot{\theta}_1\dot{\theta}_2 - k_5 \sin\theta_1 - k_6 \sin(\theta_1 + \theta_2)) \quad (8.102)$$

where

$$k_1 = m_1 l_{AS_1}^2 + m_2(l_1^2 + l_{BS_2}^2) + m_3 l_1^2 + m_6 l_{AC}^2 + \sum_{i=1}^{6} I_{S_i} \quad (8.103)$$

$$k_2 = m_2 l_1 l_{BS_2} \quad (8.104)$$

$$k_3 = m_2 l_{BS_2}^2 + I_{S_2} + I_{S_3} + I_{S_4} + I_{S_5} + I_{S_6} \quad (8.105)$$

$$k_4 = 2(k_3 - 2I_{S_6}) \quad (8.106)$$

$$k_5 = g(m_1 l_{AS_1} + m_2 l_1 + m_3 l_1 - m_6 l_{AE}) \quad (8.107)$$

$$k_6 = m_2 l_{BS_2} g \quad (8.108)$$

The static balancing of the manipulator lead to the condition:

$$k_2 = k_5 = k_6 = 0 \quad (8.109)$$

and

$$L = 0.5(k_1\dot{\theta}_1^2 + k_3\dot{\theta}_2^2 + k_4\dot{\theta}_1\dot{\theta}_2) \quad (8.110)$$

Thus, if $I_{S_6} = 0.5k_3$, i.e. $k_4 = 0$, we obtain $L = 0.5(k_1\dot{\theta}_1^2 + k_3\dot{\theta}_2^2)$ and consequently

$$\tau_1 = k_1\ddot{\theta}_1 \quad (8.111)$$

and

$$\tau_2 = k_3\ddot{\theta}_2 \quad (8.112)$$

Now let us disclose the advantages of such a design: (i) gear 5 is mounted on the frame, which allows the reduction of the balancing masses; (ii) gear 6 is mounted on the prolongation of link 1, which allows the use of the mass of gear 6 designed for dynamic linearization as a counterweight for static balancing of link 1. Such a solution allows the considerable reduction of the total masses of links of the decoupled manipulator.

As was mentioned above, the dynamic decoupling of the degrees of freedom of a manipulator can be achieved by a remote actuation, i.e. when the actuator of the second link of the examined manipulator is mounted on the base and it is coupled with the link via a belt transmission. The schematics of the suggested design concept and the design with remote actuation are almost similar. However, the manipulator with remote actuation is not effective the from point of view of the precise reproduction of the motion as it accumulates all errors due to the clearances and elasticity of the belt

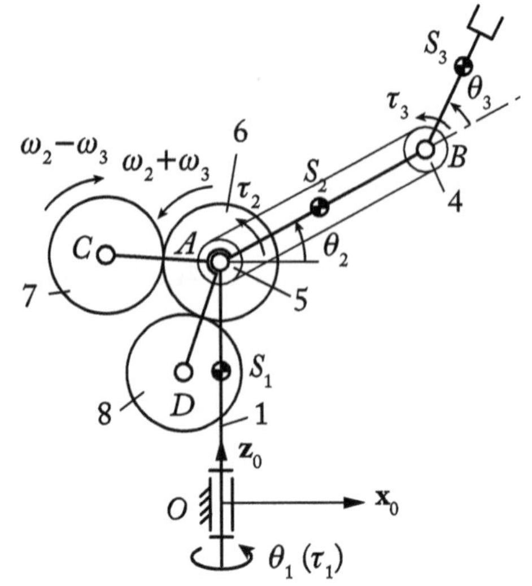

Fig. 8.21 Design of the decoupled 3-*dof* spatial serial manipulator

transmission, as well as the manufacturing and assembly errors due to the rotation transmission mechanism. In the case of the suggested design concept the actuators are located in the axes *A* and *B* but thanks to the added gears and the belt transmission the dynamic equations of the manipulator are decoupled.

It is obvious that the proposed design can be used not only for the 2-*dof* planar serial manipulators but also for any multi-degree-of-freedom spatial serial manipulator. In the case of the spatial manipulators the suggested technique will take into account that the inertia tensor of the manipulator must be diagonal and independent of arm configuration (Asada and Youcef-Toumi 1984).

Let us consider the dynamic linearization of a 3-*dof* spatial serial manipulator.

8.4.3 Design of Decoupled 3-dof Spatial Serial Manipulator

In the case of the 3-*dof* spatial serial manipulator (Fig. 8.21) two rotating gears 7 and 8 are mounted on the second link of the manipulator and coupled with gear 6 linked with third link through a belt transmission.

It should be noted that for invariant inertia of any spatial serial manipulator there is the condition concerning the identical inertias of transverse direction, i.e. $I_{xx} = I_{yy}$ (see in (Asada and Youcef-Toumi 1984)). Taking into account that in the examined spatial manipulator there are the links with masses m_2, m_3, m_7 and m_8, it is necessary that this condition will also be satisfied for these masses. Thus, the inertias of transverse direction will be identical, i.e. $I_m = I_{xx}^{(i)} = I_{yy}^{(i)}$. The kinetic energy of these masses relative to the $\mathbf{z}_1$ axis can be presented as $T_m = I_{zz}^{(i)}\dot{\theta}_1^2 = cI_m\dot{\theta}_1^2$, i.e. $I_{zz}^{(i)}$ can be replaced by the inertia I_m, where c is a constant.

Thus, the Lagrangian of the statically balanced manipulator with the mentioned identical inertias of transverse direction of masses m_2, m_3, m_7 and m_8 ($I_m = I_{xx}^{(i)} = I_{yy}^{(i)}$) (Asada and Youcef-Toumi 1984), can be written as:

$$\begin{aligned} L &= 0.5(I_{zz}^{(1)}\dot{\theta}_1^2 + cI_m\dot{\theta}_1^2 + I_m\dot{\theta}_2^2 + I_{xx}^{(2)}\dot{\theta}_1^2\sin^2\theta_2 + I_{yy}^{(2)}\dot{\theta}_1^2\cos^2\theta_2 \\ &\quad + I_{zz}^{(2)}\dot{\theta}_2^2 + I_{xx}^{(3)}\dot{\theta}_1^2\sin^2(\theta_2+\theta_3) + I_{yy}^{(3)}\dot{\theta}_1^2\cos^2(\theta_2+\theta_3) + I_{zz}^{(3)}(\dot{\theta}_2+\dot{\theta}_3)^2 \\ &\quad + (I_{zz}^{(7)} + I_{zz}^{(8)})(\dot{\theta}_2 - \dot{\theta}_3)^2) \\ &= 0.5(p_1\dot{\theta}_1^2 + p_2\dot{\theta}_2^2 + p_3\dot{\theta}_3^2 + p_4\dot{\theta}_1\dot{\theta}_2) \end{aligned} \tag{8.113}$$

with

$$\begin{aligned} p_1 &= I_{zz}^{(1)} + cI_m + I_{xx}^{(2)}\sin^2\theta_2 + I_{yy}^{(2)}\cos^2\theta_2 \\ &\quad + I_{xx}^{(3)}\sin^2(\theta_2+\theta_3) + I_{yy}^{(3)}\cos^2(\theta_2+\theta_3) \end{aligned} \tag{8.114}$$

$$p_2 = I_m + I_{zz}^{(2)} + I_{zz}^{(3)} + I_{zz}^{(7)} + I_{zz}^{(8)} \tag{8.115}$$

$$p_3 = I_{zz}^{(3)} + I_{zz}^{(7)} + I_{zz}^{(8)} \tag{8.116}$$

$$p_4 = 2(I_{zz}^{(3)} - I_{zz}^{(7)} - I_{zz}^{(8)}) \tag{8.117}$$

where $I_{zz}^{(i)}$ is the inertia of axial direction, $I_{xx}^{(i)}$ and $I_{yy}^{(i)}$ are the inertia of transverse directions with respect to the mass center of link i ($i = 1, 2, 3$); θ_1 is the angular displacement of link 1 relative to the base; θ_2 is the angular displacement of link 2 relative to link 1; θ_3 is the angular displacement of link 3 relative to link 2; $\dot{\theta}_1$ is the angular velocity of link 1 relative to the base; $\dot{\theta}_2$ is the angular velocity of link 2 relative to link 1; $\dot{\theta}_3$ is the angular velocity of link 3 relative to link 2; c is a constant due to the distribution of masses m_2, m_3, m_7 and m_8 according to the mentioned condition of identical inertia about transverse directions.

From Eqs. (8.114) and (8.117) we can see that the manipulator dynamics will be decoupled if

$$I_{xx}^{(2)} = I_{yy}^{(2)} \tag{8.118}$$

$$I_{xx}^{(3)} = I_{yy}^{(3)} \tag{8.119}$$

$$I_{zz}^{(3)} = I_{zz}^{(7)} + I_{zz}^{(8)} \tag{8.120}$$

By assuming these conditions, the kinetic energy of the manipulators will be constant and consequently

$$\tau_1 = p_1\ddot{\theta}_1 \tag{8.121}$$

$$\tau_2 = p_2\ddot{\theta}_2 \tag{8.122}$$

$$\tau_3 = p_3\ddot{\theta}_3 \tag{8.123}$$

Thus, the dynamic decoupling is again achieved.

In order to evaluate the dynamic performance of the improved design methodology let us consider a few illustrative examples.

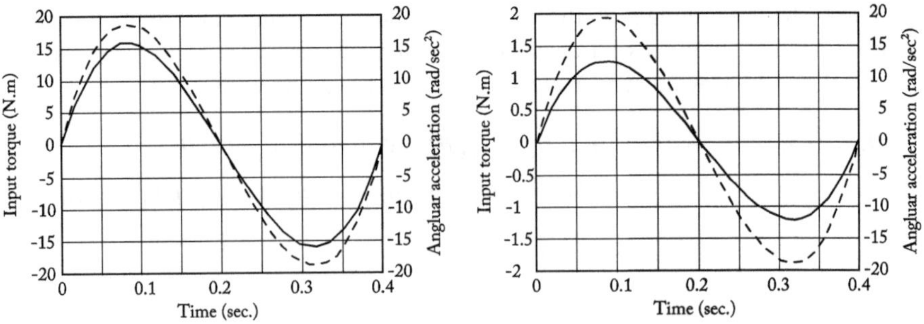

a Actuator 1 **b** Actuator 2

Fig. 8.22 Actuator torques (*solid line*) and angular accelerations (*dashed line*) for 2-*dof* linearized manipulator

8.4.4 Illustrative Examples

For illustration of the performance of the suggested design methodology, the simulations using ADAMS software have been carried out for a 2-*dof* and 3-*dof* serial manipulators. The manipulator parameters of the 2-*dof* serial manipulator (Fig. 8.20) are the following: $m_1 = 2.5$ kg, $m_2 = 2$ kg, $m_5 = 1$ kg, $m_6 = 6.5$ kg, $l_1 = 0.4$ m, $l_{AS_1} = 0.2$ m, $l_{BS_2} = 0$ m, $I_{S_1} = 0.1$ kg m^2, $I_{S_2} = 0.0275$ kg m^2, $I_{S_4} = 0.0.005$ kg m^2, $I_{S_5} = 0.0325$ kg m^2.

The generation of motions between the initial and final positions of links: $\theta_{1i} = 0$; $\theta_{1f} = 0.5236$; $\theta_{2i} = 1.0472$; $\theta_{2f} = 1.5708$, are carried out by the following fifth order polynomial laws: $\theta_j = 5.2360t^3 - 7.8540t^4 + 3.1416t^5$ ($j = 1, 2$) $0 \le t \le 0.4$ s.

Figure 8.22 shows the variations of the input torques and accelerations. We can see that the dynamic equations are linearized, i.e. $\tau_1 = k_1\ddot{\theta}_1$ and $\tau_2 = k_3\ddot{\theta}_2$.

It should be also noted that the decoupled manipulator designed according to schematics proposed in (Hess-Coelho et al. 2004) and having the same overall dimensions will be more than two times heavier.

Let us now consider the dynamic decoupling of the 3-*dof* spatial serial manipulator (Fig. 8.21). The manipulator parameters are the following: $m_1 = 4$ kg, $m_2 = 2.5$ kg, $m_3 = 2$ kg, $m_6 = 1$ kg, $m_7 = 6.5$ kg, $m_8 = 6.5$ kg, $l_{AB} = l_2 = 0.4$ m, $l_{AS_2} = 0.2$ m, $l_{BS_3} = 0$ m, $l_{AC} = l_{AD} = 0.5612$ m, $\angle BAC = \angle BAD = 125°$, $I_{zz}^{(1)} = 0.0125$ kg m^2, $I_{zz}^{(2)} = 0.1$ kg m^2, $I_{xx}^{(2)} = I_{yy}^{(2)} = 0.0175$ kg m^2, $I_{zz}^{(3)} = 0.0275$ kg m^2, $I_{xx}^{(3)} = I_{yy}^{(3)} = 0.01$ kg m^2, $I_{zz}^{(5)} = 0.005$ kg m^2, $I_{zz}^{(7)} = I_{zz}^{(8)} = 0.01625$ kg m^2.

The generation of motions between the initial and final positions of links: $\theta_{1i} = 0$; $\theta_{1f} = 0.5236$; $\theta_{2i} = 1.0472$; $\theta_{2f} = 1.5708$; $\theta_{3i} = 0$; $\theta_{3f} = 0.5236$ are carried out by the following fifth order polynomial laws: $\theta_j = 5.2360t^3 - 7.8540t^4 + 3.1416t^5$ ($j = 1, 2, 3$) $0 \le t \le 0.4$ s.

As shown in Fig. 8.23 the dynamic equations of the simulated manipulator are decoupled, i.e. $\tau_1 = p_1\ddot{\theta}_1$, $\tau_2 = p_2\ddot{\theta}_2$ and $\tau_3 = p_2\ddot{\theta}_3$.

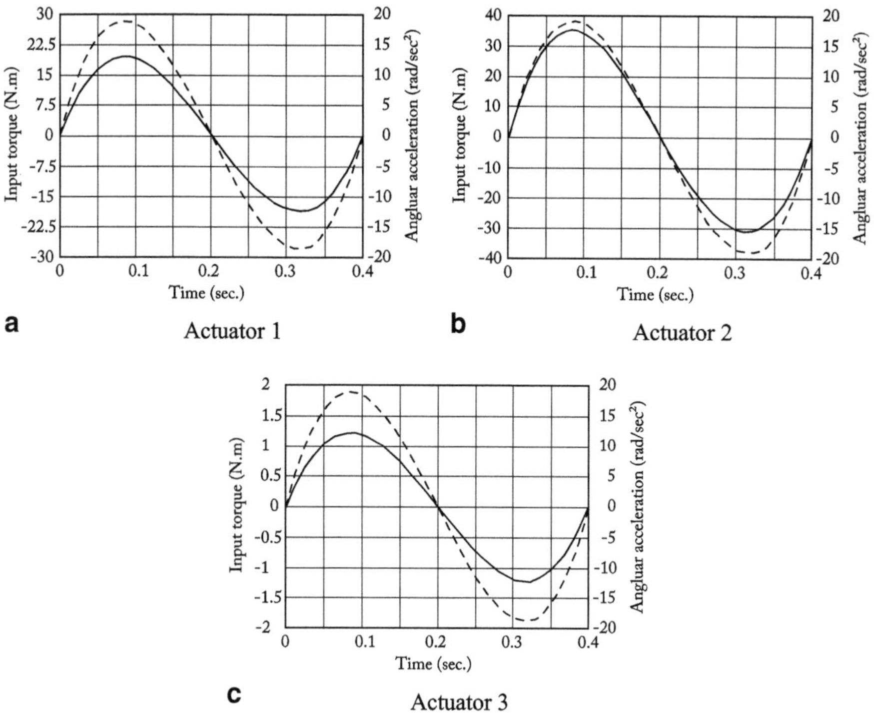

Fig. 8.23 Actuator torques (*solid line*) and angular accelerations (*dashed line*) for decoupled 3-*dof* serial manipulator

This study introduced the improved design concept for the linearization and decoupling of dynamic equations of serial manipulators. It is achieved by adding to the initial architecture of the manipulator the secondary gears having prescribed inertia parameters. The determination of the parameters of the added gears is based on eliminating coefficients of nonlinear terms in the manipulator's kinetic and potential energy equations. After such a redistribution of masses the actuator torque in each driven joint becomes a linear function of its angular acceleration. The proposed design concept permits the dynamic decoupling of the serial manipulators with a relatively small increase in the total mass of the moving links. It provides an improvement in the known design solutions, rendering them more suitable for practical applications.

References

Acevedo, M., Ceccarelli, M. Carbone, G.: Application of counter-rotary counterweights to the dynamic balancing of a spatial parallel manipulator. Appl. Mech. Mater. **162** 224–233 (2012)
Agrawal, A., Agrawal, S.: Effect of gravity balancing on biped stability. Proceedings of the IEEE International Conference on Robotics and Automation (ICRA 2004). New Orleans, USA, pp. 4228–4233 (2004)
Agrawal, A., Agrawal, S.: Design of gravity balancing leg orthosis using non-zero free length springs. Mech. Mach. Theory **40**, 693–709 (2005)
Agrawal, S., Fattah, A.: Design and prototype of a gravity-balanced leg orthosis. Int. J. Hum.-Friendly Welf. Robot. Syst. **4**(3), 13–16 (2003)
Agrawal, S., Fattah, A.: Gravity-balancing of special robotic manipulators. Mech. Mach. Theory **39**, 1331–1344 (2004a)
Agrawal, S., Fattah, A.: Reactionless space and ground robots: novel design and concept studies. Mech. Mach. Theory **39**, 25–40 (2004b)
Agrawal, S., Fattah, A.: Theory and design of an orthotic device for full or partial gravity-balancing of a human leg during motion. IEEE Trans. Neural Syst. Rehabil. Eng. **12**(2), 157–165 (2004c)
Agrawal, S., Shirumalla, S.: Planning motions of a dual-arm free-floating manipulator keeping the base inertially fixed. Mech. Mach. Theory **30**(1), 59–70 (1995)
Agrawal, S., Gardner, G., Pledgie, S.: Design and fabrication of a gravity balanced planar mechanism using auxiliary parallelograms. ASME J. Mech. Des. **123**(4), 525–528 (2001)
Agrawal, S.K., Banala, S.K., Fattah, A., Sangwan, V., Krishnamoorthy, V., Scholz, J., Hsu, W.: Assessment of motion of a swing leg and gait rehabilitation with a gravity balancing exoskeleton. IEEE Trans. Neural Syst. Rehabil. Eng. **15**(3) (2007)
Agrawal, S., Fattah, A., Banala, S.: Gravity Balanced Orthosis Apparatus. US, 7544155 (2009a)
Agrawal, S., Fattah, A., Hamnett, J.: Passive Gravity Balanced Assistive Device for Sit-to-stand Tasks. US, 7601104 (2009b)
Akeel, H.: Robot with balancing mechanism having a variable counterbalance force. US, 4659280 (1987)
Alici, G., Shirinzadeh, B.: Optimum force balancing of a planar parallel manipulator. J. Mech. Eng. Sci. **217**(5), 515–524 (2003a)
Alici, G., Shirinzadeh, B.: Optimum force balancing with mass distribution and a single elastic element for a five-bar parallel manipulator. Proceedings of the IEEE International Conference on Robotics and Automation (ICRA 2003). Taipei, Taiwan, pp. 3666–3671 (2003b)
Alici, G., Shirinzadeh, B.: Optimum dynamic balancing of planar parallel manipulators. Proceedings of the IEEE International Conference on Robotics and Automation (ICRA 2004). New Orleans, LO, USA, pp. 4527–4531 (2004)
Alici, G., Shirinzadeh, B.: Optimum dynamic balancing of planar parallel manipulators based on sensitivity analysis. Mech. Mach. Theory **41**, 1520–1532 (2006)

V. Arakelian, S. Briot, *Balancing of Linkages and Robot Manipulators*,
Mechanisms and Machine Science 27, DOI 10.1007/978-3-319-12490-2

Angeles, J., Wu, C.J.: The optimum synthesis of en elastic torque-compensating cam mechanism. Mech. Mach. Theory **36**, 245–259 (2001)

Angeles, J., Nahon, M., Thümmmel, T.: Active control for the complete dynamic balancing of linkages. Flex. Mech. Dyn. Anal. **47**, 305–310 (1992)

Arakelian, V.: Robotic technological complex. SU, 1530432 (1986)

Arakelian, V.: Manipulator. SU, 1430258 (1987)

Arakelian, V.: Manipulator. SU, 1537518 (1988)

Arakelian, V.: Balanced crank mechanism. SU, 1802244 (1993)

Arakelian, V.: Synthèse dynamique des mécanismes basée sur les méthodes d'approximation de la géométrie cinématique. Proceeding of the 9th World Congress on the Theory of Machines and Mechanisms, vol. 1. Italy, pp. 205–209 (1995)

Arakelian, V.: Structure et cinématique des mécanismes. Hermès (1997)

Arakelian, V.: équilibrage des manipulateurs manuels. Mech. Mach. Theory **33**(4), 437–442 (1998a)

Arakelian, V.: équilibrage dynamique complet des mécanismes. Mech. Mach. Theory **33**(4), 425–436 (1998b)

Arakelian, V.: An approximate method of calculating a counterweight for the optimum shaking force and shaking moment balancing of linkages. Proceedings of the 9th IFToMM International Conference on the Theory of Machines and Mechanisms. Liberec, Czech Republic, pp. 41–46 (2004a)

Arakelian, V.: The history of the creation and development of hand-operated balanced manipulators (HOBM). In: Ceccarelli, M. (ed.) Proceedings of the International Symposium on History of Machines and Mechanisms (HMM 2004), pp. 347–356. Kluwer Academic Publishers (2004b)

Arakelian, V.: Shaking moment cancellation of self-balanced slider-crank mechanical systems by means of optimum mass redistribution. Mech. Res. Commun. **33**, 846–850 (2006)

Arakelian, V.: Complete shaking force and shaking moment balancing of RSS'R spatial linkages. Proc. Inst. Mech. Eng. Part K: J. Multi-body Dyn. **221**, 303–310 (2007)

Arakelian, V., Briot, S.: Dynamic balancing of the scara robot. Proceedings of 17th CISM-IFToMM Symposium on Robot Design, Dynamics, and Control (RoManSy 2008). Tokyo, Japan (2008)

Arakelian, V., Briot, S.: Simultaneous inertia force/moment balancing and torque compensation of slider-crank mechanisms. Mech. Res. Commun. **37**(2), 265–269 (2010)

Arakelian, V., Dahan, M.: équilibrage optimal des mécanismes pour manipulateurs. Actes du 12ème Congrès Français de Mécanique. Strasbourg, France, pp. 373–376 (1995)

Arakelian, V., Dahan, M.: Dynamic balancing of mechanisms. Mech. Res. Commun. **27**(1), 1–6 (2000a)

Arakelian, V., Dahan, M.: Dynamic balancing of spatial mechanisms. C. R. l'Académie des Sci. **328**, 109–115 (2000b)

Arakelian, V., Dahan, M.: Complete shaking force and partial shaking moment balancing of planar four-bar linkages. Proc. Inst. Mech. Eng. Part K: J. Multi-body Dyn. **15**, 31–34 (2001a)

Arakelian, V., Dahan, M.: Partial shaking moment balancing of fully force balanced linkages. Mech. Mach. Theory **36**(11–12), 1241–1252 (2001b)

Arakelian, V., Dahan, M.: Balanced four-bar articulated mechanism has output member extended beyond axis of second pivot and meshing gears. FR, 2817008 (2002)

Arakelian, V., Ghazaryan, S.: Gravity balancing of the human leg taking into account the spring mass. Proceedings of the 9th International Conference on Climbing and Walking Robots. Brussels, Belgium, pp. 630–635 (2006)

Arakelian, V., Ghazaryan, S.: Improvement of balancing accuracy of robotic systems: application to leg orthosis for rehabilitation devices. Mech. Mach. Theory **43**(5), 565–575 (2008)

Arakelian, V., Makhsudyan, N.: Generalized Lanchester balancer. Mech. Res. Commun. **37**(7), 647–649 (2010)

Arakelian, V., Smith, M.: Complete shaking force and shaking moment balancing of linkages. Mech. Mach. Theory **34**(8), 1141–1153 (1999)

Arakelian, V., Smith, M.: Shaking moment minimization of fully force-balanced linkages. Proceedings of the 11th World Congress in Mechanism and Machine Science, vol. 2. Tianjin, China, pp. 633–636 (2004)

Arakelian, V., Smith, M.: Balancing method based on the properties of the Watt gear-slider mechanism. Proceedings of the International Symposium on Multibody Systems and Mechatronics (MUSME 2005). Uberlandia, Brazil (2005a)

Arakelian, V., Smith, M.: Erratum—shaking force and shaking moment balancing of mechanisms: an historical review with new examples. ASME J. Mech. Des. **127**(5), 1034–1035 (2005b)

Arakelian, V., Smith, M.: Shaking force and shaking moment balancing of mechanisms: a historical review with new examples. ASME J. Mech. Des. **127**, 334–339 (2005c)

Arakelian, V., Smith, M.: Design of planar 3-DOF 3-RRR reactionless parallel manipulators. Mechatronics **18**, 601–606 (2008)

Arakelian, V., et al.: Manipulator. SU, 1465298 (1989)

Arakelian, V., et al.: Balanced manipulator. SU, 1673432 (1991)

Arakelian, V., et al.: Slider-crank mechanism. SU, 1779853 (1992)

Arakelian, V., Dahan, M., Smith, M.: A historical review of the evolution of the theory on balancing of mechanisms. In: Ceccarelli, M. (ed.) Proceedings of the International Symposium on History of Machines and Mechanisms, pp. 291–300 (HMM 2000). Kluwer Academic Publishers, Dordrecht (2000)

Arakelian, V., Maurine, P., Briot, S., Pion, E.: Robot Parallele Comprenant des Moyens de Mise en Mouvement Décomposés en Deux Sous-Ensembles. FR, 2873317 (2006)

Arakelian, V., Sargsyan, S., Harutyunyan, M.: Friction torque compensation in the balanced leg orthosis for robotic rehabilitation. Proceedings of the 8th European Solid Mechanics Conference. Graz, Austria (2012)

Arsenault, M., Gosselin, C.: Static balancing of tensegrity mechanisms. ASME J. Mech. Des. **129**(3), 295–300 (2007)

Artobolevsky, I.: Théorie des mécanismes et des machines. Moscow (1977)

Artobolevsky, I.: Methods of balancing of inertia forces in working machines with complex kinematic schemes. Acad. Naouk SSSR. In Russian (1938)

Artobolevsky, I., Edelshtein, B.: Methods of inertia calculation for mechanisms of agricultural machines. Selkhozizdate, Moscow. In Russian (1935)

Asada, H., Youcef-Toumi, K.: Decoupling of manipulator inertia tensor by mass redistribution. Proceedings of the Design Engineering Technical Conference (DETC). Cambridge (1984)

Ashihara, J.: Body weight support device and body weight support program. US, 8177733 (2012a)

Ashihara, J.: Walking assist device. US, 8267876 (2012b)

Bagci, C.: Shaking force balancing of planar linkages with force transmission irregularities using balancing idler loops. Mech. Mach. Theory **14**, 267–284 (1979)

Bagci, C.: Complete shaking force and shaking moment balancing of link mechanisms using balancing idler loops. ASME J. Mech. Des. **104**, 482–493 (1982)

Bailey, D.: Spring counterbalance for rotating load. US, 8220765 (2012)

Balasubramanian, S., Bagci, C.: Design equations for the complete shaking force balancing of 6R 6-bar and 6-bar slider-crank mechanisms. Mech. Mach. Theory **13**, 659–674 (1978)

Banala, S., Agrawal, S., Fattah, A.: A gravity balancing leg orthosis for robotic rehabilitation. Proceedings of the IEEE International Conference on Robotics and Automation (ICRA 2004). New Orleans, USA, pp. 2427–2479 (2004)

Banala, S., Agrawal, S., Fattah, A., Krishnamoorthy, V., Hsu, W., Scholz, J., Rudolph, K.: Gravity-balancing leg orthosis and its performance evaluation. IEEE Trans. Robot. **22**(6), 1228–1237 (2006)

Banala, S., Kim, S., Agrawal, S., Scholz, J.: Robot assisted gait training with active leg exoskeleton (ALEX). IEEE Trans. Neural Syst. Rehabil. Eng. **17**(1), 2–8 (2009)

Baradat, C., Arakelian, V., Briot, S.: Design of a torque-minimizing mechanism for the delta parallel robot. Proceedings of the 20th Canadian Congress of Applied Mechanics (CANCAM2005). McGill University, Canada (2005)

Baradat, C., Arakelian, V., Maurine, P.: Parallel robot including of load-compensation system. FR, 2880575 (2006)
Baradat, C., Arakelian, V., Briot, S., Guegan, S.: Design and prototyping of a new balancing mechanism for spatial parallel manipulators. ASME J. Mech. Des. **130**(7) (2008)
Barents, R., Schenk, M., van Dorsser, W., Wisse, B., Herder, J.: Spring-to-spring balancing as energy-free adjustment method in gravity equilibrators. ASME J. Mech. Des. **133**(6) (2011)
Bartlett, D., Freed, D., Poynter, W.H.: Robot with spring pivot balancing mechanism. US, 4753128 (1988)
Bayer, A., Merk, G.: Industrial robot with a weight balancing system. EP, 2301727 (2011)
Belonovskaya, I., Dimentberg, F., Maysuk, L.: Construction of the self-balanced spatial mechanisms. Machinovedenie **1**, 3–9 (1987)
Belyanin, P.: Balanced Manipulators. Mashinostroyenie, Moscow (1988a)
Belyanin, P.: Construction des machines. Moscow (1988b)
Berestov, L.: Full dynamic balancing of pinned four-bar linkage. Izv. Vyssh. Uchebn. Zaved. Series: Machinostroenie **11**, 62–65. In Russian (1975)
Berestov, L.: Comparative analysis of the reactions in the kinematic pairs of the four-bar linkages for the different methods of balancing. J. Mech. Mach. 61–70. In Russian (1977a)
Berestov, L.: Mécanique des machines. Moscow (1977b)
Berkof, R.: Complete force and moment balancing of inline four-bar linkages. Mech. Mach. Theory **8**(3), 397–410 (1973)
Berkof, R.: Force balancing of a six-bar linkage. Proceedings of the Fifth World Congress on Theory of Machines and Mechanisms. Montreal, Canada, pp. 1082–1085 (1979a)
Berkof, R.: The input torque in linkages. Mech. Mach. Theory **14**(1), 61–73 (1979b)
Berkof, R., Lowen, G.: A new method for completely force balancing simple linkages. ASME J. Eng. Ind. **91**(1), 21–26 (1969)
Berkof, R., Lowen, G.: Theory of shaking moment optimization of force-balanced four-bar linkages. ASME J. Eng. Ind. **93**(1), 53–60 (1971)
Berkof, R., Lowen, G., Tepper, F.: Balancing of linkages. Shock Vib. Dig. **9**(6), 3–10 (1977)
Bessonov, A.: Dynamics of the Mechanisms with Variable Mass Links. Nauka, Moscow (1967)
Bessonov, A.: Balancing a planar mechanism with variable mass links. Proceedings of the Mechanisms Conference of ASME. Atlanta, Georgia, USA (1968)
Bittenbinder, W.: Lifting device for manual manipulator. DE, 4342716 (1995)
Bolotin, L.: Mechanical arm. SU, 1074709 (1982)
Briot, S., Arakelian, V.: Complete shaking force and shaking moment balancing of the position-orientation decoupled PAMINSA manipulator. Proceedings of the IEEE/ASME International Conference on Advanced Intelligent Mechatronics (AIM2009). Singapore (2009)
Briot, S., Arakelian, V.: Complete shaking force and shaking moment balancing of in-line four-bar linkages by adding a class-two RRR or RRP assur group. Mech. Mach. Theory **57**, 13–26 (2012)
Briot, S., Arakelian, V., Guegan: Design and prototyping of a partially decoupled 4-dof 3t1r parallel manipulator with high-load carrying capacity. ASME J. Mech. Des. **130** (2008a)
Briot, S., Bonev, I., Chablat, D., Wenger, P., Arakelian, V.: Self motions of general 3-RPR planar parallel robots. Int. J. Robot. Res. **7**(7), 855–866 (2008b)
Briot, S., Arakelian, V., Guégan, S.: Paminsa: a new family of decoupled parallel manipulators. Mech. Mach. Theory **44**(2), 425–444 (2009a)
Briot, S., Bonev, I., Gosselin, C., Arakelian, V.: Complete shaking force and shaking moment balancing of planar parallel manipulators with prismatic pairs. Proc. Inst. Mech. Eng. Part K: J. Multi-body Dyn. **223**(1), 43–52 (2009b)
Briot, S., Arakelian, V., Sauvestre, N., Le Baron, J.: Shaking forces minimization of high-speed robots via an optimal motion planning. Proceedings of the 18th CISM-IFToMM Symposium on Robot Design, Dynamics and Control (RoManSy 2010). Udine, Italy (2010)
Briot, S., Arakelian, V., Le Baron, J.: Shaking force minimization of high-speed robots via centre of mass acceleration control. Mech. Mach. Theory **57**, 1–12 (2012)
Brown, G., Di Guilio, A.: Support Apparatus. US, 4208028 (1980)

Brown, H., Dolan, J.: A novel gravity compensation system for space robots. Proceedings of the ASCE Specialty Conference on Robotics for Challenging Environments (Space94). pp. 250–258 (1994)

Brudniok, S., Schreiber, G., Maischberger, J.: Method and handling system for automated movement of a gravity-compensated load. EP, 2508308 (2012)

Bruzzone, L., Bozzini, G.: A statically balanced SCARA-like industrial manipulator with high energetic efficiency. Meccanica **46**, 771–784.

Buganza, A., Acevedo, M.: Dynamic balancing of a 2-dof 2RR planar parallel manipulator by optimization. Proceedings of the 13th IFToMM World Congress in Mechanism and Machine Science. Guanajuato, Mexico (2011)

Cardoso, L., Tomazio, S., Herder, J.: Conceptual design of a passive arm orthosis. Proceedings of the ASME Design Engineering Technical Conference. Montreal, Canada, pp. 1–10 (2002)

Carpenter, M., Peck, M.: Reducing base reactions with gyroscopic actuation of space-robotic systems. IEEE Trans. Robot. **25**(6), 1262–1270 (2009)

Carson, W., Stephens, J.: Feasible parameter design spaces for force and root-mean-square moment balancing on in-line 4R 4-bar linkage synthesized for kinematic criteria. Mech. Mach. Theory **13**(6), 649–658 (1978)

Carwardine, G.: Improvement in elastic force mechanisms. UK, 377251 (1932)

Carwardine, G.: Improvement in elastic force mechanisms. UK, 404615 (1934)

Carwardine, G.: Elastic Force and Equipoising Mechanism. US, 2204301 (1940)

Chaudhary, H., Saha, S.: Balancing of four-bar linkages using maximum recursive dynamic algorithm. Mech. Mach. Theory **42**(2), 216–232 (2007)

Chaudhary, H., Saha, S.: Balancing of shaking forces and shaking moments for planar mechanisms using the equimomental systems. Mech. Mach. Theory **43**(3), 310–334 (2008a)

Chaudhary, H., Saha, S.: An optimization technique for the balancing of spatial mechanisms. Mech. Mach. Theory **43**(4), 506–522 (2008b)

Chaudhary, H., Saha, S.: Dynamics and Balancing of Multibody Systems. Springer, Berlin (2009)

Chen, N.: Complete shaking force balancing of spatial linkages. Mech. Mach. Theory **19**(2) (1984a)

Chen, N.: Partial balancing of spatial force of a spatial four-bar RCCC linkage by the optimization method. Mech. Mach. Theory **19**(2), 257–265 (1984b)

Chen, N., Zhang, Q.: A method for full shaking force balancing of spatial linkages and the relevant force balancing theory. Proceedings of the Sixth World Congress on the Theory of Machines and Mechanisms. New Delhi, India, pp. 329–332 (1983)

Cheng, P., Cheng, K.: Evaluation of the dynamic performance variation of a serial manipulatot after eliminating the self-weight influence. Mechatronics **21**, 993–1002 (2011)

Chiou, S., Davies, T.: The ideal locations for the contra-rotating shafts of generalized Lanchester balancers. Proc. Inst. Mech. Eng. Part C: J. Mech. Eng. Sci. **208** (1), 29–37 (1994)

Chiou, S., Davies, T.: Partial cancellation of shaking force harmonics by cam modification. Proc. Inst. Mech. Eng. Part C: J. Mech. Eng. Sci. **211**, 253–263 (1997)

Chiou, S., Tsai, R.: The ideal shaft locations of three-rotating-mass balancers for spatial mechanisms. Mech. Mach. Theory **30**(3), 405–416 (1995)

Chiou, S., Shien, M., Tsai, R.: The two-rotating-mass balancers for partial balancing of spatial mechanisms. Mech. Mach. Theory **32**(5), 617–628 (1997)

Chiou, S., Bai, G., Chang, W.: Optimum balancing design of the drag-link drive of mechanical presses for precision cutting. Int. J. Mach. Tools Manuf. **38**(3), 131–141 (1998).

Cho, C., Kang, S.: Design of a static balancing mechanism with unit gravity compensators. Proceedings of the IEEE International Conference on Robotics and Automation (ICRA 2011). San Francisco, CA, USA, pp. 1857–1861 (2011)

Cho, C., Lee, W., Lee, J., Kang, S.: Static balancing of a manipulator with hemispherical workspace. Proceedings of the IEEE/ASME Int. Conf. on Advanced Intelligent Mechatronics (AIM 2010). Montreal, Canada, pp. 1269–1274 (2010)

Cho, C., Lee, W., Kang, S.: Design of a static balancing mechanism with unit gravity compensators. Proceedings of the IEEE/ASME Int. Conf. on Advanced Intelligent Mechatronics (AIM 2011). San Francisco, CA, USA, pp. 1857–1862 (2011)

Cho, C., Lee, W., Lee, J., Kang, S.: A 2-dof gravity compensator with bevel gears. J. Mech. Sci. Technol. **26**(9), 2913–2919 (2012)

Chung, C., Desa, S.: A global approach for using kinematic redundancy to minimize base reactions of menipulators. Proceedings of the ASME Design Technical Conference (DETC 1989), vol. 3. Montreal, Canada, pp. 297–303 (1989)

Chungand, W., Cho, H., Lee, C., Warnecke, H.: Performance of robotic manipulators with an automatic balancing mechanism. Comput.-Integr. Manuf. Robot. ASME Prod. Eng. Div. **13**, 111–121 (1984)

Ciupitu, L., Simionescu, I., Olaru, A.: Static balancing of mechanical systems used in medical engineering field—approximate balancing. Adv. Mater. **463–464**, 890–894 (2005)

Ciupitu, L., Simionescu, I., Lee, C.: Static balancing—an overview. Proceedings of the First IFToMM Asian Conference on Mechanism and Machine Sciences. Taipei, Taiwan (2010a)

Ciupitu, L., Simionescu, I., Olaru, A.: Static balancing of mechanical systems used in mechanical engineering field—continuous balancing. Proceedings of the 5th International Conference on Optimisation of the Robots and Manipulators. Calimanesti, Romania, pp. 274–278 (2010b)

Clavel, R.: Device for the movement and positioning of an element in space. US, 4976582 (1990)

Conte, F., George, G., Mayne, R., Sadler, J.: Optimum mechanisms design combining kinematic and dynamic-force consideration. ASME J. Eng. Ind. **97**(2), 662–668 (1975)

Cormac, P.: A Treatise on Engine Balance Using Exponentials. Dutton, New York (1923)

Dalby, W.: The Balancing of Engines. Arnnold, London (1923)

Davies, T.: The kinematics and design of linkages, balancing mechanisms and machines. Mach. Des. Eng. **40**, 40–51 (1968)

Davies, T., Niu, G.: On the retrospective balancing of installed planar mechanisms. Proc. Inst. Mech. Eng. Part C: J. Mech. Eng. Sci. **208**, 39–45 (1994)

Deepak, S., Ananthasuresh, G.: Static balancing of spring-loaded planar revolute-joint linkages without auxiliary links. Proceedings of the 14th National Conference on Machines and Mechanisms. NIT, Durgapur, India, pp. 37–44 (2009)

Deepak, S., Ananthasuresh, G.: Perfect static balancing of linkages by addition of springs but not auxiliary bodies. ASME J. Mech. Robot. **4** (2012a)

Deepak, S., Ananthasuresh, G.: Static balancing of a four-bar linkage and its cognates. Mech. Mach. Theory **48**, 62–80 (2012b)

Dehkordi, M., Frisoli, A., Sotgiu, E., Bergamasco, M.: Modelling and experimental evaluation of a static balancing technique for a new horizontally mounted 3-UPU parallel mechanism. Int. J. Adv. Robot. Syst. **9**, 1–12 (2012)

Delagne, G.: Certaines propriétés générales d'équilibrage des machines a piston d'après la méthode des vecteurs tournants symétriques. Comptes Rendus de l'Académie des Sciences **206**(22), 1617–1618 (1938)

Demeulenaere, B.: Dynamic Balancing of reciprocating machinery with application to weaving machines. Ph.D. thesis, K.U.Leuven, Leuven, Belgium (2004)

Demeulenaere, B., Berkof, R.: Improving machine drive dynamics: a structured design approach towards balancing. ASME J. Mech. Des. **130**(8) (2008)

Demeulenaere, B., Swevers, J., De Schutter, J.: Convex optimization framework for dynamic balancing of planar linkages. Proceedings of ISMA 2004. pp. 1997–2010 (2004a)

Demeulenaere, B., Swevers, J., De Schutter, J.: Dynamic balancing of four-bar linkages: a convex optimization framework for efficiently obtaining globally optimal counterweights. Proceedings of the ASME 2004 International Design Engineering Technical Conferences and Computers and Information in Engineering Conference (IDETC/CIE 2004). Salt Lake City, Utah, USA (2004b)

Demeulenaere, B., Aertbelien, E., Swevers, M.V.J., De Schutter, J.: Ultimate limits for counterweight balancing of crank-rocker four-bar linkages. ASME J. Mech. Des. **128**, 1272–1284 (2006)

Demeulenaere, B., Verschuure, M., Swevers, J., De Schutter, J.: Counterweight balancing for vibration reduction of elastically mounted machine frames: a second-order cone programming approach. ASME J. Mech. Des. **130** (2008)

Derkach, G.: Balanced manipulator. SU, 1826956 (1993)

Derkach, G., Umantsev, A.: Balanced manipulator. SU, 1546408 (1990)

Djavakhian, R.: Balanced manipulator. SU, 1468739 (1987)

Djavakhian, R.: Manipulator. SU, 1828803 (1990)

Djavakhian, R., Djavakhian, N.: Balanced manipulator. SU, 1521579 (1988)

Doronin, V., Pospelov, A.: Balanced slider-crank mechanism. SU, 1627769 (1991)

Doucet, E.: équilibrage dynamique des moteurs en ligne. Techniques Automobile et Arienne **37** (1946)

Dresig, H.: Schwingungen mechanischer Antriebssysteme. Springer, Berlin (2001)

Dresig, H., Holzweißig, F.: Maschinendynamik. Springer, Berlin (2004)

Dresig, H., Jacobi, P.: Vollständiger Trägheitskraftausgleich von ebenen Koppelgetrieben durch Anbringen eines Zweischlages. Maschinenbautechnik **23**(1), 5–8 (1974)

Dresig, H., Nguyen, P.: Complete shaking force and shaking moment balancing of mechanisms using a moving rigid body. Technische Mechanik **31**(2), 121–131 (2011)

Dresig, H., Schönfeld, S.: Trägheitskraftausgleich für ebene Koppelgetriebe. Wiss. Techn. Univ. Dresden **20**(5), 1341–1349 (1971)

Dresig, H., Schönfeld, S.: Rechnergestützte Optimierung der Antribs und Gestellkraftgrössen ebene. Koppelgetriebe—Teil 1. Mech. Mach. Theory **11**, 363–370 (1976a)

Dresig, H., Schönfeld, S.: Rechnergestützte Optimierung der Antribs und Gestellkraftgrössen ebene. Koppelgetriebe—Teil 2. Mech. Mach. Theory **11**, 363–370 (1976b)

Dresig, H., Naake, S., Rockausen, L.: Vollständiger und harmonischer Ausgleich ebener Mechanismen. VDI Verlag, Düsseldorf (1994)

Dubey, V., Agrawal, S.: Study of an upper arm exoskeleton for gravity balancing and minimization of transmitted forces. Proc. Inst. Mech. Eng. Part H: J. Eng. Med. **225**(11), 1025–1035 (2011)

Duning, A., Herder, J.: A review of assistive devices for arm balancing. Proceedings of the IEEE International Conference on Rehabilitation Robotics. Seattle, Washington, USA (2013)

Dunlop, G., Jones, T.: Gravity counter balancing of parallel robot for antenna aiming. Proceedings of the 6th ASRAM. pp. 153–158 (1996)

Dzhavakhyan, R., Dzhavakhyan, N.: Balanced manipulator. SU, 1521579 (1989)

Dzhavakhyan, R., Dzhavakhyan, N.: Balanced manipulator. SU, 1357218 (1987)

Ebert-Uphoff, I., Johnson, K.: Practical considerations for the static balancing of mechanisms of parallel architecture. Proc. Inst Mech. Eng. Part K: J. Multi-body Dyn. **216**(1), 73–85 (2002)

Ebert-Uphoff, I., Gosselin, C., Laliberté, T.: Static balancing of spatial parallel mechanisms—revisited. ASME J. Mech. Des. **122**, 43–51 (2000)

Eckenstein, N., Yim, M.: Modular advantages and kinematic decoupling in gravity compensated robotic systems. ASME J. Mech. Robot. **5** (2013)

Elliot, J., Tesar, D.: The theory of torque, shaking force and shaking moment balancing of four link mechanisms. ASME J. Eng. Ind. **99**(3), 715–722 (1977)

Elliot, J., Tesar, D.: A general mass balancing method for complex planar mechanisms. Mech. Mach. Theory **17**(2), 153–172 (1982)

Ellis, M., Sukal, T., DeMott, T., Dewald, J.: ACT-3D exercise targets gravity-induced discoordination and improves reaching work area in individuals with stroke. Proceedings of the 10th ICORR (2007)

Emdadi, H., Yazdanian, M., Ettefagh, M., Feizi-Derakhshi, M.: Double four-bar crank-slider mechanism dynamic balancing by meta-heuristic algorithms. Int. J. Artif. Intell. Appl. **4**(5), 1–18 (2013)

Emöd, I.: Massenausgleich am Kurbelgertiebe von Sechszylinder-viertakt-V-motoren mit 6 Kurbeln und 60deg. Zylinderwinkeln. Period. Polytech. Eng. **11**(3–4), 205–221 (1967)

Endo, G., Hirose, S.: A weight compensation mechanism with a non-circular pulley and a spring. application to a parallel link manipulator. Proceedings of the JSME annual Conference on Robotics and Mechatronics. Japan (2008)

Endo, G., Yamada, H., Yajima, A., Ogata, M., Hirose, S.: A passive weight compensation mechanism with a non-circular pulley and a spring. Proceedings of the IEEE International Conference on Robotics and Automation (ICRA 2010). Alaska, USA, pp. 3843–3848 (2010)

Erkaya, S.: Investigation of balancing problem for a planar mechanism using genetic algorithm. J. Mech. Sci. Technol. **27**(7), 2153–2160 (2013)

Esat, I., Bahai, H.: A theory of complete force and moment balancing of planar linkage mechanisms. Mech. Mach. Theory **34**(6), 903–922 (1999).

Ettefagh, M., Abbasidoust, F., Milanchian, H., Yazdanian, M.: Complex crank-slider mechanism dynamic balancing by binary genetic algorithm (BGA). Proceedings of the 2011 International Symposium on Innovations in Intelligent Systems and Applications. Istanbul, Turkey, pp. 277–281 (2011)

Fahim, A., Fernandez, M.: Performance enhancement of robot arms through active counterbalancing. Int. J. Adv. Manuf. Technol. **3**(4), 63–72 (1988)

Farmani, M., Jaamiolahmadi, A.: Optimization of force and moment balance of a four-bar linkage using multi-objective genetic algorithm. Proceedings of the ASME 2009 International Mechanical Engineering Congress and Exposition. Lake Buena Vista, Florida, USA (2009)

Fattah, A., Agrawal, S.: Design and modeling of classes of spatial reactionless manipulators. Proceedings of the IEEE International Conference on Robotics and Automation (ICRA 2003). Taipei, Taiwan, pp. 3225–3230 (2003)

Fattah, A., Agrawal, S.: Gravity balancing rehabilitative robot for human leg. Proceedings of the 26th Annual International Conference of the IEEE EMBS. San Francisco, CA, USA, pp. 2695–2698 (2004)

Fattah, A., Agrawal, S.: Design arm simulation of a class of spatial reactionless manipulators. Robotica **3**, 75–81 (2005a)

Fattah, A., Agrawal, S.: On the design of a passive orthosis to gravity balance human legs. ASME J. Mech. Des. **127**, 802–808 (2005b)

Fattah, A., Agrawal, S.: Design of a passive gravity-balanced assistive device for sit-to-stand tasks. ASME J. Mech. Des. **128**, 1122–1129 (2006a)

Fattah, A., Agrawal, S.: Gravity-balancing of classes of industrial robots. Proceedings of the IEEE International Conference on Robotics and Automation. Orlando, Florida, USA, pp. 2872–2877 (2006b)

Fattah, A., Agrawal, S.: On the design of reactionless 3-dof planar parallel mechanisms. Mech. Mach. Theory **41**(1), 70–82 (2006c)

Favorin, M.: Mass Moments of Inertia (reference book). Mashinostroenie, Moscow (1970)

Feng, B., Morita, N., Torii, T., Yoshida, S.: Optimum balancing of shaking force and shaking moment for spatial RSSR mechanism using genetic algorithm. JSME Int. J. **43**(3), 691–696 (2000)

Feng, B., Morita, N., Torii, T.: A new optimization method for dynamic design of planar linkage with clearances at joints—optimizing the mass distribution of links to reduce the change of joint forces. ASME J. Mech. Des. **124**, 68–73 (2002)

Filaretov, V., Vukobratovic, M.: Static balancing and dynamic decoupling of the motion of manipulation robots. Mech. Mach. Theory **3**(6), 767–783 (1993)

Filonov, I., Petrikovetz, I.: Balancing device of lever mechanisms. SU, 1296762 (1987)

Fischer, O.: über die reduzierten Systeme und die Hauptpunkte der Glieder eines Gelenkmechanismus. Zeit. für Math. and Phys. **47**, 429–466 (1902)

Foucault, S., Gosselin, C.: On the development of a planar 3-dof reactionless parallel mechanism. Proceedings of the ASME Design Engineering Technical Conferences and Computers and Information in Engineering Conference (IDECT/CIE 2002). Montreal, Canada (2002)

Foucault, S., Gosselin, C.: Synthesis, design, and prototyping of a planar three degree-of-freedom re-actionless parallel mechanism. ASME J. Mech. Des. **126**, 992–999 (2004)

Freudenstein, F.: Quasi lumped-parameter analysis of dynamical systems. Proceedings of the 3rd Appl. Mech. Conference (1973)
Freudenstein, F., Macey, J., Maki, E.: Optimum balancing of combined pitching and yawing moments in high-speed machinery. ASME J. Mech. Des. **103**(3), 571–577 (1981)
Frolov, K.: Theory of Mechanisms and Machines. Vishaya shkola (1987)
Frolov, K., Vorobev, E.: Mechanics of Industrial Robots. Vishaya Shkola, Moscow (1988)
Fujikoshi, K.: Balancing apparatus for jointed robot. JP, 51–12(1976)
Fukushima, E., Debenest, P., Tojo, Y., Takita, K., Frees, M., Radich, H., Hiros, S.: Teleoperated buggy vehicle and weigh balanced arm system for mechanization of mine detection and clearance tasks. Proceedings of the HUDEM 2005. Tokyo, Japan, pp. 58–63 (2005)
Gao, F.: Complete shaking force and shaking moment balancing of four types of six-bar linkages. Mech. Mach. Theory **24**(4), 275–287 (1989)
Gao, F.: Complete shaking force and shaking moment balancing of 26 types of four-, five- and six-bar linkages with prismatic pairs. Mech. Mach. Theory, **25**(2), 183–192 (1990)
Gao, F.: Complete shaking force and shaking moment balancing of 17 types of eight-bar linkages only with revolute pairs. Mech. Mach. Theory **26**(2), 197–206 (1991)
Gappoev, T.: Singularities of the balancing of the spatial mechanisms, balancing of the machines and the apparatuses. Mech. Eng. 243–251 (1979)
Gappoev, T., Salamonov, M.: Some problems of balancing of agricultural machines. Proceedings of the Conference of the Soviet Union: modern methods of balancing of the machines and the apparatuses. Moscow, Russia, pp. 49–50 (1983)
Gappoev, T., Tabouev, D.: Peculiarities of the dynamic balancing of spatial four-bar mechanisms. Proceedings of the 5th World Congress on Theory on Mechanisms and Machine. Canada, pp. 1420–1423 (1979)
Gappoev, T., Tabouev, D.: Balancing of spatial mechanisms. Dyn. Mach. 50–56 (1980)
Ghazaryan, S., Sargsyan, S., Harutyunyan, M., Arakelian, V.: The design of exoskeleton-assistant for sit-to-stand tasks. Trans. Nat. Acad. Sci. Armen. Eng. Acad. Armen. **64**(2), 121–128 (2011)
Gheronimus, Y.: An approximate method of calculating a counterweight for the balancing of vertical inertia forces. Mechanisms **3**(4), 283–288 (1968a)
Gheronimus, Y.: On the application of Chebychev's methods to the problem of balancing mechanisms. J. Mech. **3**(4), 235–281 (1968b)
Gill, G., Freudenstein, F.: Minimization of inertia-induced forces in spherical four-bar mechanisms. Part 1: the general spherical four-bar linkage. ASME J. Mech. Trans. Autom. Des. **105**(3), 471–477 (1983a)
Gill, G., Freudenstein, F.: Minimization of inertia-induced forces in spherical four-bar mechanisms. Part 2: Wobble-plate engines. ASME J. Mech. Trans. Autom. Des. **105**(3), 478–483 (1983b)
Glachet, C., Francois, D., Frioux, C., Tentelier, J.: Telescopic remote manipulator of the master-slave type and its balancing means. EP, 0084482 (1985)
Gopalswamy, A., Gupta, P., Vidyasagar, M.: A new parallelogram linkage configuration for gravity compensation using torsional springs. Proceedings of the IEEE International Conference on Robotics and Automation (ICRA 1992). Nice, France, pp. 664–669 (1992)
Goryachkin, V.: The forces of inertia and their balancing. Collection of scientific works. Kolos, pp. 283–418. In Russian (1914)
Gosselin, C.: Computational Methods in Mechanical Systems: Mechanism Analysis, Synthesis, and Optimization, pp. 68–96. (Chapter: On the design of efficient parallel mechanisms. NATO ASI). Springer, Berlin (1998)
Gosselin, C.: Adaptive robotic mechanical systems: a design paradigm. ASME J. Mech. Des. **128**(1), 192–198 (2006)
Gosselin, C.: Smart Devices and Machines for Advanced Manufacturing, pp. 27–48. (Chapter: Gravity compensation, static balancing and dynamic balancing of parallel mechanisms). Springer, London (2008)
Gosselin, C., Angeles, J.: Singularity analysis of closed-loop kinematic chains. IEEE Trans. Robot. Autom. **6**(3), 281–290 (1990)

Gosselin, C. Wang, J.: On the design of gravity-compensated six-degree-of-freedom parallel mechanisms. Proceedings of the IEEE International Conference on Robotics and Automation (ICRA 1998). Leuven, Belgium, pp. 2287–2294 (1998)

Gosselin, C., Vollmer, F., Côté, G., Wu, Y.: Synthesis and design of reactionless three-degree-of-freedom parallel mechanisms. IEEE Trans. Robot. Autom. **20**(2), 191–199 (2004)

Greene, H.: Load compensator for spring counter-weighting mechanism. US, 5400721 (1995)

Grossley, F.: Dynamics in Machines. Roland Press, New York (1954)

Gunnarsson, T.: An industrial robot with a balancing device in the form of a leaf spring. WO, 2002006018 (2002)

Gvozdev, Y.: Manipulator boom counterweighing mechanism. SU, 806405 (1981)

Gvozdev, Y.: Manipulator. SU, 1308463 (1987)

Gvozdev, Y.: Manipulator. SU, 1537512 (1990)

Gvozdev, Y.: Manipulator. SU, 1777993 (1992)

Haaker, L.: Balanced articulated manipulator. GB, 1097800 (1968)

Haines, R.: Minimum RMS shaking moment or driving torque of a force-balanced 4-bar linkage using feasible counterweights. Mech. Mach. Theory **16**, 185–190 (1981)

Han, C.: Balancing of high speed machinery. ASME J. Eng. Ind. **89**(1), 111–117 (1967)

Hank, F.: Mechanisieren von Handhabungen mit Synchon-Manipulatoren. Maschinenmarkt **84**(41), 809–810 (1978)

He, G., Lu, Z.: Optimal motion planning of parallel redundant mechanisms with shaking force reduction. Proceedings of the IMACS conference. Beijing, China, pp. 1132–1139 (2006)

Heiko, F.: Kraftsportgerät mit einem gewichtskompensierten Lastarm. DE, 19653734 (1998)

Herder, J.: Design of spring force compensation systems. Mech. Mach. Theory **33**, 151–161 (1998)

Herder, J.: Energy-free systems. Theory, conception and design of statically balanced mechanisms. Ph.D. thesis, Delft University of Technology (2001)

Herder, J.: Some considerations regarding statically balanced parallel mechanisms. Proceedings of the Workshop on Fundamental Issues and Future Research Directions for Parallel Mechanisms and Manipulators. Québec, Canada (2002)

Herder, J.: Development of a statically balanced arm support: ARMON. Proceedings of the IEEE 9th International Conference on Rehabilitation Robotics. Chicago, IL, USA (2005)

Herder, J., Gosselin, C.: A counter-rotary counterweight (CRCW) for light-weight dynamic balancing. Proceedings of the ASME Design Engineering Technical Conferences and Computers and Information in Engineering Conference (IDETC/CIE 2004). Salt Lake City, Utah, USA, pp. 659–667 (2004a)

Herder, J., Gosselin, C.: A counter-rotary counterweight for light-weight dynamic balancing. Proceedings of the ASME Design Engineering Technical Conferences and Computers and Information in Engineering Conference (IDETC/CIE 2004). Salt Lake City, Utah, USA, pp. 659–667 (2004b)

Herder, J., van der Wijk, V.: Force balanced delta robot. NL, 2002839 (2010)

Hervé, J.: Device for counter-balancing the forces due to gravity in a robot arm. FR, 2565153 (1985)

Hess-Coelho, T., Yong, L., Alves, V.: Decoupling of dynamic equations by means of adaptive balancing of 2-dof open-loop mechanisms. Mech. Mach. Theory, **39**(8), 871–881 (2004)

Hilpert, H.: Weight balancing of precision mechanical instruments. Mechanisms **3**(4), 289–302 (1968)

Hollander, K., Sugar, T.: Concept for compliant actuation in wearable robotic systems. Proceedings of the UK-Korea Conference (UKC) (2004)

Hollander, K., Sugar, T., Herring, D.: Adjustable robotic tendon using Jack spring. Proceedings of the 2005 IEEE 9th International Conference on Rehabilitation Robotics. Chicago, IL, USA, pp. 113–118 (2005)

HONDA: http://www.walkassist.honda.com (2014)

Ikeuchi, Y., Ashihara, J., Hiki, Y., Kudoh, H., Noda, T.: Walking assist device with bodyweight support systems. Proceedings of the 2009 IEEE/RSJ International Conference on Intelligent Robots and Systems (IROS 2009). St. Louis, USA, pp. 4073–4079 (2009)

Ilia, D., Sinatra, R.: A novel formulation of the dynamic balancing of five-bar linkages with applications to link optimization. Multibody Syst. Dyn., **21**, 193–211 (2009)

Ilia, D., Cammarata, A., Sinatra, R.: A novel formulation of the dynamic balancing of five-bar linkages. Proceedings of the 12th IFToMM World Congress in Mechanism and Machine Science. Besançon, France (2007)

Innocenti, C.: Harmonic balancing of planar machinery by arbitrarily placed counterweighing shafts. Proceedings of the ASME 2007 International Design Engineering Technical Conferences and Computers and Information in Engineering Conference (IDETC/CIE 2007). Las Vegas, Nevada, USA (2007)

Ishida, K., Matsuda, T.: Vibrations on vibrationless rotation-reciprocation internally geared device, and on vibrationless chain saw utilizing this device. Proceedings of the 5th ASME World Congress of Theory of Machines and Mechanisms (1977)

Ishida, K., Matsuda, T.: Performance characteristics and working comfortableness of forest workers of a new non-vibrating chain saw utilizing perfectly balanced rotation-reciprocation device. Proceedings of the fifth World Congress of Theory of Machines and Mechanisms. Montreal, Canada, pp. 951–954 (1979)

ISIS: Isis website. http://www.isis-robotics.com/ (2014)

Jacobi, P.: Vollständiger Trägheitskraftausgleich bei mehrgliedrigen Koppelgetrieben. Maschinenbautechnik **18**(11), 605–606 (1969)

Jacobi, P.: Bestimmung Optimaler Koppelgetriebe nach dem Hauptkriterium "Krefte und Moment am Gestell". Wissenschatliche Zeitshrift der Technischen Hochscule **14** (1), 45–49 (1972)

Jacobi, P., Rose, W.: Experimentelle Untersuchung Dynamisch Ausgelgichener ebener Koppelgetriebe. Maschinenbautechnik **21** (8), 354–358 (1972)

Jandrasits, W., Lowen, G.: The elastic-dynamic behavior of a counterweighted rock link with an overhanging endmass in a four-bar linkage, Part 1: Theory. ASME J. Mech. Des. **101**, 77–88 (1979a)

Jandrasits, W., Lowen, G.: The elastic-dynamic behavior of a counterweighted rock link with an overhanging endmass in a four-bar linkage, Part 2: Application and experiment. ASME J. Mech. Des. **101**, 89–98 (1979b)

Kamenski, V.: On the problem of the number of counterweights in the balancing of plane linkages. Mechanisms **3**(4), 323–333 (1968a)

Kamenski, V.: On the question of the balancing of plane linkages. Mechanisms **3**(4), 303–322 (1968b)

Karidis, J., McVicker, G., Pawletko, J., Zai, L., Goldowsky, M., Brown, R., Comulada, R.: The hummingbird minipositioner - providing three-axis motion at 50 g's with low reactions. Proceedings of the IEEE International Conference on Robotics and Automation (ICRA 1992). Nice, France, pp. 685–692 (1992)

Kaufman, R., Sandor, G.: Complete force balancing of spatial linkages. ASME J. Eng. Ind. **93**(2), 620–626 (1971)

Kazerooni, H.: Statically balanced direct drive manipulator. Robotica **7**(2), 143–149 (1989)

Kazerooni, H., Kim, S.: A new architecture for direct drive robots. Proceedings of the IEEE International Conference on Robotics and Automation (ICRA 1988). pp. 442–445 (1988)

Kfoury, F.: Load manipulator with improved balancing. EP, 70678/20121018 (2013)

Khalil, W., Dombre, E.: Modeling, Identification and Control of Robots. Hermes Penton, London (2002)

Kim, H., Song, J.: Low-cost robot arm with 3-DOF counterbalance mechanism. Proceedings of the IEEE International Conference on Robotics and Automation (ICRA 2013). Karlsruhe, Germany, pp. 4168–4173 (2013)

Kiyoaki, K.: Manual work assistant manipulator control system. JP, 11277473 (1999)

Klimchik, A., Caro, S., Wu, Y., Chablat, D., Furet, B., Pashkevich, A.: Stiffness modeling of robotic manipulator with gravity compensator. In: Thomas F., Gracia, A. P. Computational Kinematics, Mechanisms and Machine Science, vol. 15, pp. 185–192. Springer, Dordrecht (2014)

Kobayashi, A.: Analytical study of crank effort in reciprocating engines. Ryojun Coll. Eng.—Memoirs **4**(3), 127–183 (1931)

Kobayashi, K.: Comparison between spring balancer and gravity balancer in inertia force and performance. ASME J. Mech. Des. **123**, 549–555 (2001a)

Kobayashi, K.: New design method for spring balancers. ASME J. Mech. Des. **123**, 494–500 (2001b)

Kobrinskii, A., Kobrinskii, A.: Manipulation Systems of Robots. Mashinostroenie, Moscow (1989)

Kochev, I.: General method for full force balancing of spatial and planar linkages by internal mass redistribution. Mech. Mach. Theory **22**(4), 333–341 (1987)

Kochev, I.: A new general method for full force balancing of planar linkages. Mech. Mach. Theory **23**(6), 475–480 (1988)

Kochev, I.: Full shaking moment balancing of planar linkages by a prescribed input speed fluctuation. Mech. Mach. Theory **25**(4), 459–466 (1990a)

Kochev, I.: Theory of symmetrical mechanisms. Mech. Mach. Theory **25**(4), 467–478 (1990b)

Kochev, I.: Contribution to the theory of torque, shaking force and shaking moment balancing of planar linkages. Mech. Mach. Theory **26**(3), 275–284 (1991a)

Kochev, I.: Optimum balancing of a class of multiloop linkages by function cognate transformations. Mech. Mach. Theory **26**(3), 285–297 (1991b)

Kochev, I.: Active balancing of the frame shaking moment in high speed planar machines. Mech. Mach. Theory **27**(1), 53–58 (1992a)

Kochev, I.: Balancing of planar linkages for prescribed principal axes and minor/major ratio of the r.m.s. shaking force. Mech. Mach. Theory **27**(1), 45–51 (1992b)

Kochev, I.: Planar assemblies of sub-linkages with optimal dynamic characteristics. Mech. Mach. Theory **27**(3), 275–282 (1992c)

Kochev, I.: Qualitative theory of the reactions and stresses in high speed planar linkages. Mech. Mach. Theory **27**(1), 59–68 (1992d)

Kochev, I.: Root-mean-square shaking force along a given direction. Mech. Mach. Theory **27**(1), 37–43 (1992e)

Kochev, I.: General theory of complete shaking moment balancing of planar linkages: a critical review. Mech. Mach. Theory **35**, 1501–1514 (2000)

Kochev, I., Gurdev, G.: General criteria for optimum balancing of combined shaking force and shaking moment in planar linkages. Mech. Mach. Theory **23**(6), 481–489 (1988)

Kochev, I., Gurdev, G.: Balancing of linkages under the combined action of inertia and external forces. Mech. Mach. Theory **24**(2), 93–98 (1989)

Kolarski, M., Vukobratovic, M., Borovac, B.: Dynamic analysis of balanced robot mechanisms. Mech. Mach. Theory **29**(3), 427–454 (1994)

Kolotenkov, V.: Balanced mechanism. SU, 1114829 (1984)

Kondrin, A., Petrov, L., Polishchuk, N.: Pivoted arm balancing mechanism. SU, 1596154 (1990)

Korn, G., Korn, T.: Mathematical Handbook for Scientists and Engineers. McGraw-Hill Book Company, New York (1968)

Koser, K.: A cam mechanism for gravity-balancing. Mech. Res. Commun. **36**, 523–530 (2009)

Kramer, G., Romer, G., Stuyt, H.: Design of a dynamic arm support (DAS) for gravity compensation. Proceedings of the International Conference on Rehabilitation Robotics. Noodwijk, The Netherlands, pp. 1042–1048 (2007)

Kreutzinger, R.: über die bewegung des Schwerpunktes beim Kurbelgetriebe. Getriebetechnik **10**(9), pp. 397–398 (1942)

Krut, S., Benoit, M., Dombre, E., Pierrot, F.: MoonWalker, a lower limb exoskeleton able to sustain bodyweight using a passive force balancer. Proceedings of the IEEE International Conference on Robotics and Automation (ICRA 2010). Anchorage, Alaska, USA (2010)

Lacasse, M., Lachance, G., Boisclair, J., Ouellet, J., Gosselin, C.: On the design of a statically balanced serial robot using remote counterweights. Proceedings of the IEEE International Conference on Robotics and Automation (ICRA 2013). Karlsruhe, Germany, pp. 4174–4179 (2013)

Lacoste, L.: A simplification in the conditions for the zero-length-spring seismograph. Bull. Seismol. Soc. Am. **25**, 176–179 (1935)

Lakota, N., Petrov, L.: Automation of Assembly Tasks, pp. 137–153. (Chapter: Manipulators for assembly tasks). Nauka, Moscow (1985)

Lakota, N.: Vertical displacement device. SU, 1044591 (1980)

Laliberté, T., Gosselin, C., Jean, M.: Static balancing of 3-DOF planar parallel mechanisms. IEEE/ASME Trans. Mechatron. **4**(4), 363–377 (1999)

Laliberté, T., Gosselin, C., Gao, D.: Closed-loop actuation routings for cartesian scara-type manipulators. Proceedings of the 2010 International Design Engineering Technical Conferences & Computers and Information in Engineering Conference (IDETC/CIE 2010) (2010)

Lanchester, F.: Engine balancing. Horseless Age **33**(12–16) (1914)

Lauzier, N., Gosselin, C., Laliberté, T., Tremblay, P.: Adaptive gravity compensation of decoupled parallel and serial manipulators using a passive hydraulic transmission. J. Mech. Eng. Sci. **223**(12), 2871–2879 (2009)

Leblond, M., Gosselin, C.: Static balancing of spatial and planar parallel manipulators with prismatic actuators. Proceedings of the ASME 1998 DETC Conference (1998)

Lecours, A., Gosselin, C.: Reactionless two-degree-of-freedom planar parallel mechanisms with variable payload. ASME J. Mech. Robot. **2** (2010)

Lee, T., Cheng, C.: Optimum balancing of combined shaking force, shaking moment, and torque fluctuations in high-speed linkages. ASME J. Mech. Transm. Autom. Des. **106**(2), 242–251 (1984)

Leinonen, T.: Terminology for the theory of machines and mechanisms. Mech. Mach. Theory **26** (1991)

Lessard, S., Bonev, I., Bigras, P., Durand, L.G., Soulez, G., Cloutier, G., DeGuise, J.A.: Parallel robot for medical 3d-ultrasound imaging. Proceedings of the International Symposium on Industrial Electronics. École de technologie supérieure, Montréal, Canada (2006)

Lessard, S., Bigras, P., Bonev, I.: A new medical parallel robot and its static balancing optimization. ASME J. Med. Device. **1**, 272–278 (2007a)

Lessard, S., Bigras, P., Bonev, I., Briot, S., Arakelian, V.: Optimum static balancing of the parallel robot for medical 3D-ultrasound imaging. Proceedings of the 12th IFToMM World Congress. Besançon, France (2007b)

Li, C., Tso, P.: The study of dynamic balancing for high-speed presses. JSME Int. J. Serie C **79**(3), 657–662 (2006)

Lin, J.: Theoretical and experimental study on the dynamic balancing of flexible mechanisms. Master's thesis, Beijing Polytechnic University, Beijing, China (2000)

Lin, P., Shieh, W., Chen, D.: Design of perfectly statically balanced one-DOF planar linkage with revolute joints only. ASME J. Mech. Des. **131** (2009)

Lin, P., Shieh, W., Chen, D.: Design of a gravity-balanced general spatial serial-type manipulators. ASME J. Mech. Robot. **2** (2010a)

Lin, P., Shieh, W., Chen, D.: A stiffness matrix approach for the design of statically balanced planar articulated manipulators. Mech. Mach. Theory **45**, 1877–1891 (2010b)

Lin, P., Shieh, W., Chen, D.: Design of statically balanced planar articulated manipulator with spring suspension. IEEE Trans. Robotic. **28**(1), 12–21 (2012)

Lin, P., Shieh, W., Chen, D.: A theoretical study of weight-balanced mechanisms for design of spring assistive mobile arm support (MAS). Mech. Mach. Theory **61**, 156–167 (2013)

LOKOMAT: http://www.hocoma.com/ (2014)

Longman, R., Lindberg, R., Zedd, M.: Satellite-mounted robot manipulators—new kinematics and reaction moment compensation. Int. J. Robotic. Res. **6**(3), 87–103 (1987)

Lowen, G., Berkof, R.: Survey of investigations into the balancing of linkages. ASME J. Mech. **3**, 221–231 (1968)

Lowen, G., Berkof, R.: Theory of shaking moment optimization of force-balanced four-bar linkages. ASME J. Mech. Des. **12**, 53–60 (1970)

Lowen, G., Berkof, R.: Determination of force-balanced four-bar linkages with optimum shaking moment characteristics. ASME J. Eng. Ind. **93**(1), 39–46 (1971)
Lowen, G., Tepper, F., Berkof, R.: Balancing of linkages—an update. Mech. Mach. Theory **18**(3), 213–230 (1983)
Lu, Q., Ortega, C., Ma, O.: Passive gravity compensation mechanisms: technologies and applications. Recent Pat. Eng. **5**(1), 32–44 (2011)
Ma, O., Angeles, J.: Direct kinematics and dynamics of a planar 3-DOF planar manipulator. Adv. Des. Autom. **3**, 131–320 (1989)
Ma, O., Wang, J.: Apparatus and method for reduced-gravity simulation. US, 8152699 (2012)
Mahalingam, S., Sharan, A.: The optimal balance of the robotic manipulators. Proceedings of the IEEE International Conference on Robotics and Automation (ICRA 1986). San Francisco, CA, USA, pp. 828–835 (1986)
Makino, H., Furuya, N.: Scara robot and its family. Proceedings of the 3rd International Conference on Assembly Automation. Boeblingen, Germany, pp. 433–444 (1982)
Maksimov, V., Moiseenkov, V.: Manipulator pivoted arm counterbalancing mechanism. SU, 617255 (1978)
Martin, G.: Kinematics and Dynamics of Machines, 3rd edn. McGraw-Hill, Columbus (2002)
Martini, A., Troncossi, M., Rivola, A.: Elastodynamic effect of mass-balancing: experimental investigation of a four-bar linkage. Adv. Mech. Eng. (2013)
Matsumoto, R.: Load handling equipment. US, 3883105 (1975)
Maxwell, R.: Kinematics and Dynamics of Machinery. Prentice-Hall, Englewood Cliffs (1960)
Melton, B.: The Lacoste suspension—principles and practice. J. Geophys. **22**, 521–543 (1971)
Menschaar, H., Ariens, A., Herder, J., Bakker, B.: Five-bar mechanism with dynamic balancing means and method for dynamically balancing a five-bar mechanism. WO, 2006080846 (2006)
Merlet, J.: Parallel Robots, 2nd edn. Springer, Dordrecht (2006)
Mia, Y., Gao, F., Pan, D.: Mechanical design of a hybrid leg exoskeleton to augment load-carrying for walking. Int. J. Adv. Robot. Syst. **10** (2013)
Minotti, P., Pracht, P.: Ressort et mécanismes: une solution aux problèmes d'équilibrage. Mech. Mach. Theory **23**, 157–168 (1988)
Mishkind, C.: Manipulators with manual control and automatic weight balancing for engineering industry. Moscow (1981)
Mokhtarian, A., Fattah, A., Agrawal, S.: A novel passive pelvic device for assistance during locomotion. Proceedings of the IEEE International Conference on Robotics and Automation (ICRA 2010). Anchorage, Alaska, USA, pp. 2241–2246 (2010)
Moor, B., Akouna, H.: Vertical counter balanced test head manipulator. CN, 1408065 (2003)
Moore, B.: Dynamic balancing of linkages by algebraic methods. Ph.D. thesis, Research Institute for Symbolic Computation, Johannes Kepler, Linz University, Linz, Austria (2009)
Moore, B., Schicho, J.: Two methods for force balancing of Bennett linkages. Proceedings of the 5th International Workshop on Computational Kinematics. pp. 241–248 (2009)
Moore, B., Schicho, J., Gosselin, C.: Determination of the complete set of shaking force and shaking moment balanced planar four-bar linkages. Mech. Mach. Theory **44**, 1338–1347 (2009)
Moore, B., Schicho, J., Gosselin, C.: Dynamic balancing of spherical 4R linkages. ASME J. Mech. Robot. **2** (2010)
Morita, T., Kuribara, F., Shiozawa, Y., Sugano, S.: A novel mechanism design for gravity compensation in three dimensional space. Proceedings of the IEEE/ASME 2003 International Conference on Advanced Intelligent Mechatronics (AIM 2003). Kobe, Japan, pp. 163–168 (2003)
Morozovski, E.: Balanced manipulator. SU, 1813621 (1993)
Nagchaudhuri, A.: Mechatronic redesign of slider crank mechanism. Proceedings of the ASME International Mechanical Engineering Congress & Exposition. New Orleans, Louisiana, USA (2002)
Nakayama, T., Araki, Y., Fujimoto, H.: A new gravity compensation mechanism for lower limb rehabilitation. Proceedings of the IEEE International Conference on Robotics and Automation (ICRA 2009). Changchun, China, pp. 943–948 (2009)

Nathan, R.: Constant force generator mechanism and adjustable seat constructed therewith. US, 4387876 (1983)

Nathan, R.: A constant force generation mechanism. ASME J. Mech. Transm. Autom. Des. **107**(4), 508–512 (1985)

Nenchev, D.: Reaction null space of a multibody system with applications in robotics. Mech. Sci. **4**, 97–112 (2013)

Nenchev, D., Yoshida, K., Umetani, Y.: Introduction of redundant arms for manipulation in space. Proceedings of the IEEE International Workshop on Intelligent Robots, vol. 88. pp. 679–684 (1988)

Nenchev, D., Yoshida, K., Uchiyama, M.: Reaction null-space based control of flexible structure mounted manipulator systems. Proceedings of the 35th IEEE Conference on Decision and Control (CDC 1996), vol. 4. pp. 4118–4123 (1996)

Nenchev, D., Yoshida, K., Vichitkulsawat, P., Uchiyama, M.: Reaction null-space control of flexible structure mounted manipulator systems. IEEE Trans. Robot. Auto. **15**, 1011–1023 (1999)

Newman, W., Hogan, N.: The optimal control of balanced manipulators. Proceedings of the ASME winter annual meeting. CA, USA (1986)

Nguyen, V., Nguyen, P.: Balancing conditions of spatial mechanisms. Mech. Mach. Theory **42**, 1141–1152 (2007)

Offt, C.: Experimentelle Untersuchungen an Koppelgetrieben mit Leistungausgleich. Mech. Mach. Theory **9**, 239–246 (1974)

O'Leary, J., Gatecliff, G.: Computer aided balance of single-cylinder slider-crank ic engines. Computer Aided Balance of Single-Cylinder Slider-Crank IC Engines (1989)

Olsen, R.: Balanced assembly. US, 3134340 (1964)

Olsen, R.: Loading balancer assembly. US, 3259352 (1966)

Ono, Y., Morita, T.: An underactuated manipulation method using a mechanical gravity canceller. J. Robot. Mechatron. **106**(6), 563–569 (2004)

Ouyang, P., Zhang, W.: A novel force balancing method for real-time controllable mechanisms. Proceedings of the ASME Design Engineering Technical Conferences and Computers and Information in Engineering Conference (IDETC/CIE 2002). Montreal, Canada (2002)

Ouyang, P., Zhang, W.: Force balancing of robotic mechanisms based on adjustment of kinematic parameters. ASME J. Mech. Des. **127**, 433–440 (2005)

Papadopoulos, E., Abu-Abed, A.: Design and motion planning for a zero-reaction manipulator. Proceedings of the IEEE International Conference on Robotics and Automation (ICRA 1994). San Diego, CA, USA, pp. 1554–1559 (1994)

Park, C., Kwak, B.: Counterweight optimization for reducing dynamic effects of clearance at a revolute joint. Mech. Mach. Theory **22**(6), 549–556 (1987)

Perreault, S., Cardou, P.: Approximate static balancing of a planar parallel cable-driven mechanism. Proceedings of the CCToMM Symposium on Mechanism, Machines and Mechantronics. Quebec, Canada (2009)

Perrot, Y.: Ressort d'équilibrage, pour mécanisme articulé, notamment pour bras de robot. FR, 2847958 (2004)

Petrov, L., Polishchuk, N.: Vertical displacement device. SU, 643323 (1979)

Pons, J., Ceres, R., Jiménez, A.: Quasi-exact linear spring counter gravity system for robotic manipulators. Mech. Mach. Theory **33**, 59–70 (1998)

Popov, M.: Ouvrages scientifiques de VZMI. Moscow (1980)

Popov, M.: Construction des Mécanismes et dynamique des machines (1986)

Popov, M., Tyurin, V.: Balanced manipulator. SU, 1000271 (1983)

Popov, M., Tyurin, V.: Balanced manipulator. SU, 1491698 (1987)

Popov, M., Tyurin, V.: Balanced manipulator. SU, 1379105 (1988)

Popov, M., Tyurin, V.: Balanced manipulator. SU, 1664544 (1989)

Popov, M., Tyurin, V., Druyanov, B.: Counterbalanced manipulator. SU, 1065186 (1984)

Porter, B., Sandler, D.: Synthesis of dynamically optimal four-bar linkage. Proceedings of Conference on Mechanisms, Institution of Mechanical Engineers (1972) Paper C69/72, pp. 24–28, (1973)

Pracht, P., Minotti, P., Dahan, M.: Synthesis and balancing of cam-modulated linkages. Proceedings of the ASME Design and Automation Conference, vol. 10. pp. 221–226 (1987)

Previdia, F., Ficoa, F., Savaresib, S., Bellolia, D., Pesentic, I.: Direct design of a velocity controller and load disturbance estimation for a self-balancing industrial manual manipulator. Mechatronics **22**, 1177–1186 (2012)

Qi, N., Pennestri, E.: Optimum balancing of four-bar linkages. Mech. Mach. Theory **26**(3), 337–348 (1991)

Quinn, R., Chen, J., Lawrence, C.: Base reaction control for space-based robots operating in microgravity environment. J. Guid. Control Dyn. **17**(2), 263–270 (1994)

Raaijmakers, R.: Besi zoekt snelheidslimiet pakken en plaatsen op (besi attacks the speedlimit for pick and place motion). Mechatronica nieuws (Dutch Magazine), 26–31 (2007)

Raghu, E., Balasubramonian, A.: Experimental study on the elastodynamic behavior of the unbalanced and the counterweighted four bar mechanisms. ASME J. Mech. Des. **112**(3), 271–277 (1990)

Rahman, T., Ramanathan, R., Seliktar, R., Harwin, W.: A simple technique to passively gravity-balance articulated mechanisms. ASME J. Mech. Des. **117**(4), 655–658 (1995)

Rao, A.: Elastodynamic balancing of machines. Mach. Des. **6**, 92–93 (1977)

Reiner, B., Hans-Guenter, J., Otto, W.: Industrial robot with counterbalance weight attached by parallelogram linkage so that robot arm is always balanced. DE, 4014003 (1991)

Ricard, R., Gosselin, C.: On the development of reactionless parallel manipulators. Proceedings of the ASME Design Engineering Technical Conferences and Computers and Information in Engineering Conference (IDETC/CIE 2000). Baltimore, Maryland (2000)

Riele, F., Herder, J.: Perfect static balance with normal springs. Proceedings of the Design Engineering Technical Conferences. Pittsburg, New York, USA (2001)

Rizk, R., Krut, S., Dombre, E.: Design of a 3D gravity balanced orthosis for upper limb. Proceedings of the IEEE International Conference on Robotics and Automation (ICRA 2008). Pasadena, USA, pp. 2447–2452 (2008)

Root, R.: Dynamics of Engine and Shaft. Wiley, New York (1932)

Russo, A., Sinatra, R., Xi, F.: Static balancing of parallel robots. Mech. Mach. Theory **40**(2), 191–202 (2005)

Sadler, J.: Balancing of six-bar linkages by non-linear programming. Mech. Eng. **1**, 139–144 (1975)

Sadler, J., Mayne, R.: Balancing of mechanisms by non-linear programming. Proceedings of the 3rd App. Mech. Conference, vol. 29. Oklahoma State Univ. pp. 1–29 (1973)

Saravanana, R., Ramabalanb, S., Dinesh Babu, P.: Optimum static balancing of an industrial robot mechanism. Eng. Appl. Artif. Intell. **21**, 824–834 (2008)

Schenk, M., Herder, J., Guest, S.: Design of a statically balanced tensegrity mechanism. Proceedings of IDETC/CIE 2006 ASME 2006 International Design Engineering Technical Conferences & Computers and Information in Engineering Conference. Philadelphia, Pennsylvania, USA (2006)

Schmit, N., Okada, M.: Simultaneous optimization of robot trajectory and nonlinear spring to minimize actuator torque. Proceedings of the IEEE International Conference on Robotics and Automation (ICRA 2012). Saint Paul, Minnesota, USA, pp. 2806–2811 (2012)

Sconfeld, S.: Dynamische Synthese ebener Koppelgetriebe mide dun Programm—system KO-GEOP. Maschinenbautechnik **23**(3), 119–124 (1974)

Segawa, Y., Yamamoto, M., Shimada, A.: Parallel link mechanism. JP, 2000120824 (2000)

Segla, S.: Optimization of static balancing of a robot with parallelogram mechanism. Proceedings of the 9th International Conference on the Theory of Machines and Mechanisms. pp. 699–704 (2004)

Segla, S., Kalker-Kalkman, C., Schwab, A.: Statical balancing of a robot mechanism with the aide of a genetic algorithm. Mech. Mach. Theory **33**(1/2), 163–174 (1998)

Semenov, M.: Balancing of spatial mechanisms. Mechanisms **3**(4), 355–365 (1968a)
Semenov, M.: The synthesis of balanced plane mechanisms. Mechanisms **3**(4), 339–353 (1968b)
Seyferth, W.: Massenersatz duch punktmassen in räumlichen getrieben. Mech. Mach. Theory **9**, 49–59 (1974)
Shchepetilnikov, V.: Balancing of Mechanisms (1982)
Shchepetilnikov, V.: The determination of the mass centers of mechanisms in connection with the problem of mechanism balancing. Mechanisms **3**, 367–389 (1968)
Shchepetilnikov, V.: The balancing of mechanisms with unsymmetrical links. Mech. Mach. Theory **10**(6), 461–466 (1975)
Shin, E., Streit, D.: Spring equilibrator theory for static balancing of planar pantograph linkages. Mech. Mach. Theory **26**(7), 645–657 (1991)
Shirata, S., Konno, A., Uchiyama, M.: Design and evaluation of a gravity compensation mechanism for a humanoid robot. Proceedings of the 2007 IEEE/RSJ International Conference on Intelligent Robots and Systems (IROS 2007). San Diego, CA, USA, pp. 3635–3640 (2007)
Simionescu, I., Ciupitu, L.: The static balancing of the industrial arms. part i: discrete balancing. Mech. Mach. Theory, **35**, 1287–1298 (2000a)
Simionescu, I., Ciupitu, L.: The static balancing of the industrial arms. part ii: continuous balancing. Mech. Mach. Theory **35**, 1299–1311 (2000b)
Smith, M.: Dynamic analysis and balancing of linkages with interactive computer graphics. Comput. Aided Des. **7**(1), 15–19 (1975)
Smith, M., Maunder, L.: Inertia forces in a four-bar linkage. Mech. Eng. Sci. **9**(3), 218–225 (1967)
Smith, M., Walker, M.: Planar linkage. J. Eng. Mater. Des. **20**(10), 27–29 (1976)
Smith, M., Oldham, K., Walker, M.: Linkage balancing. J. Eng. Mater. Des. **21**(1), 47–50 (1977a)
Smith, M., Oldham, K., Walker, M.: Linkage design. J. Eng. Mater. Des. **21**(10), 29–31 (1977b)
Smith, R., Lobo-Prat, L., van de Kooij, H., Stienen, A.: Design of a perfect balance system for active upper-extremity exoskeletons. Proceedings of the International Conference on Rehabilitation Robots. Seattle, Washington, USA (2013)
Soethoudt, B., Herder, J.: Synthesis of perfect spring balancers with higher-order zero-free-length springs. Proceedings of the ASME 2007 International Design Engineering Technical Conferences and Computers and Information in Engineering Conference (IDETC/CIE). Las Vegas, Nevada, USA (2007)
Stevensen, E.: Balancing of machines. ASME J. Eng. Ind. **95**(2), 650–656 (1973)
Stienen, A., Hekman, E., van der Helm, F., Prange, G., Jannink, M., Aalsma, A., van der Kooij, H.: Freebal: dedicated gravity compensation for the upper extremities. Proceedings of the IEEE 10th International Conference on Rehabilitation Robotics. Noordwijk, The Netherlands, pp. 804–808 (2007)
Streit, D., Shin, E.: Equilibrators for planar linkages. Proceedings of the ASME Mechanisms Conference. Chicago, IL, USA, pp. 21–28 (1990)
Streit, D., Shin, E.: Equilibrators for planar linkages. ASME J. Mech. Des. **115**, 604–611 (1993)
Sugahara, S., Kawase, M., Mikuriya, Y., Hosobata, T., Sunazuka, H., Hashimoto, K., Lim, H., Takanishi, A.: Support torque reduction mechanism for biped locomotor with parallel mechanism. Proceedings of the 2004 IEEE/RSJ International Conference on Intelligent Robots and Systems (IROS 2004). Sendai, Japan, pp. 3213–3218 (2004)
Sugar, T., Hollander, K.: Adjustable stiffness Jack Spring actuator. US, 8322695 (2012)
Surdilovic, D., Bernhardt, R.: String-man: A novel wire robot for gait rehabilitation. Proceedings of the IEEE International Conference on Robotics and Automation (ICRA 2004). New Orleans, USA, pp. 2031–2036 (2004)
Takesue, N., Ikematsu, T., Murayama, H., Fujimoto, H.: Design and prototype of variable gravity compensation mechanism. J. Robot. Mechatron. **23**(2), 249–257 (2011)
Talbourdet, G., Shepler, P.: Mathematical solution of 4-bar linkages—IV. Balancing of linkages. Mach. Des. **13**, 73–77 (1941)

Tanaka, E., Ikehara, T., Sato, Y., Yusa, H., Sakurai, T., Saegusa, S., Ito, K., Yuge, L.: Walking assistance apparatus using special parallel link mechanism and a weight bearing lift. Proceedings of the IEEE Int. conference on rehabilitation robotics. Zurich, Switzerland (2011)

Tepper, F., Lowen, G.: General theorems concerning full force balancing of planar linkages by internal mass redistribution. ASME J. Eng. Ind. **94**(3), 789–796 (1972a)

Tepper, F., Lowen, G.: Shaking force optimization of four-bar linkage with adjustable constraints on ground bearing forces. ASME J. Eng. Ind. **97**(2) 643–651 (1972b)

Tepper, F., Lowen, G.: On the distribution of the RMS shaking moment of unbalanced planar mechanisms: theory of isomomental ellipses. ASME J. Eng. Ind. **95**(3), 665–671 (1973)

Tomas, S.: Balancing device for cable loaded with weight and spring. DE, 4420192 (1995)

Tricamo, S., Lowen, G.: A new concept for force balancing machines for planar linkages. Part 1: theory. ASME J. Mech. Des. **103**(3), 637–642 (1981a)

Tricamo, S., Lowen, G.: A new concept for force balancing machines for planar linkages. Part 2: application to four-bar linkage and experiment. ASME J. Mech. Des. **103**(4), 784–792 (1981b)

Tricamo, S., Lowen, G.: A novel method for prescribing the maximum shaking force of a four-bar linkage with flexibility in counterweight design. ASME J. Mech. Transm. Autom. Des. **105**(3), 513–522 (1983a)

Tricamo, S., Lowen, G.: Simultaneous optimization of dynamic reactions of a four-bar linkage with prescribed maximum shaking force. ASME J. Mech. Transm. Autom. Des. **105**(3), 522–528 (1983b)

Tsai, L.: Oldham-coupling second-harmonic balancer. ASME J. Mech. Transm. Autom. Des. **106**(3), 285–290 (1984)

Tsai, L., Maki, E.: Planetary-gear-type second-harmonic balancers. ASME J. Mech. Transm. Autom. Des. **11**(4), 530–536 (1989)

Tuda, G.: Gravity balancing device for rocking arm. US, 4592697 (1983)

Tuda, G., Mizuguchi, O.: Arm with gravity-balancing function. US, 4383455 (1983)

Tuijthof, G., Herder, J.: Design, actuation and control of an antropomorphic robot arm. Mech. Mach. Theory **35**, 945–962 (2000)

Turbin, B., Koropetz, A., Koropetz, Z.: The possibility of the shaking force balancing in the system with oscillating links. Russ. J. Mech. Mach. Theory **7**, 87–90 (1978)

Ulrich, N., Kumar, V.: Passive mechanical gravity compensation for robot manipulators. Proceedings of the IEEE International Conference on Robotics and Automation (ICRA 1991). Sacramento, CA, USA, pp. 1536–1540 (1991)

Urba, A.: Disposition of the correction mass for the best balancing of planar linkages. Balancing of machines and apparatuses. Moscow, Mechanical Engineering edn., pp. 226–230 (1978)

Urba, A.: Study of the elliptic harmonics and the possibility of their balancing by a counterweight. Collection of the scientific works of the Academy of agricultural science of Lithuania, **26**(3/28), 43–49 (1980)

Urba, A.: Geometric disposition of the points with r.m.s. value of shaking moment of spatial mechanisms. Collection of the scientific works of the Academy of agricultural science of Lithuania, **27**(3/89), 40–49 (1981)

van der Wijk, V.: Shaking-moment balancing of mechanisms with principal vectors and momentum. Front. Mech. Eng. **8**(1), 10–16 (2013)

van der Wijk, V., Herder, J.: Double pendulum balanced by counter-rotary counter-masses as useful element for synthesis of dynamically balanced mechanisms. Proceedings of the ASME International Design Engineering Technical Conferences and Computers and Information in Engineering Conference (IDETC/CIE 2008). New York, USA (2008a)

van der Wijk, V., Herder, J.: Dynamic balancing of mechanisms by using an actively driven counter-rotary counter-mass for low mass and low inertia. Proceedings of the Second International Workshop on Fundamental Issues and Future Research Directions for Parallel Mechanisms and Manipulators. Montpellier, France, pp. 241–251 (2008b)

van der Wijk, V., Herder, J.: Force balancing of variable payload by active force-balanced reconfiguration of the mechanism. Proceedings of the ASME/IFToMM International Conference on Reconfigurable Mechanisms and Robots. London, UK, pp. 323–329 (2009)
van der Wijk, V., Herder, J.: Dynamic balancing of Clavel's delta robot. Computational Kinematics. Springer, pp. 315–322 (2009a)
van der Wijk, V., Herder, J.: Advances in robotics research: theory, implementation, application, 21–30. Springer, Berlin (2009b)
van der Wijk, V., Herder, J.: Active dynamic balancing unit for controlled shaking force and shaking moment balancing. Proceedings of the ASME Design Engineering Technical Conferences and Computers and Information in Engineering Conference (IDETC/CIE 2010). Montreal, Canada (2010a)
van der Wijk, V., Herder, J.: Dynamic balancing of a single crank-double slider mechanism with symmetrically moving couplers. New Trends Mech. Sci., Mech. Mach. Sci. **5**, 413–420 (2010b)
van der Wijk, V., Herder, J.: Synthesis method for linkages with center of mass at invariant link point—pantograph based mechanisms. Mech. Mach. Theory **48**, 15–28 (2012)
van der Wijk, V., Herder, J.: The method of principal vectors for the synthesis of shaking moment balanced linkages. New Trends Mech. Mach. Sci., Mech. Mach. Sci. **7**, 399–407 (2013)
van der Wijk, V., Herder, J., Demeulenaere, B.: Comparison of various dynamic balancing principles regarding additional mass and additional inertia. ASME J. Mech. Robot. **1**, 21–30 (2009)
van der Wijk, V., Krut, S., Pierrot, F., Herder, J.: Generic method for deriving the general shaking force balance conditions of parallel manipulators with application to a redundant planar 4-rrr parallel manipulator. Proceedings of the 13th World Congress in Mechanism and Machine Science. Guanajuato, Mexico (2011)
van der Wijk, V., Demeulenaere, B., Gosselin, C., Herder, J.: Comparative analysis for low-mass and low-inertia dynamic balancing of mechanisms. ASME J. Mech. Robot. **4** (2012)
van der Wijk, V., Krut, S., Pierrot, F., Herder, J.: Design and experimental evaluation of a dynamically balanced redundant planar 4-RRR parallel manipulator. Int. J. Robot. Res. **32**, 744–759 (2013)
van Dijk, W., van der Kooij, H., Hekman, E.: A passive exoskeleton with artificial tendons. Proceedings of the IEEE International conference on Rehabilitation Robotics. Zurich, Switzerland (2011)
van Dorsser, W., Barents, R., Wisse, B., Herder, J.: Gravity-balanced arm support with energy-free adjustment. ASME J. Med. Device. **1**, 151–158 (2007)
van Dorsser, W., Barents, R., Wisse, B., Schenk, M., Herder, J.: Energy-free adjustment of gravity equilibrators by adjusting the spring stiffness. J. Mech. Eng. Sci. **222**, 1839–1846 (2008)
van Ninhuijs, B., van der Heide, L., Jansen, J., Gysen, B., van der Pijl, D., Lomonova, E.: Overview of actuated arm support systems and their applications. Actuators **2**(4), 86–110 (2013)
Verkhovcki, V.: Spring mechanism. SU, 932005 (1980)
Verschuure, M., Demeulenaere, B., Swevers, J., De Schutter, J.: On the benefits of partial shaking force balance in six-bar linkages. Proceedings of the 12th IFToMM World Congress. Besançon, France (2007)
Verschuure, M., Demeulenaere, B., Swevers, E.A.J., De Schutter, J.: Optimal counterweight balancing of spatial mechanisms using voxel-based discretizations. Proceedings of the international conference on noise and vibration engineering. Leuven, Belgium (2008a)
Verschuure, M., Demeulenaere, B., Swevers, J., De Schutter, J.: Counterweight balancing for vibration reduction of elastically mounted machine frames: a second-order cone programming approach. ASME J. Mech. Des. **130** (2008b)
Vladov, I., Danilevskij, V.: Loading manipulator. SU, 83113 (1981)
Vladov, I., Danilevskij, V.: Balancing manipulator. SU, 1423367 (1999)
Vladov, I., Danilevskij, V., Rassadkin, V.: Module of linear motion of industrial robot. SU, 848350 (1981)
Vrijlandt, N., Herder, J.: Seating unit for supporting a body or part of a body. NL, 1018178 (2002)
Wahl, A.: Mechanical Springs. McGraw-Hill, New-York (1963)

Walker, M., Haines, R.: An experimental study of the effects of counterweights on a six-bar chain. Mech. Mach. Theory **17**(6), 355–360 (1982a)

Walker, M., Haines, R.: A study of counterweight synthesis for a six-bar chain. Mech. Mach. Theory **17**(5), 327–334 (1982b)

Walker, M., Oldham, K.: General theory of force balancing using counterweights. Mech. Mach. Theory **13**, 175–185 (1978)

Walsh, G., Streit, D., Gilmore, B.: Spatial spring equilibrator theory. Mech. Mach. Theory **26**(2), 155–170 (1991)

Wang, J., Gosselin, C.: Static balancing of spatial three-degree-of-freedom parallel mechanisms. Mech. Mach. Theory **34**, 437–452 (1999)

Wang, J., Gosselin, C.: Static balancing of spatial four-degree-of-freedom parallel mechanisms. Mech. Mach. Theory **35**(4), 563–592 (2000)

Wang, J., Gosselin, C.: Passive mechanisms with multiple equilibrium configurations. Trans. Can. Soc. Mech. Eng. **28**, 139–151 (2004)

Wawrzecki, J.: A method of the balancing of spatial mechanisms. Mech. Mach. Theory **33**(8), 1195–1209 (1998)

Wawrzecki, J.: The conditions of optimal balancing spatial mechanisms. Proceedings of the Tenth World Congress on Theory of Machines and Mechanisms. Oulu, Finland, pp. 1589–1594 (1999)

Wiederrich, J., Roth, B.: Momentum balancing of four-bar linkages. ASME J. Mech. Des. 98B(4), 1289–1295 (1976)

Wildenberg, F.: Compensating system for a hexapod. US, 6474915 (2002)

Wisse, B., van Dorsser, W., Barents, R., Herder, J.: Energy-free adjustment of gravity equilibrators using the virtual spring concept. Proceedings of the IEEE 10th International Conference on Rehabilitation Robotics. Noordwijk, The Netherlands, pp. 742–750 (2007)

Wittenburg, J.: Dynamics of systems of rigid bodies. B. G. Teubner, Stuttgart (1977)

Wongratanaphisan, T., Chew, M.: Gravity compensation of spatial two-dof serial manipulators. J. Robot. Syst. **19**(7), 329–347 (2002)

Wu, Y., Gosselin, C.: Synthesis of reactionless spatial 3-dof and 6-dof mechanisms without separate counter-rotations. Int. J. Robot. Res. **23**(6), 625–642 (2004)

Wu, Y., Gosselin, C.: Design of reactionless 3-dof and 6-dof parallel manipulators using parallelepiped mechanisms. IEEE Trans. Robot. **21**(5), 821–833 (2005).

Wu, Y., Gosselin, C.: On the dynamic balancing of multi-dof parallel mechanisms with multiple legs. ASME J. Mech. Des. **129**, 234–238 (2007)

Xi, F.: Dynamic balancing of hexapods for high-speed applications. Robotica **17**, 335–342 (1999)

Xi, F., Sinatra, R.: Effect of dynamic balancing of four-bar linkage vibrations. Mech. Mach. Theory **32**(6), 715–728 (1997)

Yamamoto, R., Hirakawa, A., Horikawa, O.: Load balancer with automatic lifting force compensation. Proceedings of the ABCM Symposium in Mechatronics, vol. 4, pp. 580–589 (2010)

Yan, H., Soong, R.: Kinematic and dynamic design of four-bar linkages by links counterweighing with variable input speed. Mech. Mach. Theory, **36**, 1051–1071 (2001)

Yao, J., Smith, M.: An improved complex mass method for force balancing of planar linkages. Mech. Mach. Theory **28**(3), 417–425 (1993)

Ye, Z:, Smith, M.: Complete balancing of planar linkages by an equivalence method. Mech. Mach. Theory **29**(5), 701–712 (1991)

Yoshida, K., Nenchev, D., Uchiyama, M.: Moving base robotics and reaction management control. In: Giralt, G., Hirzinger, G. (eds.) Proceedings of the Seventh International Symposium on Robotics Research, pp. 101–109. Springer, Berlin (1996a)

Yoshida, K., Nenchev, D., Vichitkulsawat, P., Kobayashi, H., Uchiyama, M.: Experiments on the point-to-point operations of a flexible structure mounted manipulator system. Adv. Robot. **11**(4), 397–411 (1996b)

Yu, Y.: Optimum shaking force and shaking moment balancing of the RSS'R spatial linkage. Mech. Mach. Theory **22**(1), 39–45 (1987a)

Yu, Y.: Research on complete shaking force and shaking moment balancing of spatial linkages. Mech. Mach. Theory **22**(1), 27–37 (1987b)

Yu, Y.: Complete shaking force and shaking moment balancing of spatial irregular force transmission mechanisms using additional link. Mech. Mach. Theory **23**(4), 279–285 (1988)

Yu, Y., Jiang, B.: Analytical and experimental study on the dynamic balancing of flexible mechanisms. Mech. Mach. Theory **42**, 626–635 (2007)

Yu, Y., Lin, J.: Active balancing of a flexible linkage with redundant drives. ASME J. Mech. Des. **125**, 119–123 (2003)

Yu, Y., Smith, M.: The effect of link form on the dynamic response of flexible mechanisms. Proceedings of 9th World Congress on the Theory of Machines and Mechanisms. Milan, Italy (1995)

Yudin, V.: The balancing of machines and their stability. Edition of Academy of Red Army, Moscow (1941)

Zabalza, I., Ros, J., Gil, J., Pintor, J., Jimenez, J.: Tri-Scott. a MICABO like 6-dof quasi-decoupled parallel manipulator. Proceedings of the Workshop on Fundamental Issues and Future Researches for Parallel Mechanisms and Manipulators. Quebec, Canada, pp. 12–15 (2002)

Zang, D., Gao, F., Hu, X., Gao, Z.: Static balancing and dynamic modeling of a three-degree-of-freedom parallel kinematic manipulator. Proceedings of the IEEE International Conference on Robotics and Automation. Shanghai, China, 3211–3217 (2011)

Zayzev, C.: Manipulator. SU, 795934 (1981)

Zhang, S.: A constitutive method of objective function for the dynamic optimum balance of shaking force in linkage. Mech. Mach. Theory **29**(6), 829–835 (1994)

Zhang, S., Chen, J.: The optimum balance of shaking force and shaking moment of linkages. Mech. Mach. Theory **30**(4), 589–597 (1995)

Zhang, S., Morita, N., Torii, T.: Reduction of vibration by optimum balancing of shaking force/moment in linkage. JSME Int. J. Series C, **43**(3), 748–754 (2000)

Zhang, Y., Yao, Y., Cha, J.: An independent active balancer for planar mechanisms. Trans. CSME **31**(2), 167–190 (2007)

Zhe, L.: Sensitivity and robustness of mechanism balancing. Mech. Mach. Theory **33**(7), 1045–1054 (1998)

Zhe, L., Shixian, B.: Optimum balancing of linkages with clearances. Mech. Mach. Theory **27**(5), 535–541 (1992)

Zhu, X., Shen, H., Zhang, H., Li, J., Tao, S., Yang, T.: Finite position method for complete balancing of shaking forces of mechanisms. Proceedings of the IEEE 10th International Conference on Computer-Aided Industrial Design and Conceptual Design. Wenzhou, China (2009)

Zobairi, M., Rao, S., Sahay, B.: Kinetoelastodynamic balancing of 4R-four-bar mechanisms by internal mass redistribution. Mech. Mach. Theory. **21**(4), 317–323 (1986a)

Zobairi, M., Rao, S., Sahay, B.: Kinetoelastodynamic balancing of 4R-four-bar mechanisms combining kinematic and dynamic stress consideration. Mech. Mach. Theory. **21**(4), 307–315 (1986b)